Lecture Notes in Computer Science 16398

The series Lecture Notes in Computer Science (LNCS), including its subseries Lecture Notes in Artificial Intelligence (LNAI) and Lecture Notes in Bioinformatics (LNBI), has established itself as a medium for the publication of new developments in computer science and information technology research, teaching, and education.

LNCS enjoys close cooperation with the computer science R & D community, the series counts many renowned academics among its volume editors and paper authors, and collaborates with prestigious societies. Its mission is to serve this international community by providing an invaluable service, mainly focused on the publication of conference and workshop proceedings and postproceedings. LNCS commenced publication in 1973.

Udunna Anazodo · Dong Zhang ·
Confidence Raymond · Mehmet Kurt ·
Karim Lekadir · Alessandro Crimi
Editors

Medical Image Computing in Resource Constrained Settings

First International Workshop, MIRASOL 2025
Held in Conjunction with MICCAI 2025
Daejeon, South Korea, September 27, 2025, Proceedings

Editors
Udunna Anazodo
McGill University
Montreal, QC, Canada

Confidence Raymond
McGill University
Montreal, QC, Canada

Karim Lekadir
ICREA
Catalan, Barcelona, Spain

Dong Zhang
University of British Columbia
Vancouver, BC, Canada

Mehmet Kurt
University of Washington
Seattle, WA, USA

Alessandro Crimi
AGH University of Science and Technology
Kraków, Poland

ISSN 0302-9743 ISSN 1611-3349 (electronic)
Lecture Notes in Computer Science
ISBN 978-3-032-13653-4 ISBN 978-3-032-13654-1 (eBook)
https://doi.org/10.1007/978-3-032-13654-1

This Springer imprint is published by the registered company Springer Nature Switzerland AG
The registered company address is: Gewerbestrasse 11, 6330 Cham, Switzerland

Preface

This book is a collection of revised papers presented at the Medical Image Computing in Resource Constrained Settings Workshop & Knowledge Interchange (MIRASOL) workshop, held in conjunction with the Medical Image Computing and Computer Assisted Intervention (MICCAI) conference on September 27, 2025, in Daejeon, South Korea.

The workshop followed a double-blind peer-review process. A total of 62 submissions were received, of which 31 were accepted for publication. Each paper received an average of 3 reviews from members of the Program Committee and external reviewers. The 31 papers describe cutting-edge research from computational scientists and clinical researchers working on a variety of medical image computing challenges relevant to the low-to-middle-income countries (LMICs) and broader global contexts, as well as emerging techniques for image computing methods tailored to low-resource settings. The aim of this volume, centered on the MICCAI MIRASOL workshop, is to showcase the latest contributions from the medical imaging community in LMICs. The collection of papers is organized around three themes that describe insights into the latest advances in machine learning, medical imaging analysis, image segmentation, disease detection, diagnosis, and prognosis, with a focus on global health applications and innovations in the region for real-world clinical impact.

By bringing together researchers, clinicians, and industry stakeholders from around the world and across all career stages, this workshop fosters collaboration to tackle region-specific health challenges through the use of advanced image analysis and AI techniques. We are grateful for the incredible support of our sponsors who provided financial assistance for the workshop and travel awards to LMICs authors of the papers in this collection. This includes the International Society of Radiology, Università degli Studi di Catania, PNRR MUR (Italian Ministry of University and Research) project PE0000013-FAIR, the University of Washington, Foundation Pierre Fabre, and the Consortium for Advancement of MRI Education and Research in Africa (CAMERA).

We sincerely hope that this volume will inspire continued research and collaboration in the field of medical imaging, particularly in addressing the unique challenges faced by healthcare systems in diverse environments.

December 2025

Udunna Anazodo
Dong Zhang
Confidence Raymond
Mehmet Kurt
Karim Lekadir
Alessandro Crimi

Organization

General Chairs

Udunna Anazodo	McGill University, Canada
Karim Lekadir	Universitat de Barcelona and ICREA, Spain
Celia Cintas	IBM Research Africa, Kenya
Tinashe Mutsvangwa	IMT-Atlantique, France

Steering Committee

Bishesh Khanal	Nepal Applied Mathematics and Informatics Institute for Research (NAAMII), Nepal
Andrea Lara	Universidad Galileo, Guatemala
Farouk Dako	University of Pennsylvania, USA
Amal Saleh	Addis Ababa University, Ethiopia
Ousmane Sall	Université numérique Cheikh Hamidou Kane, Senegal
Alexander Hammers	King's College London, UK
Kiran Raj Pandey	Medharma Clinic, Nepal
William Ogallo	Google Research Africa, Kenya
Enzo Ferrante	Universidad de Buenos Aires Argentina

Program Committee

Dong Zhang	University of British Columbia, Canada
Alessandro Crimi	AGH University of Science and Technology, Poland
Mehmet Kurt	University of Washington, USA
Waqas Sultani	Information Technology University, Pakistan
Pamela Guevara	Universidad de Concepcion, Chile
Sahar Selim	Nile University, Egypt
Maruf Adewole	Medical Artificial Intelligence Laboratory, Nigeria
Deogratis Mzurikawo	Muhimbili University of Health and Allied Sciences, Tanzania

Diversity and Sponsorship Committee

Marius George Linguraru	Children's National Hospital and George Washington University, Washington, USA
Charles Delahunt	Global Health Labs, USA
Natasha Lepore	University of Southern California, USA
Esther Puyol Anton	HeartFlow, UK

Next Generation Interchange Committee

Confidence Raymond	McGill University, Canada
Mohannad Barakat	University of Erlangen-Nuremberg, Germany
Noha Magdy	University of Erlangen-Nuremberg, Germany
Oumayma Soula	University of Sfax, Tunisia
Abbas M Rabiu	Aminu Kano Teaching Hospital, Nigeria
Kelvin Murithi	Sonar Imaging, Kenya
Mubaraq Yakubu	King's College London, UK
Marawan Elbatel	Hong Kong University of Science and Technology, China

Communication Committee

Xènia Puig Bosch	Universitat de Barcelona, Spain
Juhi Tulsi	Consortium for Advancement of MRI Education and Research in Africa (CAMERA)
Ahmed Yasser	Multimedia University, Malaysia

Ambassadors

Omar Choudhry	University of Leeds, UK
Juan Pablo Rodriguez Illescas	Universidad Galileo, Guatemala
Jin Kim	University of California, Los Angeles, USA
Freedmore Sidume	Botswana Accountancy College, Botswana
Miguel López Pérez	University of Granada, Spain
Tareen Dawood	Technical University of Denmark, Denmark

Reviewers

Contents

Are ECGs Enough? Deep Learning Classification of Pulmonary Embolism Using Electrocardiograms

João D. S. Marques(✉) and Arlindo L. Oliveira

INESC-ID/Instituto Superior Técnico, Lisbon, Portugal
{joao.p.d.s.marques,arlindo.oliveira}@tecnico.ulisboa.pt

Abstract. Pulmonary embolism (PE) is a relevant cause of out-of-hospital cardiac arrest that requires fast diagnosis. While computed tomography pulmonary angiography is the standard diagnostic tool, it is not always accessible. Electrocardiography is an essential tool for diagnosing multiple cardiac anomalies, as it is affordable, fast, and available in many settings. However, the existence of public ECG datasets, especially for PE, is limited, and, in practice, these datasets tend to be small, making it essential to optimize learning strategies. In this study, we investigate the performance of multiple neural networks to assess the impact of various approaches. We further evaluate whether these practices improve generalization by transferring knowledge from larger ECG datasets (PTB-XL, CPSC18, MedalCare-XL) to a smaller dataset of PE. Our approach achieves an 8% higher AUC than clinical scores while maintaining over 80% specificity in settings where computed tomography pulmonary angiography is not available.

Keywords: Deep Learning · Electrocardiography · Anomaly detection · Pulmonary Embolism

1 Introduction

Cardiovascular diseases (CVDs) represent a significant global health challenge, being one of the leading causes of mortality and a significant contributor to the reduction in quality of life [16]. Among the most severe conditions, pulmonary embolism (PE) is the third most common CVD after stroke and heart attack [28], accounting for 2–9% of all causes of out-of-hospital cardiac arrest [12]. Due to its typically poor prognosis, fast diagnosis and treatment are critical. Although computed tomography pulmonary angiography (CTPA) is often required to confirm the diagnosis of PE, it is not feasible for all patients and may even be inaccessible in certain healthcare facilities [10].

Electrocardiography (ECG) is a widely used diagnostic tool in both pre-hospital and hospital settings due to its affordability, speed, and accessibility [25]. While improvements in ECG-based scoring systems [5,9,17] have enhanced diagnostic capabilities, they remain time-consuming and are not routinely applied

U. Anazodo et al. (Eds.): MIRASOL 2025, LNCS 16398, pp. 1–11, 2026.
https://doi.org/10.1007/978-3-032-13654-1_1

in clinical practice. Moreover, the non-specific nature of PE symptoms, which frequently overlap with other CVDs, makes ECG-based detection particularly challenging [10].

The rise of foundational AI models is having an impact on various domains, including medicine [19,22], yet anomaly detection using ECGs remains highly specialized, posing challenges for model adaptation. Although deep learning offers promising potential in this area, its effectiveness is limited by the lack of publicly available ECG datasets, particularly for PE. The few existing private datasets are typically small, containing less than 1,000 samples, and highly imbalanced, which poses significant challenges for model training and generalization.

Given these difficulties, this study aims to establish a baseline performance for deep learning models applied to PE detection using ECGs and to evaluate the impact of transfer learning in settings characterized by limited and imbalanced datasets. Our results show that, despite data scarcity, transfer learning can significantly enhance detection performance and support more effective clinical decision-making.

All code, architectures and hyperparameters are publicly available at https://github.com/joaodsmarques/Are-ECGs-enough-Deep-Learning-Classifiers.

2 Background

The application of deep learning to ECG analysis is not new. Multiple studies have explored this field over the years [6,29]. However, progress remains challenging, largely due to the limited availability of publicly accessible datasets.

A One-Dimensional Convolutional Neural Network (1D-CNN) is often the standard approach for ECG classification, as convolutional layers effectively capture local features from the signal, achieving high performance in several classification tasks [33]. However, these architectures struggle to recognize long-range dependencies. Recurrent neural networks (RNNs), particularly Long Short-Term Memory (LSTM) and Gated Recurrent Units (GRU) [4], are well-suited for capturing the temporal dependencies of ECGs. As demonstrated by Zahra Ebrahimi et al. [7], both 1D-CNNs and RNN-based models can extract relevant features for ECG classification.

Recently, Narotamo et al. [18] explored multiple deep learning architectures for detecting cardiac anomalies using the PTB-XL dataset. In their 1D experiments, they found that Convolutional Recurrent Neural Networks (CRNNs), using GRUs and LSTMs, offer an effective balance between feature extraction and capturing temporal dependencies.

Beyond traditional architectures, new transformer-inspired models have been proposed for ECG analysis. For example, ECG-Mamba [20] achieved an F1-score of 75% on PTB-XL and 79% on CPSC18 classification tasks, demonstrating the potential of new architectures in ECG classification.

In the context of PE classification, when CTPA is not available, clinicians rely on risk stratification tools such as the Wells score [17] and the PEGeD rule [9]. The Wells score estimates pre-test probability based on clinical features (e.g. signs of DVT, tachycardia), while the PEGeD rule refines this approach by combining Wells categories with age-adjusted D-dimer thresholds to reduce unnecessary imaging. In terms of deep learning, Somani et al. demonstrated that ECGs provide valuable diagnostic information for PE detection and can enhance the performance beyond what is achievable using CTPA alone [27].

In this work, we investigate several deep learning architectures applied to different ECG datasets to evaluate their effectiveness in learning directly from raw ECG signals. We then assess their performance in the context of PE detection, aiming to determine whether ECGs can contribute to improved detection in scenarios where other diagnostic tools are unavailable.

3 Methodology

Publicly available datasets for ECG anomaly detection are limited, and those that exist are often small and imbalanced. To maximize their utility, our experiments are guided by the following key questions:

- **What performance can be expected when training from scratch?** We train multiple models on two datasets to identify best practices under these settings.
- **How effective is transfer learning?** We investigate whether transfer learning from three larger datasets improves performance in a smaller PE detection setting.
- **Can our models outperform clinical approaches for PE detection?** We evaluate whether our best-performing model can surpass existing clinical procedures.

3.1 Datasets

We begin this study by focusing on 3 ECG datasets: PTB-XL [32], the China Physiological Signal Challenge 2018 (CPSC18) [15], and MedalCare-XL [8]. Insights gained from these datasets are then applied to PE using the PE-HSM dataset [30].

The selection of datasets is based on the objective of training on the largest and most complete publicly available dataset, PTB-XL, while using CPSC18 as a comparative benchmark due to its relatively large size. In addition, we assess the utility of synthetic ECGs through the MedalCare-XL dataset. Finally, we evaluate whether the insights gained from these datasets generalize to the PE-HSM setting. All selected datasets share key characteristics: 500 Hz sampling frequency, 12-lead ECG recordings, and adaptable signal lengths, ensuring consistency across experiments. An overview of these datasets is provided in Table 1.

In the PTB-XL dataset, we focus on the five diagnostic superclasses instead of the 44 subclasses, as fine-tuning on all subclasses is constrained by severe class imbalance. This imbalance can bias the learned representations toward dominant

Table 1. Overview of the datasets used for each selected task.

Dataset	Availability	Focus	Classification	ECGs	Length
PTB-XL	Public	Diagnostic	Multi-label (5)	21,837	10 s
CPSC18	Public	Arrhythmia	Multi-label (9)	6,877	6–60 s
MedalCare	Public	Morphology	Multi-class (8)	16,900	10 s
PE-HSM	Private	Acute PE	Binary (1)	927	10 s

classes, ultimately affecting transfer learning performance. Figure 1 shows the class distributions across all experimental settings. Note that the public datasets used do not include a class for PE, but contain classes that may be informative for its detection.

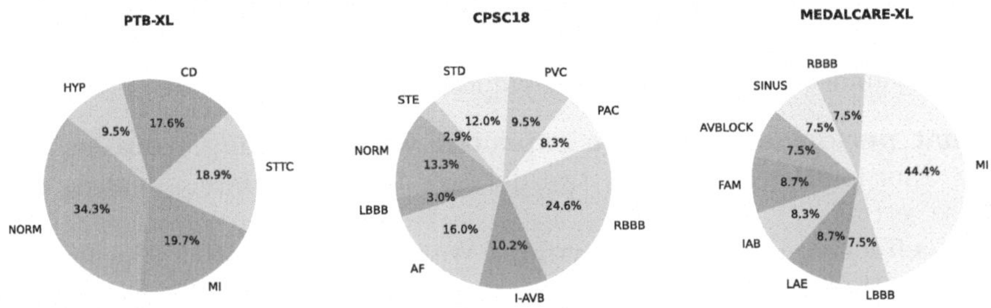

Fig. 1. Class label distributions for the PTB-XL (**left**), CPSC18 (**middle**), and MedalCare-XL (**right**) datasets.

Dataset Splits. For PTB-XL, we followed the protocol proposed by the original authors [32], using fold 9 for validation and fold 10 for testing. For CPSC18, we created 10 folds, ensuring representation of all classes in each one, and performed cross-validation. For MedalCare-XL, we combined all class-specific folders within each split into unified train, validation, and test sets. For PE-HSM, we adopted the original training and test partitions provided by the authors, comprising 824 training samples (222 positive and 602 negative) and 103 test samples (39 positive and 64 negative).

3.2 Pre-processing and Data Augmentation

Pre-processing is crucial, as it helps the models focus on the main signal and ignore noise [24]. Although specialized algorithms such as beat pickers can detect and enhance specific waveforms like R-peaks [23], we opted to evaluate neural networks directly on raw ECG signals. Our pre-processing pipeline consists of four key steps: noise removal (addressing power line interference and baseline

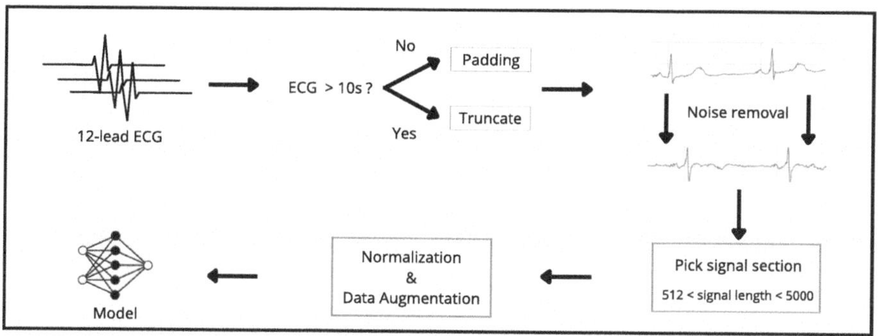

Fig. 2. Pre-processing pipeline.

drift), sequence length selection, normalization, and data augmentation. The full pipeline is illustrated in Fig. 2.

To mitigate noise, we first applied a low-pass filter at 50 Hz to remove power line interference, followed by wavelet decomposition [1]. However, applying this method during training, proved to be computationally expensive. To address this, we instead employed a second-order Butterworth bandpass filter with cut-off frequencies of 1 Hz (high-pass for baseline wander suppression) and 45 Hz (low-pass). We selected this filter and frequency range because it is widely used in the literature [18], and provided similar output quality while being more computationally efficient.

We used all 12 leads of each ECG, given that PE presents non-specific findings and clinicians typically evaluate multiple leads [30]. Empirical results further demonstrated stronger performance when all leads were included. For the selection of sequence length, to extract a segment from each ECG lead, we first randomly select a starting point s, ensuring that the extracted segment remains within the maximum available length. This process is defined in Eq. 1.

$$s \sim \mathcal{U}(0, m - l), \quad \text{with} \quad s + l \leq m \tag{1}$$

Note that s is the randomly chosen starting index (shared among all leads), l is the desired segment length, and m is the maximum length of the ECG signal (5000 samples). This ensures that the segment length is consistent across leads and does not exceed the available data range. Sequences exceeding 5000 samples (maximum available length) were truncated, while shorter sequences were zero-padded. A length of 2048 samples was chosen, as it provided similar performance to the longest sequence while being significantly less computationally demanding.

For normalization, we evaluated the impact of various methods, including MinMax, Z-score, Rscale, Logscale, and L2 normalization. Finally, we applied several data augmentation and regularization techniques, in addition to random shift, including vertical flip, random drop, lead drop, square pulse sum, and sine sum [21].

3.3 Model Architectures

We explored a range of architectures, from simpler models like AlexNet to more complex encoder-based networks. We started by using AlexNet [11] as a baseline due to its simplicity and fast training speed. Next, we investigated the VGG family [26], identifying VGG11 (with batch normalization) as the most robust contender. We then conducted experiments with ResNet architectures, where ResNet18 emerged as a strong candidate, offering a good balance between simplicity and performance, while also proving to be an effective feature extractor for other deep learning modules.

Beyond CNN-based models, we integrated GRUs and LSTMs [4] to capture temporal dependencies in ECG signals. To further enhance both deep feature extraction and temporal modeling, we employed CRNNs, where ResNet18 with GRU and ResNet18 with LSTM stood out as high-performing combinations.

We also explored representing the ECG as a 2D structured signal rather than as 12 independent 1D sequences. To this end, we applied EEGNet [13], a compact convolutional neural network that has demonstrated strong performance in the electroencephalography (EEG) domain.

Given the importance of attention mechanisms in capturing long-range dependencies [31], we studied adding attention to the ResNet18, naming this architecture AttResNet.

Finally, we explored transformer-based models [31], including transformer encoders and Residual Transformers, to assess how larger, more recent architectures perform in this setting.

3.4 Metrics

Given that our task is predominantly multi-label and that class distributions are imbalanced, relying only on accuracy can be misleading, as models may achieve high scores by prioritizing the most abundant classes. To provide a more comprehensive evaluation, we additionally report the F1-score, which balances correct detection with case coverage, and the mean average precision (MAP), which summarizes precision across all target classes and is more robust to class imbalance. Furthermore, in clinical applications, it is critical to minimize false negatives (ensuring high sensitivity) while also keeping high specificity to confirm that detected negatives are truly negative cases. To measure this balance, we also compute the G-mean (GM).

3.5 Training Setup

All models were initialized using standard procedures [3] and trained on an NVIDIA V100S GPU (32 GB) installed in a DELL PowerEdge C41402 server. We used Bayesian Optimization with Optuna [2] to determine the optimal hyperparameters for each architecture, performing 100 trials of 50 epochs each with the objective of maximizing the F1-score. For the loss function, we analyzed both Binary Cross Entropy (BCE) with class weighting, adjusted according to dataset distribution, and Focal Loss [14] with $\gamma = 2$ and $\alpha = 0.7$. Focal Loss proved to be the most effective, being used in all subsequent experiments.

4 Results and Discussion

Given their clinical relevance and frequent use in the literature, PTB-XL and CPSC18 served as benchmarks for comparing model architectures. The best results obtained from training the models from scratch are summarized in Table 2. The optimized architectures include the following configurations: (a) CNNs: logscale normalization and learning rate (lr) ≈ 0.0002; (b) CRNNs: ResNet18, L2 normalization, 2 recurrent layers, 256 hidden size and lr ≈ 0.0005; (c) AttResNet: ResNet18 combined with multihead attention layer, 512 embedding dimension, 4 heads, L2 normalization and lr ≈ 0.0002; (d) Transformer: 1 convolutional layer to increase dimensions, 4 transformer encoder layers, 512 embedding dimension, 4 heads, zscore normalization and lr ≈ 0.0001; (e) ResTransformer: ResNet18, transformer encoder, 512 embedding dimension, 4 heads, zscore normalization and lr ≈ 0.0002; (f) EEGNet: zscore normalization and lr ≈ 0.0015.

Table 2. Performance (%) of multiple models trained from scratch on PTB-XL and CPSC18 datasets.

Models	PTB-XL				CPSC18			
	Acc	F1	MAP	GM	Acc	F1	MAP	GM
AlexNet	85.9	74.3	76.1	83.8	93.2	74.8	72.4	86.2
VGG11_bn	86.2	74.1	77.3	83.2	94.0	75.0	70.9	86.1
ResNet18	86.8	75.1	**78.9**	83.9	93.9	75.3	72.0	87.2
EEGNet	83.4	62.2	70.1	75.4	91.5	54.1	59.4	74.0
$CRNN_{LSTM}$	87.1	75.8	76.7	84.6	**95.1**	79.3	76.8	88.4
$CRNN_{GRU}$	**87.2**	**76.1**	76.7	84.9	**95.1**	**79.5**	**78.3**	88.6
AttResNet	87.0	75.6	77.1	**85.0**	94.8	78.5	76.8	88.1
Transformer	75.0	50.7	44.0	64.9	87.9	57.5	51.5	74.4
ResTransformer	**87.2**	75.7	76.6	84.0	94.8	78.8	77.5	**89.0**

Based on the results obtained in Table 2, we chose 4 models to apply to the PE-HSM dataset: the best CNN (ResNet18), the best CRNN ($CRNN_{GRU}$), the Attention ResNet (AttResNet) and the ResTransformer. The performance of the four models was evaluated under both transfer learning and training from scratch settings. In the transfer learning scenario, performance improved when all weights were fine-tuned, rather than updating only the final layer. Results for these settings are shown in Table 3.

While no model dominates across all metrics, the results offer valuable insights. Transfer learning increases performance on PE detection, as evidenced

Table 3. Performance (%) of the best models on the PE-HSM dataset.

PE-HSM					
Models	Pretrain	Acc	F1	MAP	GM
ResNet18	None	58.3	56.6	42.2	59.6
	PTB-XL	**63.1**	61.2	50.8	**64.9**
	CPSC18	63.0	**62.4**	55.1	62.7
	MedalCare	59.2	58.3	**57.0**	59.3
$CRNN_{GRU}$	None	66.0	61.9	48.2	70.1
	PTB-XL	**68.9**	**65.4**	49.6	**72.7**
	CPSC18	67.0	63.2	54.1	70.81
	MedalCare	66.0	61.9	**57.2**	70.1
AttResNet	None	64.1	60.2	52.4	67.8
	PTB-XL	**68.0**	**64.1**	53.8	71.9
	CPSC18	64.1	63.4	55.7	63.5
	MedalCare	**68.0**	60.0	**56.8**	**74.2**
ResTransformer	None	66.0	57.6	49.3	**72.3**
	PTB-XL	63.1	60.8	50.3	61.8
	CPSC18	**68.0**	**66.7**	51.9	68.7
	MedalCare	64.1	59.1	**55.6**	68.6

by improvements in all pretrained settings. CRNNs offer the best balance between ease of training and performance, while ResNet18 proves to be a strong and stable feature extractor. Attention modules benefit greatly from feature extractors, whereas transformers struggle due to limited training data.

In the pre-training setting of PTB-XL, AttResNet seems close to $CRNN_{GRU}$, suggesting that with more data, the attention mechanisms may outperform RNNs. Interestingly, ResTransformer achieves better results when pretrained on CPSC18 than on PTB-XL, likely due to CPSC18's focus on arrhythmias - including a class for Right Bundle Branch Block (RBBB), a key ECG indicator of pulmonary embolism, which demonstrates effective domain adaptation.

Pretraining on MedalCare-XL yields the highest MAP across all models, improving model precision across all classes. Moreover, models generally benefit from pretraining on this dataset, showing that synthetic data can be more effective than training from scratch alone. Nevertheless, real data remains critical, and combining it with artificial data likely yields the best results.

Finally, we evaluated our best model, $CRNN_{GRU}$ (pretrained on PTB-XL) on the PE-HSM test set against two clinical scores for PE diagnosis, Wells score and PEGeD [30], as shown in Fig. 3. Although both clinical scores exhibit high sensitivity, the $CRNN_{GRU}$ achieves superior area under the curve (AUC), specificity, and positive predictive value (PPV), highlighting the potential of deep learning to enhance PE detection even in resource-limited settings.

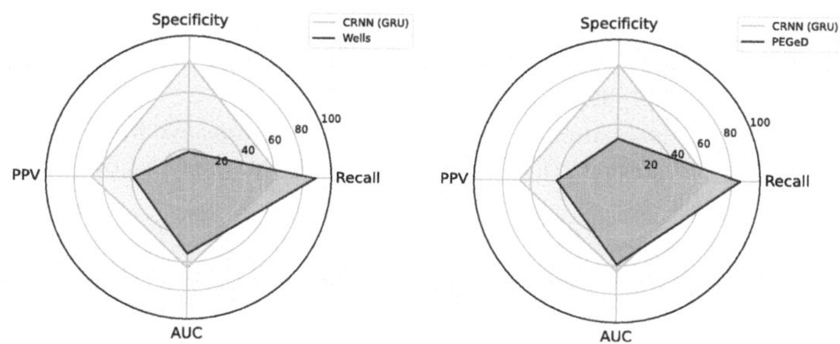

Fig. 3. CRNN_{GRU} diagnostic performance against Wells (**left**) and PEGeD (**right**).

5 Conclusions

In this study, we propose a novel methodology for PE detection using ECG signals, detailing key steps such as input length selection, model-specific normalization, and implementation best practices. Our approach is evaluated across three benchmark datasets, followed by transfer learning to enhance performance on a PE-specific dataset. Our results demonstrate that ECGs alone can provide strong classification baselines and support clinical decision-making, particularly in resource-constrained settings where CTPA is not available.

Acknowledgments. This work was supported by project n° 62 - Center for Responsible AI ref. C628696807-00454142, financed by the Recovery and Resilience Facility - Component 5, included in the NextGenerationEU funding program, by project PRELUNA, grant PTDC/CCIINF/4703/2021, and by national funds through FCT, Fundação para a Ciência e a Tecnologia, under the project UIDB/50021/2020 (DOI: 10.54499/UIDB/50021/2020). We also extend our thanks to Beatriz Valente Silva, Miguel Nobre Menezes, and Fausto J. Pinto for their research contributions and for granting us access to the PE-HSM dataset.

Disclosure of Interests. The authors have no competing interests to declare that are relevant to the content of this article.

References

1. Addison, P.: Wavelet transforms and the ECG: a review. Physiol. Meas. **26**, R155-99 (2005). https://doi.org/10.1088/0967-3334/26/5/R01
2. Akiba, T., Sano, S., Yanase, T., et al.: Optuna: a next-generation hyperparameter optimization framework. Preprint arXiv:1907.10902 (2019)
3. Boulila, W., et al.: Weight initialization techniques for deep learning algorithms in remote sensing: recent trends and future perspectives. In: Advances on Smart and Soft Computing, pp. 477–484. Springer, Singapore (2022), https://doi.org/10.1007/978-981-16-5559-3_39

4. Cahuantzi, R., et al.: A comparison of LSTM and GRU networks for learning symbolic sequences. In: Intelligent Computing, pp. 771–785. Springer, Cham (2023). https://doi.org/10.1007/978-3-031-37963-5_53
5. Daniel, K.R., et al.: Assessment of cardiac stress from massive pulmonary embolism with 12-lead ECG. Chest **120**(2), 474–481 (2001). https://doi.org/10.1378/chest.120.2.474
6. Ding, Z., et al.: Cross-modality cardiac insight transfer: a contrastive learning approach to enrich ECG with CMR features. In: Medical Image Computing and Computer Assisted Intervention MICCAI 2024, pp. 109–119. Springer, Cham (2024). https://doi.org/10.1007/978-3-031-72384-1_11
7. Ebrahimi, Z., Loni, M., Daneshtalab, M., Gharehbaghi, A.: A review on deep learning methods for ECG arrhythmia classification. Expert Syst. Appl. X **7**, 100033 (2020). https://doi.org/10.1016/j.eswax.2020.100033
8. Gillette, K., Gsell, M.A.F., Nagel, C., et al.: Medalcare-xl: 16,900 healthy and pathological synthetic 12-lead ECGs from electrophysiological simulations. Sci. Data **10**, 531 (2023). https://doi.org/10.1038/s41597-023-02416-4
9. Kearon, C., de Wit, K., Parpia, et al.: Diagnosis of pulmonary embolism with d-dimer adjusted to clinical probability. New Engl. J. Med. **381**(22), 2125–2134 (2019). https://doi.org/10.1056/NEJMoa1909159
10. Konstantinides, S.V., et al.: 2019 ESC guidelines for the diagnosis and management of acute pulmonary embolism developed in collaboration with the european respiratory society (ERS). Eur. Heart J. **41**(4), 543–603 (2019)
11. Krizhevsky, A., et al.: ImageNet classification with deep convolutional neural networks. In: Advances in Neural Information Processing Systems, vol. 25. Curran Associates, Inc. (2012). https://doi.org/10.1145/3065386
12. Laher, A.E., Richards, G.: Cardiac arrest due to pulmonary embolism. Indian Heart J. **70**(5), 731–735 (2018). https://doi.org/10.1016/j.ihj.2018.01.014
13. Lawhern, V.J., Solon, A.J., Waytowich, N.R., Gordon, S.M., Hung, C.P., Lance, B.J.: EEGNet: a compact convolutional neural network for EEG-based brain-computer interfaces. J. Neural Eng. **15**(5), 056013 (2018). https://doi.org/10.1088/1741-2552/aace8c
14. Lin, T.Y., Goyal, P., Girshick, R., et al.: Focal Loss for Dense Object Detection (2018). https://doi.org/10.48550/arXiv.1708.02002, preprint arXiv:1708.02002 [cs]
15. Liu, F., Liu, C., Zhao, L., et al.: An open access database for evaluating the algorithms of electrocardiogram rhythm and morphology abnormality detection. J. Med. Imaging Health Inf. **8**(7), 1368–1373 (2018)
16. Mensah, G.A., Roth, G.A., Fuster, V.: The global burden of cardiovascular diseases and risk factors. J. Am. Coll. Cardiol. **74**(20), 2529–2532 (2019). https://doi.org/10.1016/j.jacc.2019.10.009
17. Modi, S., et al.: Wells criteria for DVT is a reliable clinical tool to assess the risk of deep venous thrombosis in trauma patients. World J. Emerg. Surg. **11**, 24 (2016). https://doi.org/10.1186/s13017-016-0078-1
18. Narotamo, H., Dias, M., Santos, R., et al.: Deep learning for ECG classification: a comparative study of 1d and 2d representations and multimodal fusion approaches. Biomed. Signal Process. Control **93**, 106141 (2024). https://doi.org/10.1016/j.bspc.2024.106141
19. Nori, H., King, N., McKinney, S.M., et al.: Capabilities of gpt-4 on medical challenge problems (2023). https://doi.org/10.48550/arXiv.2303.13375. arXiv:2303.13375 [cs.CL]

20. Qiang, Y., Dong, X., Liu, X., Yang, Y., Hu, F., Wang, R.: Ecgmamba: towards ECG classification with state space models. In: 2024 IEEE International Conference on Bioinformatics and Biomedicine (BIBM), pp. 6498–6505 (2024). https://doi.org/10.1109/BIBM62325.2024.10822754
21. Raghu, A., Shanmugam, D., Pomerantsev, E., et al.: Data Augmentation for Electrocardiograms (2022). https://doi.org/10.48550/arXiv.2204.04360. arXiv:2204.04360 [cs]
22. Saab, K., Tu, T., Weng, W.H., et al.: Capabilities of gemini models in medicine (2024). https://doi.org/10.48550/arXiv.2404.18416. arXiv:2404.18416 [cs.AI]
23. Sadhukhan, D., Mitra, M.: R-peak detection algorithm for ECG using double difference and RR interval processing. Procedia Technol. **4**, 873–877 (2012)
24. Safdar, M.F., Nowak, R.M., Pałka, P.: Pre-Processing techniques and artificial intelligence algorithms for electrocardiogram (ECG) signals analysis: a comprehensive review. Comput. Biol. Med. **170**, 107908 (2024). https://doi.org/10.1016/j.compbiomed.2023.107908
25. Sattar, Y., Chhabra, L.: Electrocardiogram, 1st edn. StatPearls Publishing, Treasure Island (2023)
26. Simonyan, K., Zisserman, A.: Very deep convolutional networks for large-scale image recognition (2015). preprint arXiv:1409.1556 [cs.CV]
27. Somani, S.S., et al.: Development of a machine learning model using electrocardiogram signals to improve acute pulmonary embolism screening. Eur. Heart J. Digital Health **3**(1), 56–66 (2021). https://doi.org/10.1093/ehjdh/ztab101
28. Stevens, S.M., et al.: Executive summary: antithrombotic therapy for VTE disease: second update of the chest guideline and expert panel report. CHEST **160**(6), 2247–2259 (2021). https://doi.org/10.1016/j.chest.2021.07.056
29. Tang, X., Berquist, J., Steinberg, B.A., Tasdizen, T.: Hierarchical transformer for electrocardiogram diagnosis (2024). https://arxiv.org/abs/2411.00755
30. Valente Silva, B., Marques, J., et al.: Artificial intelligence-based diagnosis of acute pulmonary embolism: development of a machine learning model using 12-lead electrocardiogram. Portuguese J. Cardiol. **42**(7), 643–651 (2023)
31. Vaswani, A., et al.: Attention is all you need. In: Advances in Neural Information Processing Systems, vol. 30, pp. 6000–6010. Curran Associates, Inc. (2017)
32. Wagner, P., et al.: PTB-XL, a large publicly available electrocardiography dataset. Sci. Data **7**(1), 154 (2020). https://doi.org/10.1038/s41597-020-0495-6
33. Xu, X., Liu, H.: ECG heartbeat classification using convolutional neural networks. IEEE Access **8**, 8614–8619 (2020). https://doi.org/10.1109/ACCESS.2020.2964749

Brain Tumor Segmentation in Sub-Sahara Africa with Advanced Transformer and ConvNet Methods: Fine-Tuning, Data Mixing and Ensembling

Toufiq Musah[1,19(✉)], Chantelle Amoako-Atta[2], John Amankwaah Otu[3], Lukman E. Ismaila[6,19], Swallah Alhaji Suraka[4], Oladimeji Williams[5], Isaac Tigbee[8], Kato Hussein Wabbi[7], Samantha Katsande[9], Kanyiri Ahmed Yakubu[8], Adedayo Kehinde Lawal[10], Anita Nsiah Donkor[8], Naeem Mwinlanaah Adamu[8], Adebowale Akande[11], John Othieno[12], Prince Ebenezer Adjei[1,18], Zhang Dong[13], Confidence Raymond[15], Udunna C. Anazodo[14,15], Abdul Nashirudeen Mumuni[8], Adaobi Chiazor Emegoakor[16], Chidera Opara[14], Maruf Adewole[14], and Richard Asiamah[17]

[1] Department of Computer Engineering, Kwame Nkrumah University of Science and Technology, Kumasi, Ghana
[2] Department of Mathematics, University of Ghana, Accra, Ghana
[3] Department of Physics, Kwame Nkrumah University of Science and Technology, Kumasi, Ghana
[4] Department of Medical Imaging, University of Health and Allied Sciences, Ho, Ghana
[5] Department of Electrical and Electronics Engineering, University of Lagos, Akoka, Nigeria
[6] F.M. Kirby Research Center for Functional Brain Imaging at Kennedy Krieger Institute, Johns Hopkins University, Baltimore, MD, USA
[7] School of Biomedical Sciences, College of Health Sciences, Makerere University Biomedical Research Centre, Makerere, Uganda
[8] Department of Medical Imaging, University for Development Studies, Tamale, Ghana
[9] Action on Preeclampsia Ghana (APEC-GH), Accra, Ghana
[10] Department of Radiology, Lagos State University Teaching Hospital, Ikeja, Nigeria
[11] Department of Electronic and Electrical Engineering, Obafemi Awolowo University, Ile-Ife, Osun State, Nigeria
[12] Department of ICT and Biostatistics, Ernest Cook UltraSound Research and Education Institute, Mengo, Kampala, Uganda
[13] Department of Electrical and Computer Engineering, University of British Columbia, Vancouver, BC, Canada
[14] Medical Artificial Intelligence Lab, Lagos, Nigeria
[15] Department of Biomedical Engineering, McGill University, Montreal, QC, Canada
[16] Nnamdi Azikiwe University Teaching Hospital, Nnewi, Nigeria
[17] Department of Biomedical Engineering, University of Ghana, Accra, Ghana
[18] Global Health and Infectious Disease Group, Kumasi Centre for Collaborative Research in Tropical Medicine, Kumasi, Ghana
[19] Johns Hopkins University, Baltimore, MD, USA

U. Anazodo et al. (Eds.): MIRASOL 2025, LNCS 16398, pp. 12–21, 2026.
https://doi.org/10.1007/978-3-032-13654-1_2

Abstract. Brain tumors are among the deadliest cancers worldwide, with particularly devastating impact in Sub-Saharan Africa (SSA) where limited access to medical imaging infrastructure and expertise delays diagnosis and treatment planning. Accurate brain tumor segmentation is critical for treatment planning, surgical guidance, and monitoring disease progression, yet manual segmentation is time-consuming and subject to inter-observer variability. Recent advances in deep learning, based on Convolutional Neural Networks (CNNs) and Transformers have demonstrated significant potential in automating this task. This study evaluates three state-of-the-art architectures, SwinUNETR-v2, nnUNet, and MedNeXt for automated brain tumor segmentation in multi-parametric Magnetic Resonance Imaging (MRI) scans. We trained our models on the BraTS-Africa 2024 and BraTS2021 datasets, and performed validation on the BraTS-Africa 2024 validation set. We observed that training on a mixed dataset (BraTS-Africa 2024 and BraTS2021) did not yield improved performance on the validation set in all tumor regions compared to training solely on SSA data with well-validated methods. Ensembling predictions from different models also lead to notable performance increases. Our best-performing model, a finetuned MedNeXt, achieved an average lesion-wise Dice score of 0.84, with individual scores of 0.81 (enhancing tumor), 0.81 (tumor core), and 0.91 (whole tumor). While further improvements are expected with extended training and larger datasets, these results demonstrate the feasibility of using deep learning methods for reliable tumor segmentation in resource-limited settings.

Keywords: Brain Tumor Segmentation · Magnetic Resonance Imaging · Deep Learning · Ensembling

1 Introduction

Glioblastomas, commonly referred to as brain tumors, are among the most aggressive and proliferative forms of cancer with a five-year survival rate of only 6.9% as reported by the National Brain Tumor Society [1]. In Sub-Saharan Africa (SSA), where cases were once considered rare, recent findings have challenged this notion, suggesting that the perceived rarity is largely due to underreporting [2,3]. This under-reporting has been linked to the lack of tumor diagnostic tools centralized tumor boards [4,5], a consequence of resource limitations in brain tumor research across the continent [6,7]. In 2022 alone, about 10,000 new glioblastoma cases were recorded in the SSA region, with a mortality rate of 84%, according to the World Health Organization's International Agency for Research on Cancer (IARC) [8]. This high mortality rate reflects the gaps in early detection, access to diagnostic tooling, and effective treatment, highlighting the need for context-specific research and clinical strategies in the region [5,9].

Accurately delineating brain tumor regions on MRI provides vital information on tumor size, location, and extent, directly informing surgical approach and radiation treatment plans [10]. In current practice, tumor regions in neuroimages are outlined manually by expert radiologists. While considered the reference standard, this process is time-consuming, and may suffer from inter- and intra-observer variability based on the annotating radiologist's level of expertise, and other compounding factors [11]. Automated methods, including deep learning, have the potential to address the stated challenges in the segmentation of brain tumors [12]. However, resource constraints in SSA and other LMICs introduce additional hurdles, including suboptimal image quality and resolution [13] due to a lack of onsite imaging expertise and low-quality imaging equipment [9]. A contributing constraint is the lack of sufficient annotated datasets for training robust deep learning algorithms designed to support diagnosis in SSA. For instance, in the Brain Tumor Segmentation challenge, the primary task of automated segmentation of glioma (BraTS-GLI) includes over 1,250 training cases [14], whereas the Africa-specific task (BraTS-Africa) has only 60 training cases [13].

Deep learning-based methods have demonstrated superior performances in biomedical image segmentation, with the U-Net architecture in [15] being the most widely adopted. Recently, transformer based methods have gained traction, especially SwinUNETR [16] which composed of an shifting window encoder with self-attention mechanism to capture both local and global contextual information [17], and a ConvNet decoder. An updated Swin UNETR-v2 built on this, introducing stagewise convolutions within the transformer encoder to improve upon the model's inductive bias. MedNeXt [18], built on the ConvNeXt architecture [19], is a fully convolutional paradigm, adopting design principles from the shifting window transformer.

In this work, we applied state-of-the-art transformer- and ConvNet- based methods to automatically segment brain tumors on MRI scans of SSA patients. Taking into account region-specific challenges, such as suboptimal imaging quality, we evaluate these methods to identify the most suitable approaches for addressing these constraints. We further investigate the impact of training with non-homogeneous datasets from diverse sources, including BraTS 2021 [14], and assess the effectiveness of fine-tuning models pretrained on these external datasets. We employed the STAPLE (Simultaneous Truth and Performance Level Estimation) ensembling technique on selected predictions, demonstrating that the best-performing outputs from multiple models can be effectively combined to achieve improved segmentation results on limited training datasets.

2 Materials and Methods

2.1 Data

The primary dataset utilized in this study is the BraTS-Africa 2024 dataset, which includes multi-parametric MRI scans from 95 individuals consisting of diverse cases relevant to the target population in SSA [13]. 60 samples are designated for training and 35 samples for validation. Data preprocessing by the challenge organizers prior to releasing the dataset included skull stripping, and co-registering the MRI scans to a common anatomical template.

2.2 Model Selection

SwinUNETR-v2. SwinUNETR-v2 (Shifting Window U-Net Transformer) [20] is built upon the original SwinUNETR [16], introducing architectural changes that led to slight benchmark improvements in various biomedical imaging segmentation tasks. The SwinUNETR architecture consists of a transformer-based encoder that performs downsampling and feature extraction at multiple stages, connected to a convolutional decoder that concatenates and upsamples features to produce the final segmentation output. A purely transformer-based encoder may offer the advantage of global context awareness through the attention mechanism, but lacks the inductive bias of convolutions, which has been shown to be beneficial for medical image segmentation tasks, as demonstrated in models like the U-Net [15]. SwinUNETR-v2 introduces stagewise convolutional modules after each transformer downsampling stage in the encoder, thereby reintroducing inductive bias into the feature extraction process. It was trained using the AdamW optimizer [21] with a cosine annealing learning rate scheduler initialized at $1 \times 10^{-}3$, and optimized using a Dice Cross-Entropy loss function.

nnUNet. The nnUNet (no new U-Net) framework provides an out-of-the-box tool for biomedical image segmentation [22], which has been previously demonstrated to perform admirably in neuroimaging segmentation tasks [23]. In this study, we specifically make use of the new Residual Encoder presets [24]. It introduced a stronger residual encoder for better feature extraction in the earlier stages of the U-Net. All preprocessing steps, hyperparameter selection, and training procedures followed the standardized nnUNet pipeline [22].

MedNeXt. MedNeXt is a transformer-inspired fully-ConvNeXt encoder-decoder U-Net, specially made for 3D biomedical image segmentation [18]. Building on the ConvNeXt architecture introduced by Liu et al. [19], it employs architectural elements originally devised for Shifting Window Transformers, including larger convolutional kernel sizes and an inverted bottleneck design, improving the network's ability to capture global representation. We specifically selected the medium variant of MedNeXt with a kernel size of $3 \times 3 \times 3$ and trained it using the standardized nnUNet preprocessing and training pipeline [22].

2.3 STAPLE Ensembling

Simultaneous Truth and Performance Level Estimation (STAPLE) was explored in a limited capacity as an ensembling strategy, where predictions from both nnUNet and MedNeXt were combined. STAPLE works by estimating a probabilistic ground truth segmentation from a collection of candidate segmentations, weighting each one based on its estimated performance [25]. In our case, STAPLE was applied to segmentations from the validation split, using outputs from both models to generate a consensus segmentation that showed improved performance over individual predictions. The algorithm estimates the true label T_v

at each voxel v, by maximizing the posterior probability:

$$P(T_v = 1 \mid D_v)$$

where D_v represents the set of predicted segmentations from MedNeXt and nnUNet at voxel, producing a voxel-wise probability map estimating the likelihood of each voxel being part of a tumor, based on inter-model agreement.

2.4 Experiments

We conducted several experiments utilizing the BraTS-Africa dataset, alongside the publicly available BraTS2021 dataset, to explore different training strategies and evaluate their impact on segmentation performance. Initially, each model was trained from scratch exclusively on the BraTS-Africa dataset.

We also explored a *data mixing* approach, combining the original 60 BraTS-Africa samples with 200 randomly selected samples from the BraTS2021 dataset. This experiment aimed to assess whether incorporating additional BraTS2021 data could improve segmentation performance. Contrary to expectations, initial evaluations suggested a slight performance drop, likely due to differences in data quality and domain shifts between the two datasets. Such quality issues in the BraTS-Africa dataset have been previously noted [9,13].

For the SwinUNETR model, we evaluated the effect of *pretraining*, where the model was initially trained on the BraTS2021 dataset and subsequently fine-tuned on BraTS-Africa data. For nnUNet and MedNeXt which have demonstrated strong out-of-the-box performance without extensive external pretraining [18,22], after initial training on the BraTS-Africa dataset (i.e. nn_{SSA}, M_{SSA}), we conducted an additional round of training by iteratively fine-tuning the models, initializing this step from the model weights obtained in the initial training run (i.e. nn_{SSA-FT}, M_{SSA-FT}).

3 Results

3.1 Segmentation Performance

Our solution is evaluated using the Dice Similarity Coefficient (DSC), Hausdorff Distance at the 95th percentile (HD95), as well as lesion-wise Dice and lesion-wise HD95 [26]. Lesion-wise metrics quantify segmentation performance at the individual lesion level rather than the entire image, reducing bias toward larger lesions[1]. These metrics are computed across the following tumor subregions: tumor core (TC), enhancing tumor (ET), and whole tumor (WT).

Tables 1 and 2 summarize the segmentation performance of the nnUNet and MedNeXt models across different training setups. Results are reported for both lesion-wise and legacy variants of the Dice score and HD95 metric, measured on the tumor sub-regions, along with the corresponding averages (Fig. 1).

[1] https://github.com/rachitsaluja/BraTS-2023-Metrics.

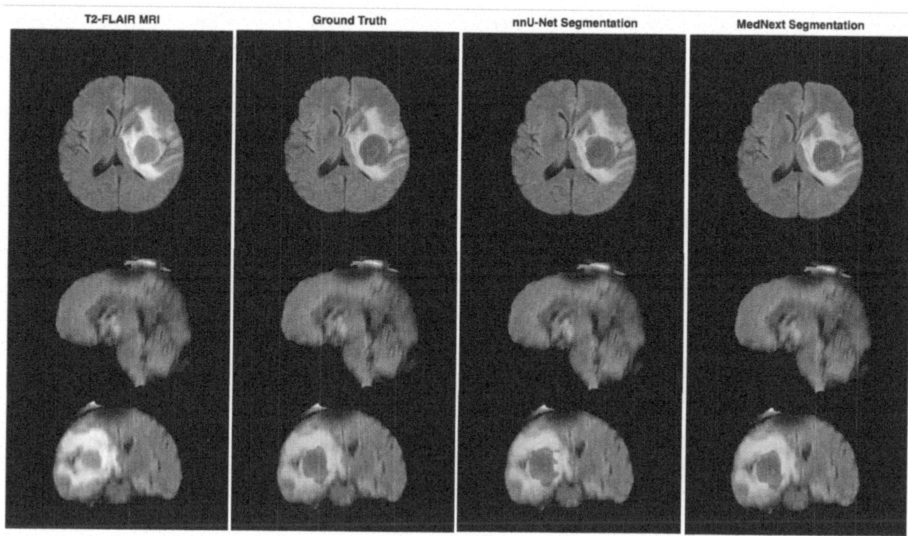

Fig. 1. Predicted tumor sub-regions from the nnUNet and MedNeXt models, shown in case 055. (Tumor Core-Red, Enhancing Tumor-Blue, and Whole Tumor-Green. (Color figure online)

For nnUNet, three configurations were evaluated: training on BraTS-Africa (nn_{SSA}), fine-tuning (further training of nn_{SSA}) on BraTS-Africa (nn_{SSA-FT}), and data mixing with BraTS2021 and BraTS-Africa(nn_{Mix}). MedNeXt was evaluated using the same BraTS-Africa data (M_{SSA}) and its fine-tuned variant (M_{SSA-FT}). SwinUNETR-v2 was evaluated only through the a held out subset of 12 cases from the training dataset. A complete Synapse submission could not be submitted within the project timeline. The resulting dice scores are summarized in Table 3.

Table 1. Dice scores for nnUNet (nn) and MedNeXt (M) models. Lesion-wise and legacy variants are reported per region and as an average. Best shown in **bold**.

Model	Lesion-wise Dice				Legacy Dice			
	ET	TC	WT	Avg	ET	TC	WT	Avg
nn_{SSA}	0.797	0.786	0.846	0.810	0.850	0.853	0.913	0.872
nn_{SSA-FT}	**0.805**	**0.794**	0.853	**0.817**	**0.871**	**0.873**	**0.923**	**0.889**
nn_{Mix}	0.759	0.766	**0.872**	0.799	0.836	0.844	0.904	0.861
M_{SSA}	0.793	0.786	0.893	0.824	**0.878**	**0.876**	0.924	**0.893**
M_{SSA-FT}	**0.811**	**0.806**	**0.907**	**0.841**	0.869	0.869	**0.924**	0.887

Table 2. HD95 scores for nnUNet (nn) and MedNeXt (M) models. Lesion-wise and legacy variants are reported per region and as an average. Best shown in **bold**.

Model	Lesion-wise HD95 (↓)				Legacy HD95 (↓)			
	ET	TC	WT	Avg	ET	TC	WT	Avg
nn_{SSA}	42.717	49.288	39.053	43.686	24.944	26.819	6.743	19.502
nn_{SSA-FT}	**37.213**	**43.913**	35.188	**38.771**	**14.262**	**16.187**	7.422	**12.623**
nn_{Mix}	53.574	54.706	**14.953**	41.078	26.088	27.091	**5.768**	19.649
M_{SSA}	**45.470**	**48.889**	**14.157**	**36.172**	**13.615**	**15.797**	**5.329**	**11.580**
M_{SSA-FT}	46.344	49.539	19.325	38.402	24.012	26.068	14.485	21.521

Table 3. In-training validation Dice scores for SwinUNETR-v2 on 12 Cases.

Dataset	ET	TC	WT	Avg
BraTS-Africa	0.623	0.680	0.752	0.720
BraTS-Mix	0.796	0.848	0.874	0.839
BraTS2021	**0.847**	**0.917**	0.888	**0.884**
BraTS-Africa-FT on BraTS2021	0.813	0.881	**0.923**	0.872

STAPLE Ensemble. We further explored a STAPLE ensemble combining predictions from nn_{SSA}, nn_{SSA-FT}, and M_{SSA}. The resulting lesion-wise dice scores are 0.805, 0.791, 0.895, for ET, TC and WT respectively. The ensemble achieved improved overall performance, with an average Dice score of 0.830 across the tumor sub-regions, outperforming each of the individual models in isolation.

4 Discussion

Fine-tuning with SSA-specific data consistently improved performance across all tested models. Both MedNeXt and nnUNet showed notable gains after fine-tuning (further training on the same BraTS-Africa dataset), with M_{SSA-FT} and nn_{SSA-FT} achieving the highest average Dice scores and the lowest HD95 distances in their respective model categories. Interestingly, M_{SSA} attained the lowest average HD95, despite having a lower Dice score than its further trained counterpart. This discrepancy may be attributed to the fact that most models are explicitly optimized for region-based overlap metrics such as Dice loss, rather than boundary-sensitive metrics like Hausdorff distance. As a result, improvements in Dice score may not always correspond to improved HD95. Performance on WT segmentation remained strong across several configurations, including nn_{Mix}, suggesting that WT may be less sensitive to domain-specific variation than subregions like ET and TC.

The nn_{Mix} model, while competitive on WT segmentation, underperformed on ET and TC compared to its fine-tuned SSA-only counterpart. This result attributed domain shift and the potential negative effects of mixing heterogeneous datasets. Specifically, the presence of varying image quality, annotation styles, and scanner characteristics in SSA data may introduce inconsistencies that disrupt learned feature representations when directly combined with higher-quality, curated datasets like BraTS2021. These findings suggest that datasets like BraTS2021 may be better leveraged for pretraining, allowing models to capture generalizable features before adapting to a specific domain via targeted fine-tuning.

MedNeXt also demonstrated strong baseline performance, with the out-of-the-box model (M_{SSA}) achieving competitive legacy Dice and HD95 scores, particularly for WT. However, fine-tuning (M_{SSA-FT}) further enhanced lesion-wise performance, reaffirming the value of domain adaptation even for architectures designed with cross-domain robustness in mind.

Ensembling via the STAPLE method yielded an average lesion-wise Dice of 0.830, surpassing the performance of each model individually. This highlights the advantage of ensemble strategies in data-constrained environments, where combining model outputs can compensate for individual weaknesses and enhance overall robustness.

Collectively, these findings emphasize the importance of domain-specific fine-tuning, the risks of direct dataset mixing across heterogeneous sources, and the utility of lightweight ensembling methods in low-resource environments. This work contributes actionable insights for enhancing the robustness, generalizability, and practical deployment of segmentation models in SSA and similarly underrepresented medical imaging settings.

5 Limitations

This study presents several limitations that should be acknowledged to contextualize its findings. First, due to resource constraints, the SwinUNETR-v2 model was evaluated solely on an internal subset of 12 training cases, without external validation on the BraTS-Africa validation set. As such, its performance cannot be directly or fairly compared to nnUNet and MedNeXt, which were evaluated on the full validation set. Consequently, any architectural insights regarding SwinUNETR-v2 should be treated as preliminary. Although we report both standard and lesion-wise segmentation metrics, we did not perform statistical significance testing to assess the robustness of the observed performance differences between models. Future work should incorporate statistical tests to better assess the reliability of performance differences.

6 Conclusion

This study emphasizes the importance of developing segmentation solutions using data collected from the environments in which they are intended to be applied. Models trained or fine-tuned on Sub-Saharan African MRI data outperformed those relying on mixed datasets due to domain shifts. While established architectures like nnUNet and MedNeXt demonstrated strong baseline performance, further improvements were observed through fine-tuning and lightweight ensembling. These findings support the view that robust and context-relevant solutions requires direct engagement with local data, rather than relying exclusively on datasets curated in unrelated settings.

Acknowledgement. The authors would like to thank all faculty and instructors of the Sprint AI Training for African Medical Imaging Knowledge Translation (SPARK) Academy (https://www.cameramriafrica.org/spark) 2024 Summer School on Deep Learning in Medical Imaging for providing valuable insights into brain tumor pathology, which informed the research presented here. Special thanks to Linshan Liu for administrative support throughout the SPARK Academy training and capacity-building activities from which the authors greatly benefited. The authors also acknowledge the contributions of Akwasi Asare, Edward Amenyaglo, and Edifon Jimmy. Computational infrastructure support was provided by the Digital Research Alliance of Canada (The Alliance), with additional knowledge translation support from the McGill University Doctoral Internship Program through the student exchange component of the SPARK Academy. The authors are further grateful to McMedHacks for delivering foundational training in Python programming for medical image analysis as part of the 2024 SPARK Academy curriculum. This research was supported by the Lacuna Fund for Health and Equity (PI: Udunna Anazodo, grant number 0508-S-001) and the Natural Sciences and Engineering Research Council of Canada (NSERC) Discovery Launch Supplement (PI: Udunna Anazodo, grant number DGECR-2022-00136).

References

1. National Brain Tumor Society. About glioblastoma (2025). Accessed 27 Apr 2025
2. Kanmounye, U.S., Karekezi, C., Nyalundja, A.D., Awad, A.K., Laeke, T., Balogun, J.A.: Adult brain tumors in Sub-Saharan Africa: a scoping review. Neuro Oncol. **24**(10), 1799–1806 (2022)
3. Odeku, E.L., Osuntokun, B.O., Adeloye, A., Williams, A.O.: Tumors of the brain and its coverings: An African series. Int. Surg. **57**(10), 798–801 (1972)
4. Mbi Feh, M.K.N., Lyon, K.A., Brahmaroutu, A.V., Tadipatri, R., Fonkem, E.: The need for a central brain tumor registry in Africa: a review of central nervous system tumors in Africa from 1960 to 2017. Neuro-Oncol. Pract. **8**(3), 337–344 (2021)
5. Odukoya, L.A., et al.: Establishment of a brain tumor consortium of Africa: advancing collaborative research and advocacy for brain tumors in Africa. Neuro-Oncol. Adv. **6**(1), vdae198 (2024)
6. Ngulde, S.I., et al.: Improving brain tumor research in resource-limited countries: a review of the literature focusing on West Africa. Cureus **7**(11), e372 (2015)
7. Ismaila, L.E., Turki, H., Frikha, M., Weinstein, T., Hunja, F., Fourie, C., Adeshina, S.A.: Afribiobank: empowering Africa's medical imaging research and practice through data sharing and governance. In: Meets Africa Workshop, pp. 189–198. Springer, Heidelberg (2024)
8. International Agency for Research on Cancer. Global Cancer Observatory: Cancer Today. Global Cancer Observatory (2019)
9. Anazodo, U.C., et al.: A framework for advancing sustainable magnetic resonance imaging access in Africa. NMR Biomed. **36**(3), e4846 (2023)

10. Kouli, O., Hassane, A., Badran, D., Kouli, T., Hossain-Ibrahim, K., Steele, J.D.: Automated brain tumor identification using magnetic resonance imaging: a systematic review and meta-analysis. Neuro-Oncol. Adv. **4**(vdac1), 081 (2022)
11. Bø, H.K., Solheim, O., Jakola, A.S., Kvistad, K.A., Reinertsen, I., Berntsen, E.M.: Intra-rater variability in low-grade glioma segmentation. J. Neurooncol. **131**, 393–402 (2017)
12. Dorfner, F.J., Patel, J.B., Kalpathy-Cramer, J., Gerstner, E.R., Bridge, C.P.: A review of deep learning for brain tumor analysis in MRI. NPJ Precis. Oncol. **9**(1), 2 (2025)
13. Adewole, M., et al.: The brain tumor segmentation (brats) challenge 2023: glioma segmentation in Sub-Saharan Africa patient population (brats-africa). ArXiv, pages arXiv–2305 (2023)
14. Baid, U., et al.: The RSNA-ASNR-MICCAI brats 2021 benchmark on brain tumor segmentation and radiogenomic classification. arXiv preprint arXiv:2107.02314 (2021)
15. Ronneberger, O., Fischer, P., Brox, T.: U-net: convolutional networks for biomedical image segmentation. In: Medical Image Computing and Computer-Assisted Intervention–MICCAI 2015: 18th International Conference, Munich, Germany, 5–9 October 2015, Proceedings, Part III 18, pp. 234–241. Springer, Heidelberg (2015)
16. Hatamizadeh, A., Nath, V., Tang, Y., Yang, D., Roth, H.R., Xu, D.: Swin unetr: swin transformers for semantic segmentation of brain tumors in MRI images. In: International MICCAI Brainlesion Workshop, pp. 272–284. Springer, Heidelberg (2021)
17. Liu, Z., et al.: Swin transformer: hierarchical vision transformer using shifted windows. In: Proceedings of the IEEE/CVF International Conference on Computer Vision, pp. 10012–10022 (2021)
18. Roy, S., et al.: Mednext: transformer-driven scaling of convnets for medical image segmentation. In: International Conference on Medical Image Computing and Computer-Assisted Intervention, pp. 405–415. Springer, Heidelberg (2023)
19. Liu, Z., Mao, H., Wu, C.Y., Feichtenhofer, C., Darrell, T., Xie, S.: A convnet for the 2020s. In: Proceedings of the IEEE/CVF Conference on Computer Vision and Pattern Recognition, pp. 11976–11986 (2022)
20. He, Y., Nath, V., Yang, D., Tang, Y., Myronenko, A., Xu, D.: Swinunetr-v2: stronger swin transformers with stagewise convolutions for 3d medical image segmentation. In: International Conference on Medical Image Computing and Computer-Assisted Intervention, pp. 416–426. Springer, Heidelberg (2023)
21. Loshchilov, I., Hutter, F.: Decoupled weight decay regularization. arXiv preprint arXiv:1711.05101 (2017)
22. Isensee, F., Jaeger, P.F., Kohl, S.A.A., Petersen, J., Maier-Hein, K.H.: nnu-net: a self-configuring method for deep learning-based biomedical image segmentation. Nat. Methods **18**(2), 203–211 (2021)
23. Musah, T., Adjei, P.E., Otoo, K.O.: Automated segmentation of ischemic stroke lesions in non-contrast computed tomography images for enhanced treatment and prognosis. In: Meets Africa Workshop, pp. 73–80. Springer, Heidelberg (2024)
24. Isensee, F., et al.: nnu-net revisited: a call for rigorous validation in 3d medical image segmentation. arXiv preprint arXiv:2404.09556 (2024)
25. Warfield, S.K., Zou, K.H., Wells, W.M.: Simultaneous truth and performance level estimation (staple): an algorithm for the validation of image segmentation. IEEE Trans. Med. Imaging **23**(7), 903–921 (2004)
26. Menze, B.H., et al.: The multimodal brain tumor image segmentation benchmark (brats). IEEE Trans. Med. Imaging **34**(10), 1993–2024 (2015)

Multimodal Fusion for Melanoma Classification Using Dermoscopic Images and Clinical Metadata

Aby Diallo[1(✉)], Mouhamad M. Allaya[1], Dame Samb[1], Marawan Elbatel[2], Serigne Lo[3], and Mamadou Bousso[1]

[1] Iba Der Thiam University, Thiès, Senegal
abydiallo456@gmail.com
[2] The Hong Kong University of Science and Technology, Hong Kong, China
[3] Sydney University, Sydney, Australia

Abstract. Melanoma diagnosis, a particularly aggressive form of skin cancer, remains challenging due to the variability of lesions. To improve classification accuracy, we propose a **novel multimodal approach** combining dermoscopic images and clinical metadata (sex, age, anatomical location). Our architecture relies on different **visual backbones** (Vision Transformer, ResNet, EfficientNet) for image feature extraction, **coupled with a multilayer perceptron (MLP)** dedicated to metadata encoding. The two branches are fused to produce a joint prediction. Results show that the **MultimodalViTModel** achieves the best overall performance (F1-score of 0.8897, accuracy of 0.9955), closely followed by the MultimodalResNet. The MultimodalEfficientNet model exhibits greater variability on minority classes. In comparison, a baseline model relying solely on images performs worse, confirming the **added value of integrating clinical metadata**. Finally, the complexity analysis reveals that ResNet represents an **excellent trade-off between efficiency and accuracy**, making it suitable for constrained clinical settings. Our code will be publicly available at: https://github.com/Aby1diallo/Skin_lesion.

Keywords: classification · deep learning · melanoma · Nevus · Seborrheic Keratosis

1 Introduction

Melanoma is a particularly aggressive skin cancer, responsible for the majority of skin cancer-related deaths despite its low incidence [1,7]. Its diagnosis is difficult due to the high morphological variability of lesions, which increases the risk of error, even among experienced clinicians [8]. Artificial intelligence, particularly deep learning, is emerging as a promising solution for automating medical image analysis. Convolutional networks (CNNs), such as ResNet, VGGNet or DenseNet, have shown excellent results for melanoma classification, particularly when pre-trained on large ensembles such as ImageNet [9,10]. Assemblistic

U. Anazodo et al. (Eds.): MIRASOL 2025, LNCS 16398, pp. 22–30, 2026.
https://doi.org/10.1007/978-3-032-13654-1_3

approaches such as Efat et al. [11] have even achieved accuracies in excess of 94%. Nevertheless, these models remain sensitive to image quality and lack robustness in the clinical context.

More recently, Transformers, notably Vision Transformers (ViT), have been explored for skin lesion classification and segmentation [12,13]. Despite promising performance, they still have limitations, such as local feature extraction or high computational complexity [15]. Several studies have attempted to improve these approaches through hybrid variants or oversampling strategies [14,16,17], but remain image-centric only. The integration of multimodal image data and clinical metadata is a promising avenue that has yet to be explored. Vachmanus et al. [18] have shown significant gains by merging these two modalities. However, the majority of recent work [19–22] is still limited to an image-centric approach.

These observations motivate the exploration of new strategies capable of exploiting complementary sources of information. In this context, we propose a truly innovative approach based on multimodal fusion of dermoscopic and clinical data for melanoma classification. In contrast to conventional image-only approaches, our method simultaneously integrates two distinct modalities: images are processed by different visual backbones (ViT or CNN), while metadata are encoded using a dedicated multilayer perceptron. Representation fusion takes place at an intermediate level, enabling the model to extract enriched and complementary diagnostic signals. This strategy compensates for the known limitations of Transformers in terms of local feature extraction [13], while enhancing the contextualization of predictions, as suggested by the work of Yacob et al. [14]. It also keeps computational complexity under control, unlike heavier architectures such as the XBound-Former [15]. By combining performance, interpretability and personalization, our multimodal approach opens up a promising yet little-explored avenue for more reliable, contextually-informed automatic melanoma diagnosis.

2 Methodology

In this work, we introduce a novel and still underexplored approach by developing a multimodal architecture called **MultimodalFusionModel**, designed to jointly integrate visual data (dermoscopic images) and tabular data (clinical metadata) for the task of melanoma classification. Unlike traditional methods that process either modality separately or exclusively, our approach relies on a direct and simultaneous integration. This design highlights the intrinsic contribution of multimodality through a simple yet effective architecture. The proposed architecture is composed of two independent branches. The first is dedicated to visual feature extraction using a pretrained backbone, while the second encodes the clinical metadata through a series of linear layers followed by normalization. The representations from both branches are then concatenated to form a single multimodal representation, which is directly used for final classification via fully connected layers. Three different models were tested using three distinct visual backbones in this study.

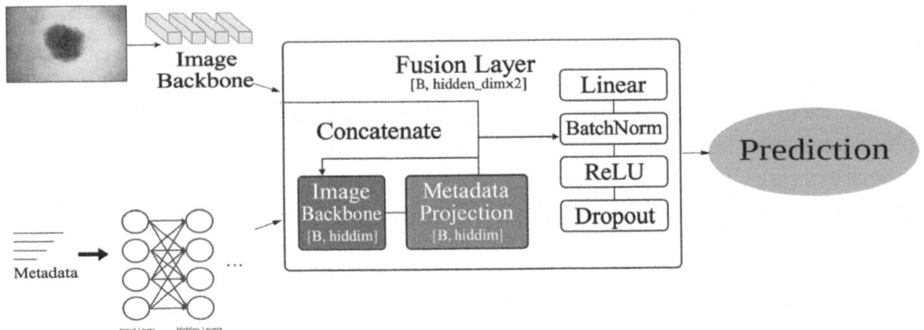

Fig. 1. General architecture of the different multimodal models.

MultimodalViTModel. It uses a Vision Transformer (ViT) model with fine-tuning to extract image features, removing its original classification head to directly obtain visual embeddings. These embeddings are then projected into a latent space of dimension `hidden_dim` through a linear block with batch normalization and `ReLU` activation. In parallel, the metadata is processed by a similar network to obtain a vector representation of the same dimension. The two vectors derived from the images and metadata are concatenated and passed through a fusion layer composed of linear layers, normalization, `dropout`, and a non-linear activation, in order to produce the final output of the model with dimension `output_dim` Fig. 1.

MultimodalEfficientNetModel. It is based on a pretrained *EfficientNetV2* backbone, from which the classification head is removed and all parameters are made trainable. Visual features are extracted using the `forward_features` method, then reduced to a global vector through adaptive pooling, before being projected into a latent space of dimension `hidden_dim` via a linear block followed by batch normalization and a `ReLU` activation. The processing of metadata, modality fusion, and the classification head follow the same structure as described in the ViT-based model Fig. 1.

MultimodalResNetModel. It is based on a pretrained *ResNet18* backbone from which the classification head has been removed to extract high-level visual features. To reduce overfitting, only the final layers (`layer4`) are unfrozen for fine-tuning, while the rest of the network is frozen. The extracted image features are projected into a latent space of dimension `hidden_dim` through a linear block with batch normalization and `ReLU` activation.The processing of metadata, modality fusion, and the classification head follow the same structure as described in the ViT-based model Fig. 1.

Unimodal Model. To quantify the actual contribution of clinical information, we systematically compare each multimodal model to a control version referred to as the **image-only** or **baseline** model using ResNet, in which only the image branch is used for classification. This reference model shares the same optimization parameters, except for the absence of the tabular branch. This comparison highlights the added value of multimodality in melanoma prediction.

The use of models pretrained on large datasets such as ImageNet accelerates convergence and allows leveraging robust visual representations from the early stages of training. This simple yet innovative methodology constitutes a first step toward the rigorous evaluation of fusion strategies in real-world medical contexts.

3 Experimental Results

3.1 Dataset and Experimental Strategy

Dataset. This study is based on the **ISIC 2020** dataset [5], provided by the International Skin Imaging Collaboration. It includes approximately 33,000 annotated dermoscopic images, accompanied by clinical metadata such as age, sex, and anatomical location. To focus on visually similar but clinically distinct diagnoses, only three classes were retained: *melanoma*, *nevus*, and *seborrheic keratosis*. In the training set used, the class distribution is as follows: melanoma – 575 images, nevus – 5,147 images, and seborrheic keratosis – 130 images.

Preprocessing. The data preprocessing pipeline combines dermoscopic images and clinical metadata, which are first cleaned by removing missing values and then numerically encoded using `LabelEncoder`, before being converted to the `float32` format compatible with PyTorch. Images are linked to their corresponding metadata via their identifier and processed through two distinct transformation pipelines: a basic transformation (resizing, tensor conversion, and normalization) applied to majority classes, and an augmented transformation including the same steps as the basic one, enriched with random flips, rotations, color variations (`ColorJitter`), and random cropping, specifically applied to minority classes to mitigate class imbalance. Imagemetadata pairs are encapsulated in a custom `CustomDataset` object, which also returns the label, and the entire dataset is loaded into a `DataLoader` optimized for training.

Implementation Details. Training follows a **K-fold cross-validation** strategy, with the best model saved for each fold. Several metrics are used to evaluate performance under class imbalance: *accuracy*, *precision*, *recall*, *specificity*, *F1-score*, and *AUC-ROC*. These indicators provide a comprehensive and balanced assessment of the model's ability to detect and distinguish skin lesions.

A phase of **hyperparameter optimization** was conducted using `Optuna`, targeting the following parameters:

- `hidden_dim`: size of the latent representation,
- `lr`: learning rate,
- `weight_decay`: regularization factor,

– `batch_size`: batch size.

Training relies on the `Adam` optimizer, a `CrossEntropy` loss function with *label smoothing* (0.1), and the use of **mixed precision** (AMP) to accelerate computation. **Early stopping** is applied to prevent overfitting. Each model is trained for 60 epochs. ROC-AUC curves and classification reports are generated at each fold for thorough evaluation.

To assess the effectiveness of the proposed architectures for skin lesion classification, we conducted a series of experiments involving three multimodal models: ViT, EfficientNet, and ResNet. For each, we compare the performance of models integrating clinical metadata to that of the image-only model. Performance is measured using macro-averaged metrics, which are particularly well-suited for imbalanced datasets as they assign equal weight to each class, regardless of its frequency.

3.2 Results

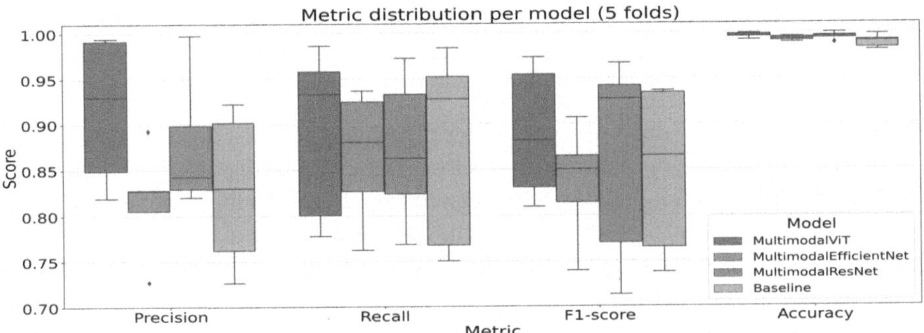

Fig. 2. Comparative distribution of performance metrics across the four evaluated models.

MultimodalViTModel. The results obtained with the MultimodalViTModel show excellent stability and overall performance across the 5 folds Fig. 2. The distribution of metrics across the validation folds highlights the strong and consistent performance of the model. Accuracy is particularly high and stable, consistently exceeding 0.99, indicating excellent predictive capacity across most samples. Macro-averaged metrics, which are more sensitive to class balance, also show strong performance. Precision shows some variability (ranging from 0.82 to 0.99), indicating fluctuations in the model's ability to avoid false positives across folds. In contrast, recall and F1-score are more consistent, often reaching or exceeding 0.95, reflecting the model's good sensitivity to all classes, including minorities.

MultimodalEfficientNetModel. For the MultimodalEfficientNetModel, accuracy across the five validation folds remains very high and relatively stable (between 0.9887 and 0.9948), reflecting strong overall performance across classes Fig. 2. On the other hand, certain metrics, particularly accuracy and F1-score, show greater variability, reflecting a certain instability in the way minority classes are taken into account. This dispersion suggests that the model may struggle to capture the characteristics of rare samples, despite a satisfactory overall performance.

MultimodalResNetModel. The MultimodalResNetModel demonstrates very strong and stable performance across all metrics, especially in accuracy, which consistently exceeds 0.99 across several folds Fig. 2. This indicates the model's consistent ability to classify data accurately. Metrics (precision, recall, F1-score) are also strong, with high averages and less variability than observed with EfficientNet, suggesting that MultimodalResNet is more robust to class imbalance. Notably, F1-score and precision scores are among the highest across all models tested, reflecting a strong balance between precision and recall, even for minority classes.

Baseline. The **Baseline** Fig. 2 model reveals overall good performance of the baseline model, based solely on image data, with an average accuracy of 0.9881, confirming that a well-trained model on dermoscopic images can already offer reliable results. However, metrics, which are more sensitive to the representation of minority classes, show greater variability, with an average Accuracy of 0.8286, an average Recall of 0.8703 and an average F1-score of 0.8474. This increased dispersion, particularly in precision and recall, suggests that the model struggles to maintain balanced performance across all classes, especially underrepresented ones.

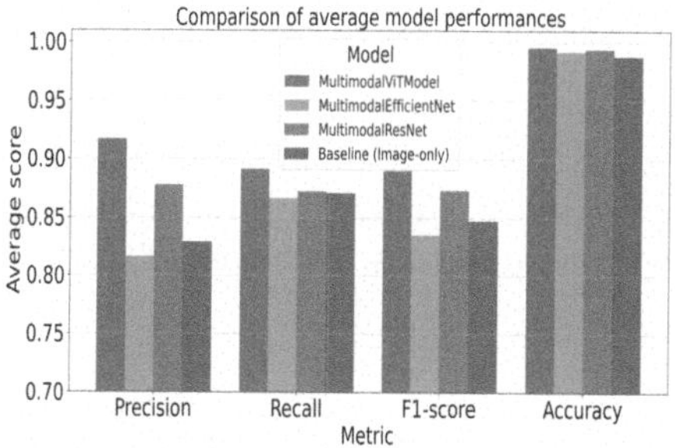

(a) Comparison of average model performance

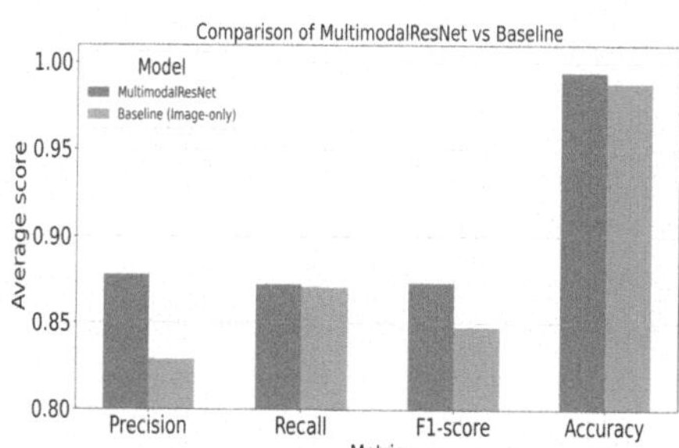

(b) Multimodal vs. Unimodal Learning with ResNet Backbone

Fig. 3. Comparison of classification performance between different multimodal models and unimodal baseline.

Table 1. Average performance comparison (5 folds) of the different models

Model	Precision	Recall	F1-score	Accuracy
ResNet (Image-only)	0.8286	0.8703	0.8474	0.9881
Multimodal-ViT	**0.9166**	**0.8910**	**0.8897**	**0.9955**
Multimodal-EfficientNet	0.8158	0.8660	0.8349	0.9920
Multimodal-ResNet	0.8774	0.8719	0.8729	0.9941

Table 2. Summary comparison of models in terms of complexity and performance

Criterion	ViT	EfficientNet	ResNet
FLOPs Complexity	Very High (17.61 GMac)	Medium (12.02 GMac)	Very Low (1.83 GMac)
Number of Parameters	High (86M)	Very High (117M)	Very Low (9M)
Performance/Cost Trade-off	Low (very expensive)	Good (improved performance)	Excellent (light and accurate)

Comparaison. Among the evaluated models, the *MultimodalViTModel* stands out with the best overall performance, showing a good balance between precision, recall, and F1-score in the context of imbalanced data Fig. 3a. The *Multimodal-ResNet* also achieves very good results, especially in precision and F1-score, although with a slightly lower recall. The *MultimodalEfficientNet*, on the other hand, shows more modest results, suggesting a less effective fusion strategy. Finally, the *Baseline* model, which relies solely on images, performs respectably but remains inferior to all multimodal models, confirming the added value of integrating clinical metadata. Figure 3b **clearly illustrates that multimodal models outperform the image-only model across all evaluation metrics** Table 1

Complexity. Table 2 also compares the complexity and performance of the three models. MultimodalViTModel presents a very high computational cost (17.61 GMac) and a large number of parameters (86M), making it expensive to deploy. MultimodalEfficientNet offers a good trade-off with medium complexity (12.02 GMac) and a high number of parameters (117M). MultimodalResNet is the most lightweight, with low complexity (1.83 GMac) and few parameters (9M), while still maintaining strong accuracy. Thus, MultimodalResNet appears best suited for resource-constrained environments, whereas ViT, while performant, is more resource-intensive.

4 Discussion

The results show that multimodal integration (dermoscopic images + clinical metadata) significantly improves melanoma classification compared with image-only models. The *MultimodalViT* model shows the best overall performance, while *MultimodalResNet* offers a good compromise between efficiency and lightness, suitable for resource-limited environments. However, the evaluation is limited to a single dataset, and it would be necessary to use an independent test set to verify generalizability. Moreover, the addition of more clinically relevant metadata, such as medical history or previous diagnoses, could enhance the performance and reliability of the models by providing additional context for decision-making. Finally, adapting the models to the local context, particularly in Senegal, by collecting specific data and adjusting the models, could enhance their relevance and impact in real-life conditions. This study confirms the potential of multimodal approaches and opens up prospects for improving the robustness, diversity, interpretability and contextualization of artificial intelligence systems in dermatology.

References

1. World Health Organization: Ultraviolet (UV) radiation and skin cancer. WHO QA (2023). https://www.who.int/fr/news-room/questions-and-answers/item/ultraviolet-(uv)-radiation-and-skin-cancer
2. Wu, Y., Chen, B., et al.: Ultraviolet radiation and skin cancer: epidemiology, mechanisms and preventive strategies. Front. Oncol. **12**, 893972 (2022). https://www.frontiersin.org/articles/10.3389/fonc.2022.893972/full
3. Liu, Y., Jain, A., Eng, C., et al.: A deep learning system for differential diagnosis of skin diseases. Sci. Transl. Med. **12**(563), eabb3652 (2020). https://www.science.org/doi/10.1126/scitranslmed.abb3652
4. Jojoa Acosta, M.F., Caballero Tovar, L.Y., Garcia-Zapirain, M.B., Percybrooks, W.S.: Melanoma diagnosis using deep learning techniques on dermatoscopic images. BMC Med. Imaging **20**(1), 65 (2021). https://bmcmedimaging.biomedcentral.com/articles/10.1186/s12880-020-00534-8
5. International Skin Imaging Collaboration (ISIC): ISIC Challenge (2020). https://challenge.isic-archive.com/data/#2020
6. Wu, Y., Chen, B., Zeng, A., Pan, D., Wang, R., Zhao, S.: Skin cancer classification with deep learning: a systematic review. Front. Oncol. **12**, 893972 (2022)
7. Société canadienne du cancer: Mélanome de la peau : diagnostic (2024). Consulté en mai 2025. https://cancer.ca/fr/cancer-information/cancer-types/melanoma-skin/diagnosis
8. Dinnes, J., et al.: Dermoscopy, with and without visual inspection, for the diagnosis of melanoma in adults. Cochrane Datab. Syst. Rev. **2018**(12), CD011902 (2018). https://doi.org/10.1002/14651858.CD011902.pub2. https://www.researchgate.net/publication/326460219_Dermoscopy_with_and_without_visual_inspection_for_the_diagnosis_of_melanoma_in_adults
9. Brinker, T.J., et al.: Deep learning outperformed 11 pathologists in the classification of histopathological melanoma images. J. Med. Internet Res. **20**(10), e11936 (2018)
10. Stofa, M.M., Zulkifley, M.A., Zainuri, M.A.A.M.: Skin lesions classification and segmentation: a review. Int. J. Adv. Comput. Sci. Appl. **12**(10) (2021). https://doi.org/10.14569/ijacsa.2021.0121060. https://thesai.org/Publications/ViewPaper?Code=IJACSA&Issue=10&SerialNo=60&Volume=12
11. Efat, A.H., Hasan, S.M.M., Uddin, M.P., Mamun, M.A.: A multi-level ensemble approach for skin lesion classification using customized transfer learning with triple attention. PLoS ONE **19**(10), e0309430 (2024)

12. Viso.ai: Vision Transformer (ViT): How It Works and Why It Matters in Deep Learning (2024). Consulté le 2026/03/28. https://viso.ai/deep-learning/vision-transformer-vit/
13. Cirrincione, G., et al.: Transformer-based approach to melanoma detection. Sensors **23**(12), 5677 (2023)
14. Yacob, F., et al.: Weakly supervised detection and classification of basal cell carcinoma using graph-transformer on whole slide images. Sci. Rep. **13**, 7555 (2023). https://doi.org/10.1038/s41598-023-33863-z. https://www.nature.com/articles/s41598-023-33863-z
15. Wang, X., Li, Y., Zhang, Z., Chen, Y.: A novel approach for melanoma detection utilizing GAN synthesis and BatchFormer vision transformer model. Comput. Methods Prog. Biomed. (2024). https://www.sciencedirect.com/science/article/abs/pii/S0010482524006577
16. Roy, V.K., Thakur, V., Baliyan, N., Goyal, N., Nijhawan, R.: A framework for seborrheic keratosis skin disease identification using vision transformer. In: Malik, P., Nautiyal, L., Ram, M. (eds.) Machine Learning for Cyber Security, pp. 117–128. De Gruyter (2023). https://doi.org/10.1515/9783110766745-006
17. Yang, G., Luo, S., Greer, P.: A novel vision transformer model for skin cancer classification. Neural Process. Lett. **55**, 9335–9351 (2023). https://doi.org/10.1007/s11063-023-11204-5
18. Vachmanus, S., Noraset, T., Piyanonpong, W., Rattananukrom, T., Tuarob, S.: DeepMetaForge: a deep vision-transformer metadata-fusion network for automatic skin lesion classification. IEEE Access **11** (2023). https://doi.org/10.1109/ACCESS.2023.3345225. https://www.researchgate.net/publication/376674813_DeepMetaForge_A_Deep_Vision-Transformer_Metadata-Fusion_Network_for_Automatic_Skin_Lesion_Classification
19. Khan, S., Khan, A.: SkinViT: a transformer based method for Melanoma and Nonmelanoma classification. PLoS ONE **18**(12), e0295151 (2023)
20. Wang, R., et al.: A novel approach for melanoma detection utilizing GAN synthesis and vision transformer. Comput. Biol. Med. **176**, 108572 (2024). https://doi.org/10.1016/j.compbiomed.2024.108572. https://www.sciencedirect.com/science/article/abs/pii/S0010482524006577
21. Catal Reis, H., Turk, V.: Fusion of transformer attention and CNN features for skin cancer detection. Appl. Soft Comput. **164**, 112013 (2024). https://doi.org/10.1016/j.asoc.2024.112013. https://www.sciencedirect.com/science/article/abs/pii/S1568494624007877
22. Dai, W., Liu, R., Wu, T., Wang, M., Yin, J., Liu, J.: Deeply supervised skin lesions diagnosis with stage and branch attention. IEEE J. Biomed. Health Inf. **28**(2), 719–729 (2024)

An Empirical Study on Liver Volumetry: Deep Learning-Based Estimation of Tumor and Remnant Liver Volumes for Preoperative Planning

Metou Sanghe[1](✉), Mouhamad M. Allaya[1], Dame Samb[1], Marawan Elbatel[2], Serigne Lo[4], Balla Diop[3], and Mamadou Bousso[1]

[1] University Iba Der Thiam, Thies, Senegal
metou.sanghe@univ-thies.sn
[2] The Hong Kong University of Science and Technology, Sai Kung District, China
[3] General Surgery Service, HMO, Dakar, Senegal
[4] University of Sydney, Camperdown, Australia

Abstract. Hepatocellular carcinoma, a common liver cancer with a high mortality rate, requires accurate estimation of tumor and liver volumes to optimize surgical planning. Current manual methods are time-consuming and prone to errors. This study evaluates several convolutional neural network architectures for liver and tumor segmentation, including U-Net, ResU-Net, Fully Convolutional Network (FCN), Attention U-Net, and U-Net combined with ResNet50 and CBAM (Convolutional Block Attention Module). The U-Net model achieved the best performance, with a Dice score of 0.98 for liver segmentation and 0.90 for tumor segmentation. In comparison, FCN reached 0.87 for tumor segmentation, while U-Net+ResNet50+CBAM achieved 0.95 for liver but only 0.77 for tumor segmentation. Regarding volumetric accuracy, U-Net showed a root mean square error (RMSE) of $6.41\,\text{cm}^3$ for tumors and $20.95\,\text{cm}^3$ for the residual liver. These results demonstrate that U-Net is a reliable and efficient solution for automatic segmentation and volumetric analysis. Its robust performance reduces practitioners' workload and enhances clinical decision-making by providing accurate, consistent, and reproducible measurements. Furthermore, the simplicity and flexibility of U-Net make it well-suited for integration into real-world medical imaging workflows, potentially accelerating preoperative planning and improving patient outcomes. Future work will focus on adapting the model to locally annotated data from underrepresented regions to improve robustness and generalization. Code is available at: GitHub Repository

Keywords: Liver volumetry · Segmentation · Deep learning

1 Introduction

Hepatocellular carcinoma (HCC) is the most common primary liver cancer and one of the leading causes of cancer-related deaths worldwide. It is often associated

U. Anazodo et al. (Eds.): MIRASOL 2025, LNCS 16398, pp. 31–40, 2026.
https://doi.org/10.1007/978-3-032-13654-1_4

with chronic liver diseases such as hepatitis B, hepatitis C, or cirrhosis, which are particularly prevalent in regions of Africa and Asia. For many patients, hepatic resection–surgically removing part of the liver–remains one of the main curative treatments. To ensure the safety of this procedure, accurate preoperative and postoperative liver volume estimation is critical to prevent severe complications, such as postoperative liver failure.

Traditionally, these estimations are performed using medical imaging modalities such as computed tomography (CT) or magnetic resonance imaging (MRI), in combination with methods like Fuzzy C-Means, active contours, or region growing. Although commonly used, these approaches are time-consuming, heavily dependent on operator expertise, and sometimes lack reliability, especially in the presence of complex tumors or noisy images [11,13,19]. Moreover, their limited reproducibility hampers integration into standardized clinical protocols.

The emergence of deep learning, particularly convolutional neural networks (CNNs), has opened new avenues for automating medical image segmentation. Architectures such as U-Net [6], ResU-Net, and Fully Convolutional Network (FCN) [9] have achieved significant progress, especially on public datasets like LiTS [10,14]. These models offer promising results in terms of accuracy and speed, but still face challenges, particularly in capturing the 3D complexity of liver structures [15,16] and the anatomical variability across patients.

In this study, we propose a deep learning-based approach to automatically segment the liver and tumors from DICOM images and perform volumetric estimation. Several CNN architectures were explored, including U-Net, ResU-Net [21], FCN, Attention U-Net [23], and a U-Net enhanced with a ResNet-50 encoder combined with the CBAM attention module [24]. These variants incorporate advanced mechanisms to improve the localization of structures of interest while capturing contextual and spatial information.

Our objective is to have an innovative, robust, and clinically relevant tool based on deep learning, designed for automatic liver segmentation and volumetric estimation to support preoperative decision-making. Beyond optimizing clinical workflows, our approach aims to reduce human error and improve the accuracy of residual liver volume predictions, which is a key factor in surgical planning. The novelty of our approach lies in the intention to adapt and validate the method on locally annotated medical data from Senegal, in order to extend its generalizability to underrepresented populations and to address the specific challenges of resource-limited environments. This cross-context adaptability positions our tool as a step toward more inclusive and equitable intelligent medical solutions on a global scale.

2 Methodology

2.1 Overall Framework

Our method is based on an automated pipeline for processing DICOM medical images to segment and estimate liver and tumor volumes. The medical images are first converted and preprocessed, then analyzed by a deep learning model that generates segmentation masks. These masks, combined with DICOM metadata, are subsequently used to automatically compute the volumes. This process standardizes and enhances the reliability of quantitative analysis in clinical settings (Fig. 1).

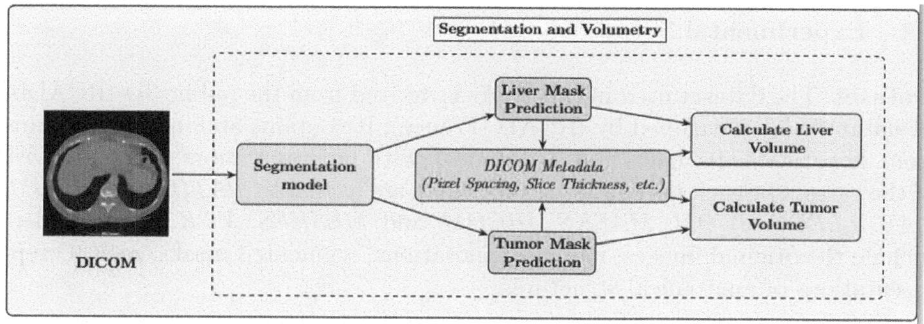

Fig. 1. Overall Framework

2.2 Automatic Segmentation

The first core component of our framework is the automatic segmentation step. Traditional methods either lack precision or are time-consuming when performed manually. To address these limitations, we implemented and compared several deep segmentation architectures, including U-Net, ResUNet, FCN, Attention U-Net, as well as an advanced version combining U-Net with ResNet50 as encoder and the CBAM attention module.

2.3 Hepatic Volume

The second crucial component is the integration of DICOM metadata for volumetric computation. To calculate volumes from the segmented masks, we compared two approaches:

$$Volume = \sum_i voxel_i \times voxel\_volume \quad [2] \tag{1}$$

Where

$$voxel\_volume = voxel\_size_x \times voxel\_size_y \times voxel\_size_z$$

$$Volume = \sum_i A_i \times d \quad [1] \tag{2}$$

$voxel\_size_x$ and $voxel\_size_y$ represent the voxel dimensions along the x and y axes, while $voxel\_size_z$ corresponds to the slice thickness. These values are hyperparameters extracted from the DICOM metadata.

Although formulated differently, both methods lead to the same final result, confirming the validity and robustness of the volumetric computation in our pipeline.

3 Experiments

3.1 Experimental Setup

Dataset. The dataset used in this study is derived from the public 3D-IRCADb-01 database [8], developed by IRCAD (France). It contains abdominal CT scans from 20 patients (10 males and 10 females), with hepatic tumors present in 75% of the cases. For each patient, four subfolders are available: *PATIENT_DICOM*, *LABELLED_DICOM*, *MASKS_DICOM*, and *MESHES_VTK*. These folders include the original images, manual annotations, segmented masks, and 3D representations of anatomical structures.

Preprocessing. The DICOM volumes were converted into 2D JPEG image series to be compatible with deep learning models. This 2D approach was chosen due to limited computational resources and to facilitate potential deployment in clinical environments, where processing full 3D volumes in real time may be challenging. An intensity normalization process was applied to standardize pixel value scales. The data were then split into three sets – training (70%), validation (15%), and test (15%) – using a random shuffle method [7] to avoid any bias related to data ordering.

Baseline. To evaluate the effectiveness of our approach, several segmentation models were compared:

- Standard U-Net,
- Attention U-Net,
- U-Net with ResNet50 encoder + CBAM,
- ResUNet,
- FCN (Fully Convolutional Network),

Implementation Details. Training was performed using the Adam optimizer, with early stopping and learning rate reduction to prevent overfitting. To address the limited dataset size, various data augmentation techniques were applied, including rotations, zooms, flips, and brightness/contrast variations. Hyperparameters (learning rate, batch size, etc.) were automatically optimized using Optuna [3], an efficient Bayesian optimization framework. The best configuration obtained for the U-Net model includes a learning rate of 9.78×10^{-4}, a dropout rate of 0.402, 64 base filters, and a batch size of 8.

Metrics. Model performance was assessed using accuracy, the Dice coefficient, and loss values, with Binary Cross-Entropy (BCE) serving as the primary loss function. The best results obtained on the validation set were confirmed on the independent test set. Volumetric accuracy, which is critical for surgical planning, was also evaluated to ensure reliable estimation of both liver and tumor volumes.

3.2 Quantitative Analysis

The quantitative results obtained by the different models are presented in Table 1 for liver segmentation and tumor segmentation, respectively (Fig. 2).

Table 1. Performance comparison of liver and tumor segmentation models. Best values per task are highlighted in bold.

Model	Liver Segmentation			Tumor Segmentation		
	Dice Score	Accuracy (%)	Loss	Dice Score	Accuracy (%)	Loss
FCN [4]	0.95	99.74	0.007	0.87	99.97	0.001
Attention U-Net [23]	0.97	99.80	0.005	0.80	99.96	0.001
ResNet50 + CBAM [24]	0.95	99.65	0.010	0.77	99.96	0.002
U-Net [6]	**0.98**	99.88	**0.003**	0.90	**99.98**	**0.0006**
ResU-Net [5]	**0.98**	**99.89**	0.004	**0.90**	**99.98**	0.0007

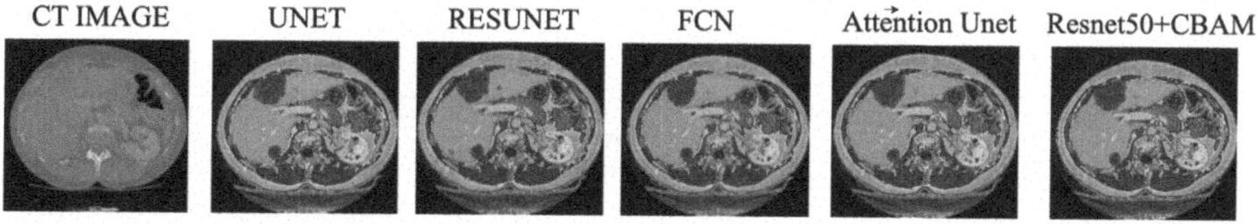

Fig. 2. Predicted liver (green) and tumor (red) masks overlaid on the input image (Color figure online)

Segmentation Accuracy. Table 1 reports Dice, Accuracy, and Loss for liver and tumor segmentation. U-Net and ResU-Net achieve the highest Dice scores for liver (0.98) and tumor (0.90). FCN and Attention U-Net show slightly lower performances.

Volumetric Accuracy. Table 2 presents RMSE and Bland-Altman metrics comparing predicted and ground truth volumes. The U-Net model was used for this analysis. Figures 3 and 4 illustrate volumetric agreement visually.

Table 2. Precision measurements of the volumetry obtained from the U-Net model

Metrics	Tumor (cm^3)	Remaining liver (cm^3)
RMSE	6.41	20.95
Bland-Altman Mean	−3.29	−15.39
Bland-Altman Interval	[−14.09, 7.51]	[−43.26, 12.48]

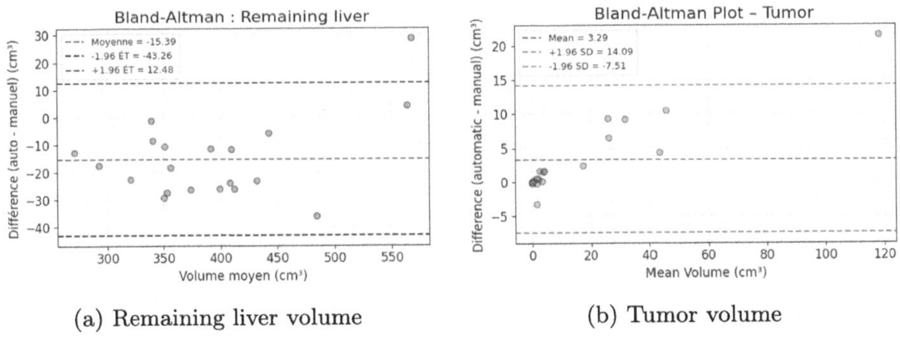

(a) Remaining liver volume

(b) Tumor volume

Fig. 3. Bland-Altman plots comparing manual and automatic volumes

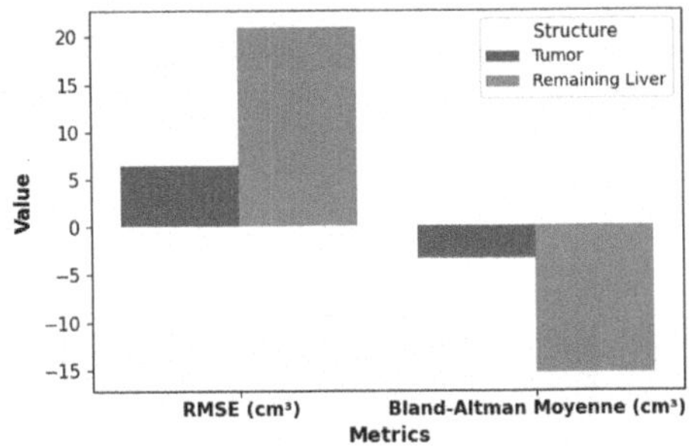

Fig. 4. Comparison of volumetric metrics for tumor and remaining liver structures

The results (Table 2, Fig. 3, 4) show a low mean error (RMSE) of 6.41 cm^3 for the tumor and 20.95 cm^3 for the residual liver, indicating good overall agreement between predicted and actual volumes. Bland-Altman analysis reveals small biases: $-3.29\,\text{cm}^3$ for the tumor and $-15.39\,\text{cm}^3$ for the residual liver, reflecting a slight underestimation of volumes. The confidence intervals remain acceptable in both cases, confirming the reliability of volumetric estimation by the U-Net model.

These results highlight the importance of accurate and precise segmentation to achieve reliable volumetric quantification, particularly in the context of surgical planning or therapeutic monitoring.

4 Discussion

4.1 Interpretation of Results

The U-Net model achieves the best performance in both liver and tumor segmentation as well as volumetric estimation. Its robustness is likely due to its simple yet effective architecture, capable of capturing both global and local anatomical structures efficiently.

Models with added complexity, such as Attention U-Net or U-Net+ResNet50 +CBAM, do not improve results significantly, suggesting that overfitting or reduced generalization may occur with limited data.

Tumor segmentation remains more challenging due to high variability in shape and size. Nonetheless, U-Net maintains stable performance across the dataset. Volumetric analysis indicates slight underestimation for residual liver, likely caused by imaging artifacts or less well-defined boundaries. However, errors remain clinically acceptable, particularly for preoperative planning.

From an algorithmic perspective, U-Net contains approximately 7.78 million parameters and requires 15.56 million FLOPs per inference. Average inference time per image is 110.72 ms, and memory usage is estimated at 29.7 MB per image or 237.5 MB for a batch of eight, supporting near real-time deployment.

These results demonstrate that well-trained conventional architectures can outperform more complex models when data is limited. U-Net thus provides a lightweight, fast, and easily deployable solution suitable for clinical environments.

Additionally, although Zhang et al. [25] employed a 3D U-Net with dual attention for liver segmentation on the LiTS dataset, their reported Dice scores provide a useful reference for qualitative comparison with our 2D U-Net results. This comparison highlights that our simpler 2D approach achieves competitive performance while requiring fewer computational resources.

4.2 Clinical Implications

The outcomes of this study could have a significant impact on clinical practice by providing a faster and potentially more accurate alternative to manual methods for measuring tumor and adjacent organ volumes. Reliable volumetric segmentation would improve decision-making in several medical contexts, notably hepatic surgery, radiotherapy, and transplantation.

In tumor resection, precise estimation of tumor volume and residual liver is crucial to assess surgical feasibility and minimize postoperative complications such as liver failure.

In radiotherapy, where targeting accuracy is paramount, automated segmentation could help refine treatment margins and better preserve surrounding healthy tissues, thereby reducing side effects. Moreover, integrating these models into clinical workflows could optimize radiologists' and surgeons' time, allowing them to focus on higher-value tasks.

However, the observed variability in residual liver segmentation underscores the need for further improvements before full clinical integration. Such refinements could include better modeling of interpatient anatomical variation, fine hyperparameter calibration, and validation on larger, more diverse patient cohorts.

Prospective studies will also be necessary to evaluate the real-world impact of this approach on clinical outcomes and patient management.

4.3 Study Limitations

This study has several limitations that should be acknowledged. First, the relatively small sample size may limit the generalizability of the results to a broader population. Validation on a larger patient cohort is needed to confirm model robustness.

Although the U-Net model demonstrated good performance, segmentation errors remain. These may stem from anatomical variability among patients or the quality of CT images. Finally, our approach relies on preprocessed data and does not yet address certain artifacts or noise present in real clinical images, which could affect accuracy in practice.

4.4 Future Research Directions

Future work will focus on integrating new local medical data, which is currently in the collection phase. After rigorous annotation, this data will allow fine-tuning of the model to better capture regional anatomical and imaging characteristics. Validation on larger, more heterogeneous patient cohorts will ensure clinical reliability. Architectural improvements, including attention mechanisms, 3D networks, stronger encoders, and hybrid convolution–Transformer designs, will also be explored to enhance segmentation accuracy and robustness.

Acknowledgements. This work was partly supported by the Italian Ministry of University and Research (MUR) under project PE0000013 – Future of Artificial Intelligence Research (FAIR)

References

1. Kormano, M., Dean, P.B.: Volume determinations using computed tomography. Am. J. Roentgenol. **138**(2), 329–333 (1982). https://doi.org/10.2214/ajr.138.2.329
2. Wendler, T., Kreissl, M.C., Schemmer, B., Rogasch, J.M.M., De Benetti, F.: Artificial Intelligence-powered automatic volume calculation in medical images – available tools, performance and challenges for nuclear medicine. Nuklearmedizin **62**(6), 343–353 (2023). https://doi.org/10.1055/a-2200-2145
3. Akiba, T., Sano, S., Yanase, T., Ohta, T., Koyama, M.: Optuna: a next-generation hyperparameter optimization framework. In: Proceedings of the 25th ACM SIGKDD International Conference on Knowledge Discovery & Data Mining, pp. 2623–2631. ACM (2019)
4. Singh, A., Kaur, A., Kaur, M.: Fully convolutional network for the semantic segmentation of medical images. Biomed. Sig. Process. Control **71**, 103122 (2022). https://doi.org/10.1016/j.bspc.2021.103122
5. Jha, D., Smedsrud, P.H., Riegler, M.A., et al.: ResUNet++: an advanced architecture for medical image segmentation. In: 2019 IEEE International Symposium on Multimedia (ISM), pp. 225–2255. IEEE (2019). https://doi.org/10.1109/ISM46123.2019.00049

6. Ronneberger, O., Fischer, P., Brox, T.: U-Net: convolutional networks for biomedical image segmentation. In: MICCAI 2015, LNCS, vol. 9351, pp. 234–241. Springer, Cham (2015). https://doi.org/10.1007/978-3-319-24574-4_28
7. GeeksforGeeks: Why Should the Data Be Shuffled for Machine Learning Tasks? (2024). https://datascience.stackexchange.com/questions/24511/why-should-the-data-be-shuffled-for-machine-learning-tasks
8. IRCAD France: 3D-IRCADb-01: 3D Image Reconstruction for Comparison of Algorithm Database (2010). https://www.ircad.fr/research/data-sets/liver-segmentation-3d-ircadb-01/
9. Christ, P., Elshaer, M.E.A., Ettlinger, F., et al.: Automatic liver and lesion segmentation in CT using cascaded FCNs and 3D CRFs. arXiv preprint arXiv:1610.02177 (2016)
10. Venkatesh, P., Bharath, A.B.V., Livingston, J.J.: ResNet50-Boosted UNet for improved liver segmentation accuracy. J. Artif. Intell. Clin. Diag. **1**(1), 006 (2024). https://doi.org/10.36548/jaicn.2024.1.006
11. Sridhar, K., Kavitha, C., Lai, W.-C., Kavin, B.P.: Detection of liver tumour using deep learning based segmentation with coot extreme learning model. Biomedicines (2024)
12. Heimann, T., van Ginneken, B., Styner, M.A., Arzhaeva, Y., et al.: Comparison and evaluation of methods for liver segmentation from CT datasets. IEEE Trans. Med. Imaging **28**(8), 1251–1265 (2009). https://doi.org/10.1109/TMI.2009.2013851
13. Kumar, S.S., Moni, R.S., Rajeesh, J.: Automatic liver and lesion segmentation: a primary step in diagnosis of liver diseases. Signal, Image Video Process. **7**, 163–172 (2011)
14. Bilic, P., Christ, P.F., Vorontsov, E., et al.: The liver tumor segmentation benchmark (LiTS). arXiv preprint arXiv:1901.04056 (2019)
15. Milletari, F., Navab, N., Ahmadi, S.-A.: V-Net: fully convolutional neural networks for volumetric medical image segmentation. In: 3DV 2016, pp. 565–571. IEEE (2016). https://doi.org/10.1109/3DV.2016.79
16. Zhou, Z., Sodha, V., Pang, J., Shen, W., Fishman, E., Yuille, A.: Models genesis: generic autodidactic models for 3D medical image analysis. Med. Image Anal. **67**, 101840 (2021). https://doi.org/10.1016/j.media.2020.101840
17. Frid-Adar, M., Klang, E., Amitai, M., Goldberger, J., Greenspan, H.: Synthetic data augmentation using GAN for improved liver lesion classification. IEEE Trans. Med. Imaging **38**(3), 677–685 (2019). https://doi.org/10.1109/TMI.2018.2867380
18. Bai, W., Oktay, O., Sinclair, M., et al.: Self-supervised learning for cardiac MR image segmentation by anatomical position prediction. In: MICCAI 2019, pp. 541–549. Springer, Cham (2019). https://doi.org/10.1007/978-3-030-32245-8_60
19. Shimizu, A., Narihira, T., Kobatake, H., Nawano, S.: Automated liver segmentation from 3D CT images using probabilistic atlas and multilevel statistical shape model. Medical Imaging 2010: Image Process. **7623**, 76231E (2010)
20. Ronneberger, O., Fischer, P., Brox, T.: U-Net: convolutional networks for biomedical image segmentation. In: International Conference on Medical Image Computing and Computer-Assisted Intervention (MICCAI), pp. 234–241. Springer, Cham (2015)
21. Zhang, Z., Liu, Q., Wang, Y.: Road extraction by deep residual U-Net. IEEE Geosci. Remote Sens. Lett. **15**(5), 749–753 (2018)
22. Long, J., Shelhamer, E., Darrell, T.: Fully convolutional networks for semantic segmentation. In: Proceedings of the IEEE Conference on Computer Vision and Pattern Recognition (CVPR), pp. 3431–3440 (2015)

23. Oktay, O., Schlemper, J., Le Folgoc, L., et al.: Attention U-Net: learning where to look for the pancreas. arXiv preprint arXiv:1804.03999 (2018)
24. Woo, S., Park, J., Lee, J.-Y., Kweon, I.S.: CBAM: convolutional block attention module. In: Proceedings of the European Conference on Computer Vision (ECCV), pp. 3–19 (2018)
25. Zhang, X., Li, Y., Wang, H., et al.: Dual-attention 3D U-Net for liver segmentation. Bioengineering **11**(7), 737 (2024). https://doi.org/10.3390/bioengineering11070737

Towards Trustworthy Breast Tumor Segmentation in Ultrasound Using Monte Carlo Dropout and Deep Ensembles for Epistemic Uncertainty Estimation

Toufiq Musah[1,2(✉)], Chinasa Kalaiwo[3], Maimoona Akram[4], Ubaida Napari Abdulai[1], Maruf Adewole[5], Farouk Dako[7], Adaobi Chiazor Emegoakor[8], Udunna C. Anazodo[5,6], Prince Ebenezer Adjei[1,2], and Confidence Raymond[6]

[1] Department of Computer Engineering, Kwame Nkrumah University of Science and Technology, Kumasi, Ghana
toufiqmusah32@gmail.com
[2] Global Health and Infectious Disease Group, Kumasi Centre for Collaborative Research in Tropical Medicine, Kumasi, Ghana
[3] Department of Radiology, National Hospital Abuja, Abuja, Nigeria
[4] Computer Science Department, FAST National University of Computer and Emerging Sciences, Lahore, Pakistan
[5] Medical Artificial Intelligence Lab, Lagos, Nigeria
[6] Department of Biomedical Engineering, McGill University, Montreal, Canada
[7] Perelman School of Medicine, University of Pennsylvania, 3400 Spruce Street, Philadelphia, PA 19104, USA
[8] Nnamdi Azikiwe University Teaching Hospital, Nnewi, Nigeria

Abstract. Automated segmentation of BUS images is important for precise lesion delineation and tumor characterization, but is challenged by inherent artifacts and dataset inconsistencies. In this work, we evaluate the use of a modified Residual Encoder U-Net for breast ultrasound segmentation, with a focus on uncertainty quantification. We identify and correct for data duplication in the BUSI dataset, and use a deduplicated subset for more reliable estimates of generalization performance. Epistemic uncertainty is quantified using Monte Carlo dropout, deep ensembles, and their combination. Models are benchmarked on both in-distribution and out-of-distribution datasets to demonstrate how they generalize to unseen cross-domain data. Our approach achieves state-of-the-art segmentation accuracy on the Breast-Lesion-USG dataset with in-distribution validation, and provides calibrated uncertainty estimates that effectively signal regions of low model confidence. Performance declines and increased uncertainty observed in out-of-distribution evaluation highlight the persistent challenge of domain shift in medical imaging, and the importance of integrated uncertainty modeling for trustworthy clinical deployment (Code available at: https://github.com/toufiqmusah/nn-uncertainty.git).

Keywords: Breast Ultrasound Segmentation · Uncertainty Estimation · Out-of-Distribution Data · Deep Learning

U. Anazodo et al. (Eds.): MIRASOL 2025, LNCS 16398, pp. 41–51, 2026.
https://doi.org/10.1007/978-3-032-13654-1_5

1 Introduction

Breast tumors are masses resulting from abnormal cellular proliferation within breast tissues, encompassing a broad range of pathologies, the most clinically significant of which is breast cancer. Breast cancer remains highly prevalent, and was reported as the most common cancer among females in 157 out of 185 countries [1], resulting in approximately 670,000 deaths in 2021, with projections indicating a constant increase in cases past 2050, especially impacting low- and middle-income regions of the world [2]. Early detection and accurate diagnosis are fundamental strategies for improving survival outcomes [3,4].

Multiple medical imaging techniques are employed in the detection and diagnosis of breast cancers, including mammography, breast ultrasonography (BUS), and magnetic resonance imaging (MRI). Breast ultrasonography serves as an essential complement to mammography, as it is particularly valuable for early scanning, follow-ups, and treatment monitoring [5,6]. It offers several practical advantages, including real-time imaging without exposure to ionizing radiation, suitability for repeated examinations, and particular effectiveness in imaging dense breast tissue commonly found in younger populations. It plays a central role in clinical workflows, especially in low- and middle-income settings where mammography or MRI may be inaccessible [5].

Accurate segmentation aids in reporting tumor features with BI-RADS [7], and clinical decision-making by improving lesion characterization, radiation therapy planning, response monitoring, and surgical preparation [8–10]. Though ultrasound-based segmentation faces notable challenges due to inherent imaging artifacts, including low contrast, speckle noise, blurred lesion boundaries, and significant operator dependence. The diverse morphological presentation of breast tumors and limitations arising from inadequate and imbalanced datasets further complicate downstream tasks including segmentation [11].

In this body of work, we explore the use of deep learning methods in the segmentation of breast ultrasound images, and further estimate the uncertainty in model predictions using various combinations of epistemic methods. The widely used Breast Ultrasound Images (BUSI) dataset is shown to have unreliable segmentation performance due to data duplication and inconsistent annotations across the same subject images, resulting in data leakage between training and validation sets. We provide code for the deduplication of the dataset as was done in this work. Epistemic uncertainty is estimated via Bayesian inference approximation using Monte Carlo dropout, followed by deep ensembling. We also experiment with a combined Monte Carlo dropoutâĂŞdeep ensembling approach. These methods are evaluated on an out-of-distribution test set to emulate real-world deployment scenarios.

2 Related Works

The paradigm shifted with encoderâĂŞdecoder deep neural network architectures such as U-Net [12] and its improved variants, including UNet++ [13] and

Attention-UNet [14], which significantly advanced segmentation accuracy and robustness in breast ultrasound segmentation [15–17]. Recent innovations include AAU-Net [18], HCTNet [19], and LightBTSeg [20]. Despite these advances, most methods focus on in-distribution data without robust cross-domain validation.

Epistemic uncertainty can be approximated via MC dropout [21] or deep ensembles [22]. Combining these approaches produces well-calibrated uncertainty maps aligned with areas of anatomical ambiguity [23]. In a study by Marisa et al. [24], epistemic uncertainty was quantified in the classification of breast tumor sub-types, exploring approximated Bayesian inference in MC dropout, and deep ensembles. These models produce coherent uncertainty estimates across anatomical regions or entire structures, enabling more meaningful confidence assessment along boundaries and complex areas. Such structured uncertainty frameworks improve interpretability, and can better support clinical decision-making where trustworthiness is necessary.

3 Methods

3.1 Dataset

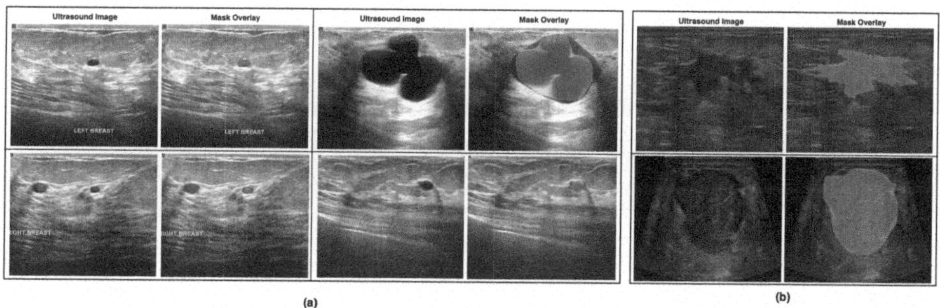

Fig. 1. (a) Sample cases with duplicate masks from the training and validation sets. (*Annotator-1* - Red | *Annotator-2* - Blue | Overlap - Magenta) (b) Sample cases from the out-of-distribution testing dataset. (Color figure online)

This study utilizes two datasets; the **Breast UltraSound Images (BUSI)** (Fig. 1(a)) dataset [25] for model training and validation, and the **Breast-Lesions-USG** dataset [26] (Fig. 1(b)) for out-of-distribution testing and uncertainty quantification. This is to enable both in-distribution performance assessment and evaluation of model generalizability. The original BUSI dataset contained several discrepancies identified by [27], such as duplicated images, and the inadvertent inclusion of non-breast images (maxilla ultrasound scans), which were not explicitly stated in the dataset publication [25]. We further note that the duplicated sets were of varying annotations, which led us to systematically deduplicate the dataset in three distinct ways:

1. ***BUSI-A1:*** Removed the first occurrence of each duplicate pair.
2. ***BUSI-A2:*** Removed the second occurrence of each duplicate pair.
3. ***BUSI-A3:*** Retained the duplicate deemed most accurate by a radiologist.

3.2 Modelling

We employ a modified Residual Encoder U-Net with dropout layers, trained with the nnUNet framework [28]. It follows an identical setup as described in previous work [29] including 8 encoder stages and 7 decoder stages with increasing feature sizes per stage. A typical residual block in the modified setup comprises 6 layers;

$$Conv2D \rightarrow Dropout \rightarrow InstanceNorm \rightarrow LeakyReLU \rightarrow Conv2D \rightarrow InstanceNorm$$

The models were trained with deep supervision and optimized using stochastic gradient descent with a batch dice loss. We used a patch size of 512 × 512 for the input of 2D breast ultrasound images, and a batch size of 13. By default, we train all folds for 250 epochs, and further train *BUSI-A3* for another 750 epochs before applying it on the test dataset for out-of-distribution evaluation.

3.3 Uncertainty Estimation

We quantify uncertainty using three complementary methods: Monte Carlo (MC) dropout, Deep Ensembles, and a combined Deep Ensemble-MC dropout approach.

MC Dropout. Estimates epistemic uncertainty via multiple stochastic forward passes with dropout active at inference [21]. In Eq. 1, for input x, predictions are averaged as:

$$p(y|x) \approx \frac{1}{T}\sum_{t=1}^{T} f_{\theta_t}(x), \tag{1}$$

where $f_{\theta_t}(x)$ is the prediction with dropped weights at iteration t. Uncertainty is the variance across these predictions.

Deep Ensembles. Aggregate predictions from K independently trained models [22]:

$$p(y|x) \approx \frac{1}{K}\sum_{k=1}^{K} f_{\theta^{(k)}}(x). \tag{2}$$

Variability captures uncertainty from data splits and initialization.

Combined Approach. Each ensemble member performs T stochastic passes:

$$p(y|x) \approx \frac{1}{K}\sum_{k=1}^{K}\frac{1}{T}\sum_{t=1}^{T} f_{\theta_t^{(k)}}(x). \tag{3}$$

This jointly captures intra- and inter-model epistemic uncertainty. Aleatoric uncertainty is not modeled.

Uncertainty Evaluation. To quantify epistemic uncertainty at the pixel level, we evaluate on the following metrics:

Predictive Entropy. For pixel (i, j), mean predicted probability over T stochastic forward passes is

$$\bar{p}_{ij} = \frac{1}{T} \sum_{t=1}^{T} \hat{p}_{ij}^{(t)}, \quad \mathcal{H}(\bar{p}_{ij}) = -\bar{p}_{ij} \log \bar{p}_{ij} - (1 - \bar{p}_{ij}) \log(1 - \bar{p}_{ij}). \tag{4}$$

Mutual Information. Epistemic uncertainty is

$$\mathcal{I}(y, \theta | x_{ij}) = \mathcal{H}(\bar{p}_{ij}) - \frac{1}{T} \sum_{t=1}^{T} \mathcal{H}(\hat{p}_{ij}^{(t)}). \tag{5}$$

Expected Calibration Error (ECE). ECE quantifies the alignment between predicted confidence and actual accuracy [30,31]. To compute it, pixels are grouped into M bins according to their confidence score, defined here as $\max(\bar{p}_{ij}, 1-\bar{p}_{ij})$. For each bin B_m, we calculate the average accuracy and the average confidence, and ECE is then given by

$$\text{ECE} = \sum_{m=1}^{M} \frac{|B_m|}{N} \left| \text{acc}(B_m) - \text{conf}(B_m) \right|, \tag{6}$$

where $|B_m|$ is the number of pixels in bin m, N is the total number of pixels, and $\text{acc}(B_m)$, $\text{conf}(B_m)$ are the average accuracy and confidence for that bin. A lower ECE indicates better calibration (i.e., confidence values are well-matched to empirical accuracy). In our experiments, we compute ECE using 30 bins across all 83 million pixels, of which 5.8 million correspond to foreground regions.

4 Results and Discussion

4.1 Segmentation Performance

We conducted 5-fold cross-validation on four variations of the BUSI dataset: *BUSI-Full*, which includes the complete dataset with duplicates, resulting in overlapping cases between training and validation sets, and *BUSI-A1, A2 and A3*, curated subsets containing only unique annotated cases.

Table 1 presents the Dice scores across folds. The *BUSI-Full* dataset achieved a higher average Dice score of 0.7512, compared to 0.7144, 0.7179, and 0.7211 for *A1*, *A2*, and *A3*, respectively. This difference may be attributed to data leakage between the training and validation sets in the original dataset. The performance on *BUSI-A1, A2 and A3* are therefore considered more indicative of true generalization. Lower scores are expected when data leakage and redundancy

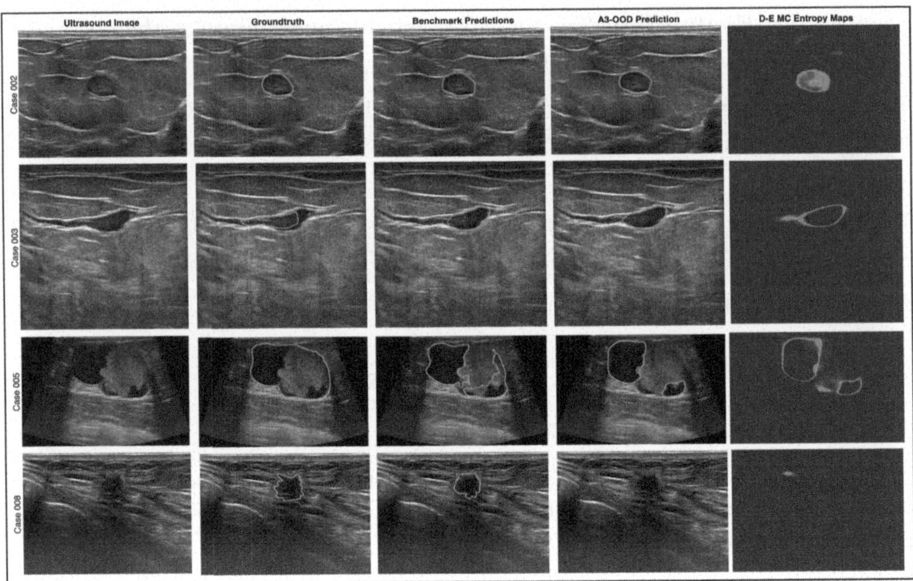

Fig. 2. Qualitative examples of segmentation and uncertainty entropy maps on the Breast-Lesion-USG dataset. Columns show the original ultrasound image, ground truth annotation (red), benchmark (in-domain) and A3-OOD (out-of-distribution) predictions (green), and the corresponding deep ensemble Monte Carlo (D-E MC) entropy map. Higher entropy (yellow) highlights regions of increased model uncertainty. (Color figure online)

are corrected, as evaluation is no longer artificially inflated by overlaps between training and validation sets.

We further benchmarked our method on the Breast-Lesion-USG dataset, training for 250 epochs using 5-fold cross-validation to provide a measure of segmentation performance when the model is trained and validated on the same data distribution. We achieved an average Dice score of 0.7726 ± 0.0212 across folds.

To evaluate the out-of-distribution performance of the model, we selected the *BUSI-A3* subset, which we recommend as the most representative, as it was deduplicated by a trained radiologist. We compare this performance against other methods reported in [17], evaluating on Dice scores, and Intersection over Union (IoU) summarized in Table 2.

Table 1. 5-Fold Cross-Validation Dice Scores for BUSI Datasets

Fold	BUSI-Full	BUSI-A1	BUSI-A2	BUSI-A3
Fold 0	0.7478	0.7048	0.6927	0.7732
Fold 1	0.7147	0.7092	0.7366	0.6657
Fold 2	0.7769	0.6815	0.7324	0.7084
Fold 3	0.7667	0.7512	0.7260	0.7670
Fold 4	0.7509	0.7253	0.7016	0.6911
Average	**0.7514**	**0.7144**	**0.7179**	**0.7211**

Table 2. Comparison of State-of-the-Art methods on Breast-Lesion-USG vs. our methods using in-distribution validation (**Benchmark**) and OOD validation (*A3-OOD*).

Model	Dice	IoU
ResUNet	0.4563	0.3444
UNet++	0.3734	0.2564
Attention-UNet	0.4764	0.3000
SwinUNet	0.4436	0.3331
D-DDPM [17]	0.7104	0.6140
$Ours_{Benchmark}$	**0.7726**	**0.6801**
$Ours_{A3-OOD}$	*0.4855*	*0.4309*

4.2 Uncertainty Estimation Results

Table 3. Summary of Uncertainty and Calibration Metrics Across Methods

Method	Entropy (↓)	MI (↓)	ECE (↓)	Pixel-wise Acc. (↑)
Monte Carlo (MC) Dropout	0.009	0.002	0.0367	0.9595
Deep-Ensemble (D-E)	0.021	0.013	0.0300	0.9607
D-E MC Dropout	0.031	0.019	0.0303	0.9604

We evaluated model uncertainty using three strategies: Monte Carlo (MC) dropout, deep ensembles, and a combined deep ensemble MC dropout approach. All analyses were conducted on the 256 cases of the Breast-Lesions-USG dataset [26], comprising a total of 83,099,921 analyzed pixels, summarized in Table 3.

Monte Carlo Dropout. MC dropout yielded a mean predictive entropy (total uncertainty) of 0.009 (range: [0.000, 0.693]; average standard deviation within cases: 0.046), and a mean epistemic uncertainty (mutual information) of 0.002 (range: [0.000, 0.488]; average within-case standard deviation: 0.010) on 10 stochastic forward passes. The median entropy and mutual information across cases were both near zero, indicating that most pixels were predicted with high confidence.

Deep Ensemble. Using a deep ensemble, the model exhibited a mean predictive entropy of 0.021 (range: $[0.000, 0.693]$; average standard deviation: 0.076) and a mean epistemic uncertainty of 0.013 (range: $[0.000, 0.673]$; average standard deviation: 0.050).

Deep Ensemble Monte Carlo Dropout. For the combined approach (5 ensemble members $\times$ 3 MC dropout samples each; 15 samples per case), the mean predictive entropy increased to 0.031 bits, and mean mutual information to 0.019 bits.

5 Discussion and Conclusion

We recommend *BUSI-A3* as the preferred subset for future studies, as it provides a more reliable basis for evaluating generalization. Benchmarking on the Breast-Lesion-USG dataset using standard in-domain cross-validation showed that our method achieves state-of-the-art Dice and IoU scores, outperforming models including ResUNet, UNet++, SwinUNet, and D-DDPM [17]. However, training on the strictly deduplicated *BUSI-A3* subset and testing out-of-distribution on Breast-Lesion-USG results in a significant drop in segmentation accuracy. Case 008 (Fig. 2, bottom) illustrates such a challenging example. This shows the persistent challenge of domain shift due to the inherent distributional differences between the BUSI and Breast-Lesion-USG datasets, and the need for models that generalize reliably across diverse datasets [11].

All uncertainty quantification methods and their combination, yielded low average predictive entropy and mutual information, indicating high confidence in the prediction of most pixels. D-E and combined D-E MC showed better Expected Calibration Error (ECE) [30,31] and pixel-wise accuracy than MC alone. On out-of-distribution data, higher uncertainty values corresponded with decreased accuracy, with entropy and mutual information effectively highlighting unreliable prediction regions. Clinically, these findings emphasize the importance of robust dataset preparation to avoid optimistic generalization estimates, and highlight uncertainty quantification as an important safeguard in decision support, enabling practitioners to recognize and manage predictions under uncertainty or domain shift.

6 Limitations and Future Work

While entropy maps may provide intuitive qualitative uncertainty visualization, quantitative calibration via ECE depends on binning schemes that can be sensitive and may not generalize across different data distributions. Pixel-wise accuracy tends to be inflated because of the imbalance between foreground and background pixels. Future work should employ segmentation-aware calibration methods [32] to obtain more realistic estimates. Further, epistemic uncertainty quantification can be computationally expensive. MC dropout is T times more computationally demanding than single forward pass (Eq. 1, deep ensembles are K times more expensive, and DE-MC $K \times T$ times more, posing challenges for real-time clinical applications.

In datasets like *BUSI-Full* with multiple annotator segmentations, human uncertainty calibration could align model confidence with clinical opinion variability rather than relying on majority votes. Collecting such datasets with intentional design is valuable. Although our methods advance ultrasound breast lesion segmentation, they expose limitations in cross-domain deployment. Future work should focus on adaptive models to bridge generalization gaps and refine uncertainty estimation for safer and more transparent clinical integration.

Acknowledgments. This work was supported by the Lacuna Fund on Sexual, Reproductive and Maternal Health and Rights (SRMHR) for the African Breast imaging dataset for equitable cancer care (ABreast data) Project and completed as part of the 2024 Precision Cancer Care in Africa (PRECISE) Symposium Hackathon in collaboration with the University of Pennsylvania Center for Global and Population Health Research in Radiology, the Medical Artificial Intelligence Laboratory (MAI Lab), the National Institute for Cancer Research and Treatment (NICRAT) Nigeria, Ernest Cooke Ultrasound Research and Education Institute Uganda, Consortium for Advancement of MRI Education and Research in Africa (CAMERA).

References

1. World Health Organization. Breast cancer (2022). https://www.who.int/newsroom/fact-sheets/detail/breast-cancer. Accessed 17 May 2025
2. Deng, T., et al.: Global, regional, and national burden of breast cancer, 1990–2021, and projections to 2050: a systematic analysis of the global burden of disease study 2021. Thoracic Cancer **16**(9), e70052 (2025)
3. Ginsburg, O., et al.: Breast cancer early detection: a phased approach to implementation. Cancer **126**, 2379–2393 (2020)
4. Malakouti, S.M., Menhaj, M.B., Suratgar, A.A.: ML: early breast cancer diagnosis. Curr. Probl. Cancer: Case Rep. **13**, 100278 (2024)
5. Iacob, R., et al.: Evaluating the role of breast ultrasound in early detection of breast cancer in low-and middle-income countries: a comprehensive narrative review. Bioengineering **11**(3), 262 (2024)
6. Pan, H.-B.: The role of breast ultrasound in early cancer detection. J. Med. Ultras. **24**(4), 138–141 (2016)
7. Spak, D.A., Plaxco, J.S., Santiago, L., Dryden, M.J., Dogan, B.E.: Bi-rads® fifth edition: a summary of changes. Diagnost. Intervent. Imag. **98**(3), 179–190 (2017)
8. Porrath, S., Avallone, L.T.: Radiation therapy planning using ultrasound. In: Ultrasound in Medicine: Volume 2 Proceedings of the 20th Annual Meeting of the American Institute of Ultrasound in Medicine, pp. 165–172. Springer (2012)
9. Labora, A.N., Kapoor, N.S.: The evolution of breast ultrasound in surgical practice: current applications, missed opportunities, and future directions. Surg. Oncol. Insight **1**(3), 100084 (2024)
10. Bennett, I.C., Biggar, M.A.: The role of ultrasound in the management of breast disease. Aust. J. Ultras. Med. **14**(2), 25–28 (2011)

11. Huang, Q., Luo, Y., Zhang, Q.: Breast ultrasound image segmentation: a survey. Int. J. Comput. Assist. Radiol. Surg. **12**, 493–507 (2017)
12. Ronneberger, O., Fischer, P., Brox, T.: U-net: convolutional networks for biomedical image segmentation (2015)
13. Zhou, Z., Siddiquee, M.M.R., Tajbakhsh, N., Liang, J.: UNet++: a nested U-net architecture for medical image segmentation (2018)
14. Oktay, O., et al.: Attention U-net: learning where to look for the pancreas (2018)
15. Sharma, A., Mishra, P.K.: Inception UNet architecture for breast tumor segmentation and detection using hybrid deep learning approach. Multimed. Tools Appl. 1–39 (2024)
16. Anari, S., Sadeghi, S., Sheikhi, G., Ranjbarzadeh, R., Bendechache, M.: Explainable attention based breast tumor segmentation using a combination of UNet, ResNet, DenseNet, and EfficientNet models. Sci. Rep. **15**(1), 1027 (2025)
17. Alblwi, A., Makkawy, S., Barner, K.E.: D-DDPM: deep denoising diffusion probabilistic models for lesion segmentation and data generation in ultrasound imaging. IEEE Access (2025)
18. Chen, G., Li, L., Dai, Y., Zhang, J., Yap, M.H.: AAU-net: an adaptive attention U-net for breast lesions segmentation in ultrasound images. IEEE Trans. Med. Imaging **42**(5), 1289–1300 (2022)
19. He, Q., Yang, Q., Xie, M.: HCTNet: a hybrid CNN-transformer network for breast ultrasound image segmentation. Comput. Biol. Med. **155**, 106629 (2023)
20. H Guo, et al.: LightBTSeg: a lightweight breast tumor segmentation model using ultrasound images via dual-path joint knowledge distillation. In: 2023 China Automation Congress (CAC), pp. 3841–3847 (2023)
21. Gal, Y., Ghahramani, Z.; Dropout as a Bayesian approximation: representing model uncertainty in deep learning. In: International Conference on Machine Learning, pp. 1050–1059. PMLR (2016)
22. Lakshminarayanan, B., Pritzel, A., Blundell, C.: Simple and scalable predictive uncertainty estimation using deep ensembles. In: Advances in Neural Information Processing Systems, vol. 30 (2017)
23. Kurz, A., et al.: Uncertainty estimation in medical image classification: systematic review. JMIR Med. Inform. **10**(8), e36427 (2022)
24. Wodrich, M., Karlsson, J., Lång, K., Arvidsson, I.: Trustworthiness for deep learning based breast cancer detection using point-of-care ultrasound imaging in low-resource settings. In: Meets Africa Workshop, pp. 42–51. Springer (2024)
25. Al-Dhabyani, W., Gomaa, M., Khaled, H., Fahmy, A.: Dataset of breast ultrasound images. Data Brief **28**, 104863 (2020)
26. Pawłowska, A., et al.: Curated benchmark dataset for ultrasound based breast lesion analysis. Sci. Data **11**(1), 148 (2024)
27. Pawłowska, A., Karwat, P., Żołek, N.: Re: [dataset of breast ultrasound images by W. Al-dhabyani, M. Gomaa, H. Khaled & A. Fahmy, data in brief, 28, 104863]. Data Brief **48**(109247), 2023 (2020)
28. Isensee, F., et al.: nnU-net revisited: a call for rigorous validation in 3d medical image segmentation. In: International Conference on Medical Image Computing and Computer-Assisted Intervention, pp. 488–498. Springer (2024)
29. Musah, T., Adjei, P.E., Otoo, K.O.: Automated segmentation of ischemic stroke lesions in non-contrast computed tomography images for enhanced treatment and prognosis. In: Meets Africa Workshop, pp. 73–80. Springer (2024)
30. Guo, C., Pleiss, G., Sun, Y., Weinberger, K.Q.: On calibration of modern neural networks. In: International Conference on Machine Learning, pp. 1321–1330. PMLR (2017)

31. Kull, M., Nieto, M.P., Kängsepp, M., Filho, T.S., Song, H., Flach, P.: Beyond temperature scaling: obtaining well-calibrated multi-class probabilities with Dirichlet calibration. In; Advances in Neural Information Processing Systems, vol. 32 (2019)
32. Zeevi, T., Lieffrig, E.V., Staib, L.H., Onofrey, J.A.: Spatially-aware evaluation of segmentation uncertainty. arXiv preprint arXiv:2506.16589 (2025)

Lightweight 3D U-Net for Brain Tumor Segmentation on CPUs: Enabling Deep Learning in Low-Resource Environments

Ayomide B. Oladele[1], Raymond Confidence[1,2,3], Dong Zhang[1,4], Charity Umoren[1], Aondana M. Iorumbur[5], Ifeoluwa Oladeji[1], Tolulope Olusuyi[1], Gamaliel Adebayo[1], Anu Gbadamosi[1], Oluyemisi Toyobo[1], Farouk Dako[1,6], Abiodun Fatade[1], Maruf Adewole[1,6], and Udunna Anazodo[1,2,3](✉)

[1] Medical Artificial Intelligence Laboratory, Lagos, Nigeria
udunna.anazodo@mcgill.ca
[2] Montreal Neurological Institute, McGill University, Montreal, Canada
[3] Department of Biomedical Engineering, McGill University, Montreal, Canada
[4] Department of Electrical and Computer Engineering, University of British Columbia, Vancouver, Canada
[5] Department of Physics, Federal University of Technology, Minna, Nigeria
[6] Department of Radiology, University of Pennsylvania, Philadelphia, USA

Abstract. The application of deep learning (DL) methods in diagnostic imaging has gained significant popularity and relevance in research and clinical settings, leading to discoveries that are improving human health. However, the implementation of conventional DL models for diagnostic tasks often requires high computational resources, limiting their use in resource-constrained settings (RCS). This computational burden also limits skills training in DL methods in RCS. To bridge this gap, we implemented a lightweight 3D U-Net architecture that can be trained and deployed entirely on CPU-only systems, lowering hardware barriers. Our aim was to 1) assess the feasibility of a CPU-based DL model in performing common diagnostic tasks, and 2) use it as a teaching tool for medical image analysis skills training in RCS. Multiparametric brain magnetic resonance imaging (MRI) scans (146) from the Brain Tumor Segmentation Africa (BraTS-Africa) dataset was used to implement the model and evaluate its performance. The lightweight model achieved remarkable success in the segmentation of the background as well as the three tumor subregions-enhancing tumor (ET), non-enhancing tumor core (NETC), and surrounding non-enhancing FLAIR hyperintensity (FLAIR). The pipeline was externally validated by 53 participants of a workshop from low-and middle-income countries who have a wide range of AI skillsets and CPU configurations. By equipping local researchers and clinicians with accessible, reproducible tools, our approach revealed the potential of CPU-based deep learning models for democratizing AI-powered medical imaging solutions and supports equitable innovation and sustainable research in RCS.

Keywords: Brain Tumor Segmentation · AI Democratization · BraTS-Africa · CPU-only deployment · Resource-Constrained Settings · SPARK Academy

U. Anazodo et al. (Eds.): MIRASOL 2025, LNCS 16398, pp. 52–61, 2026.
https://doi.org/10.1007/978-3-032-13654-1_6

1 Introduction

Every year, brain tumors affect over 300,000 people worldwide with survival outcomes highly dependent on early and accurate diagnosis. In many low-resource settings, however, diagnosis is often delayed or inaccessible due to a shortage of skilled radiologists and limited access to advanced imaging technologies. Notably, cancer mortality in sub-Saharan Africa (SSA) has increased by 45% since 2000, in contrast to declining rates in high-income countries [1]. Brain tumor segmentation plays a critical role in the diagnostic pathway, enabling precise identification and delineation of tumors on magnetic resonance imaging (MRI) scans [2]. Accurate segmentation directly informs treatment planning, surgical intervention, and patient monitoring. Yet, manual segmentation remains labor-intensive, subjective, and reliant on specialized expertise—resources that are scarce in many regions most affected by brain tumors.

Several deep learning methods have been applied for automated brain tumor segmentation. In particular, 3D convolutional neural networks, such as U-Net [3] and its 3D variants (e.g., [4]), have achieved state-of-the-art results in benchmark challenges like the Brain Tumor Segmentation (BraTS) competitions [5]. However, these models typically require substantial computational power, including GPUs with 8–16 GB memory, multi-core processors, and large RAM—resources rarely available in low-resource environments, where standard CPU-based systems are common.

This study addresses this gap by demonstrating the feasibility of training and deploying a 3D U-Net model for brain tumor segmentation entirely on CPUs. Using the Brain Tumor Segmentation Africa (BraTS-Africa) 2024 dataset, a lightweight deep learning pipeline tailored for low-resource environments was developed. All processes—training, inference, and evaluation—were conducted without GPU support in short duration. To support reproducibility and encourage adoption, the complete protocol was published on protocols.io, offering step-by-step guidance for researchers and students, particularly in underrepresented regions. This approach was also validated through deployment in a training workshop, where participants from resource-constrained settings successfully implemented the lightweight method on CPU-only machines.

2 Related Works

Traditional brain tumor segmentation methods use classical image processing techniques, but they are not very robust and require manual adjustments. Early attempts, like Prastawa et al. [6] outlier detection method, showed that rule-based approaches struggle with the complexity and variability of brain tumors. The introduction of U-Net by Ronneberger et al. [3] in biomedical image segmentation marked a paradigm shift, combining encoder-decoder architectures with skip connections to preserve contextual and spatial resolution. This architecture has since been adapted to 3D volumes for tasks involving volumetric brain imaging. Variants like 3D U-Net and frameworks such as nnU-Net, and Swin UNETR, have demonstrated state-of-the-art (SOTA) performance in global challenges such as BraTS, leveraging GPU acceleration to handle large-scale data with improved segmentation accuracy (i.e. Dice Score) and boundary precision (HD95) [7, 8].

The recognition of computational limitations has led to recent studies on lightweight segmentation architectures that are suited for resource-constrained deployment [9]. Recent surveys of lightweight segmentation methods revealed that the use of model optimization techniques such as model pruning, weight quantization, and knowledge distillation can be useful for efficient deployment in constrained environments [10]. Recent efforts include Mizanu et al. [11], who proposed a VGG-infused U-Net for low-resource settings; Sharma et al. [12], who created a lightweight U-Net (~35 MB) that runs well on modest hardware. Beyond computational limitations, another major challenge in low- and middle-income countries (LMICs) is the quality and availability of annotated medical imaging datasets. Studies by Kanmounye [13] and Zhang [14] emphasize how low-quality MRIs and the scarcity of structured annotations in LMICs hinder the development of robust, generalizable brain tumor segmentation models.

3 Methods

3.1 Dataset

The BraTS-Africa dataset, a publicly available repository specifically curated to represent brain tumor cases from SSA was used in this study [15, 16]. This dataset comprises 146 multi-contrast magnetic resonance imaging (MRI) volumes, of which 95 are glioma cases and 51 are other types of brain tumor. All images are stored in Neuroimaging Informatics Technology Initiative (NIfTI) format (. Nii.gz) and include four distinct MRI modalities alongside an expert-annotated segmentation mask for each case. The four modalities are: T1-weighted native (T1n), T1-weighted contrast-enhanced (T1c), T2-weighted Fluid Attenuated Inversion Recovery (FLAIR), and T2-weighted (T2w). The accompanying segmentation mask (seg.nii.gz) adheres to standard BraTS conventions by labeling three tumor sub-regions as: the non-enhancing tumor core (NETC), enhancing tumor (ET), and surrounding non-enhancing FLAIR hyperintensity (SNFH) (Fig. 1).

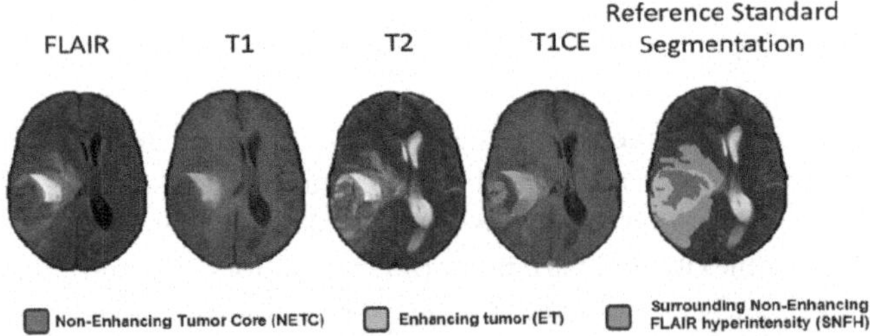

Fig. 1. Illustration of the BraTS-Africa Dataset showing brain slices of the four MRI image contrasts and Segmentation masks of the three tumor sub-regions in one representative patient. Adapted and modified from Adewole M. et al. [16].

Our study focused on glioma cases because the model was intended to tackle segmentation of glioma-specific tumor sub regions, hence the 95 glioma cases from the dataset were used, given that gliomas are the most common primary brain tumors and partitioned into three subsets for model development and validation. Specifically, 66 cases were used for the training set, 13 cases for the validation set, and the remaining 9 cases for the test set.

3.2 Dataset Preparation and Processing

The dataset preparation and preprocessing were carefully designed to enable execution of this study on CPU-only systems, prioritizing minimal memory consumption. The NiBabel library (version 3.2.1) was utilized to load each volumetric (.nii.gz) file into memory. Initially, each 3D MRI volume (corresponding to T1n, T1c, FLAIR, and T2w) was reshaped into two-dimensional arrays of size (H × W, D) to allow a slice-wise intensity normalization, scaling voxel intensities within each axial slice to the [0, 1] range. Then, each volume was uniformly cropped to a cubic size of 128 × 128 × 128 voxels. The cropping process was centered on the brain to maximize inclusion of relevant anatomical structures and tumor regions, thereby reducing redundant background information. Subsequently, the four preprocessed MRI modalities for each case were concatenated along a dedicated channel axis, resulting in a four-dimensional NumPy array of shape (128, 128, 128, 4).

3.3 3D U-Net Model

A lightweight 3D U-Net architecture (Fig. 2) was adopted to address the challenges of volumetric brain tumor segmentation while maintaining computational feasibility within an 8 GB RAM constraint. Building upon established 3D U-Net implementations the network architecture features an encoder with progressively increasing filter counts, from 16 to 128 across four levels, incorporating 3D convolutions, dropout, and max-pooling. A bottleneck layer with 256 filters captures high-level features. The decoder reverses this process, using 3D transposed convolutions for upsampling and concatenating with encoder features, while reducing filter counts back to 16. A final (1 × 1 × 1) convolution with softmax activation generates the voxel-wise segmentation probabilities.

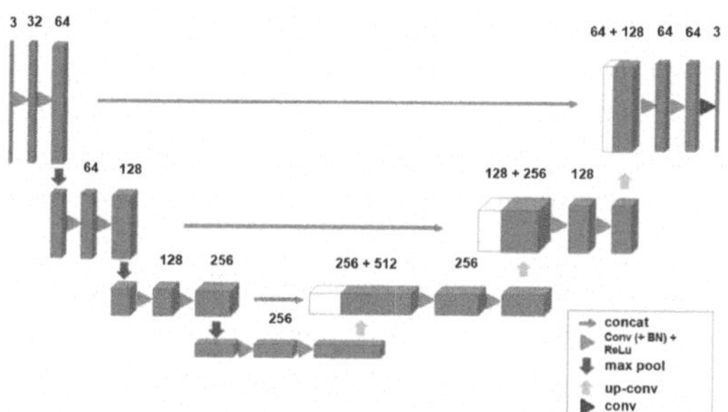

Fig. 2. U-shaped architecture of the original 3D U-Net model for volumetric medical image segmentation with the default configuration [19].

Our study's approach incorporates several key modifications aimed at reducing memory usage and computational overhead; The filter depth was reduced throughout the network, starting from 16 filters in the first layer, which significantly decreased both memory usage and computational complexity. Additionally, dropout rates were carefully tuned within the range of 0.1 to 0.3 to ensure effective regularization while maintaining computational efficiency. Instead of processing the entire (128^3) volume, we adopted a patch-based strategy by cropping to (94^3) sub-volumes, thereby dramatically reducing data size and enabling processing on standard CPU-based systems. Consistent use of ($3 \times 3 \times 3$) kernel sizes provided a balanced trade-off between local feature extraction and computational demand. The output is a segmentation mask with the same spatial shape, ($96 \times 96 \times 96 \times 4$), delineating four classes: background, NETC, ET, SNFH.

3.4 3D U-Net Model Training

This study trained the model on a CPU-only workstation to ensure accessibility in low-resource settings. The setup included an AMD Ryzen 5 5500U (6 cores, 12 threads), 32 GB RAM, Windows 11 Pro, Python 3.8, and PyTorch 1.13 (CPU). Without a GPU, careful tuning of hyperparameters ensured feasible training on this limited hardware.

To enhance segmentation performance, the training employed a combination of Dice Loss and Categorical Focal Loss. Dice Loss handled class imbalance, while Focal Loss focused on difficult tumor voxels. Model performance was measured using Dice Similarity Coefficient, 95th-percentile Hausdorff Distance (HD95), and mean Intersection over Union (IoU). Training used a batch size of one patch, an initial learning rate of 1×10^{-4}, and the Nadam optimizer. A ReduceLROnPlateau scheduler lowered the learning rate as needed, and early stopping (patience of 15 epochs) avoided overfitting. Training ran for 100 epochs, but stopped at epoch 44 after ~ 3 h due to early stopping criteria.

3.5 Real-World Validation: SPARK Academy 2025 Workshop

For reproducibility, the full pipeline including dataset access, preprocessing, augmentations, architecture, and evaluation scripts, was openly shared on protocols.io[1] to support further research, replication, and deployment in low-resource settings. The generalizability and practical deployment of the pipeline was further evaluated during workshop at the Sprint AI Training for African Medical Imaging Knowledge Translation (SPARK) Academy 2025 program[2]. The workshop engaged 53 participants from low-resource settings in Africa, Southeast Asia and Latin America, including students, clinicians, and early-career researchers. Participants were given 1 h overview of the pipeline and 1 week to implement the pipeline published on protocols.io using their local CPU systems and submit their predictions. A performance evaluation survey was conducted before the start of the SPARK Academy and after participants implemented the pipeline, to assess the feasibility of replicating the pipeline, its performance on types of computer hardware in these regions, and the ease of using this pipeline as a teaching tool for enhancing computational imaging knowledge and skills in resource-constrained settings.

[1] https://dx.doi.org/10.17504/protocols.io.dm6gpdwmdgzp/v3.

[2] https://event.fourwaves.com/spark

3.6 Performance Evaluation

The performance of the lightweight 3D U-Net model and its reliability as implemented through the pipeline was evaluated on the BraTS-Africa 2024 dataset using standard segmentation metrics: Mean Intersection Over-Union (IoU), Dice Similarity Coefficient (DSC) and the 95th percentile Hausdorff Distance (HD95) across the tumor sub-regions (NT, ET, and NC). The evaluation was conducted through lesion-wise comparison of predicted segmentation masks against the ground truth annotations.

4 Results

4.1 Training Metrics

A primary objective of this study was demonstrating the feasibility of a lightweight 3D medical image segmentation on resource-constrained hardware. The lightweight model was trained for an average epoch time of 314.16 ± 20.68 s, a total training time for 44 epochs of 13835.50 s (230.59 min, 3.84 h), using a peak memory of 1647.62 MB/1.6 GB and average CPU of 81.86% ± 2.60%.

4.2 Performance Evaluation

Visual inspection of the test sets revealed consistent boundary delineation for large tumor sizes, where voxel-level precision becomes challenging in smaller lesions (Fig. 3). The mean IoU, DSC and HD95 scores are reported for the 13 validation and the 9 test sets in Table 1.

Table 1. Evaluation of the Model on the Validation and Test datasets.

Region	Validation IoU	Validation Dice	Validation HD95	Test IoU	Test Dice	Test HD95
ET	0.57	0.84	23.9 ± 11.5	0.63	0.79	18.0 ± 11.71
NETC	0.67	0.84	26.36 ± 15.7	0.71	0.85	20.6 ± 11.15
SNFH	0.61	0.75	18.42 ± 14.0	0.73	0.73	18.9 ± 14.38
WT	0.76	0.75	17.54 ± 14.6	0.79	0.74	16.4 ± 12.14

For more concept on the performance of our model, we compared it with recent lightweight 3D segmentation models reported in the literature (Table 2). Compared to CPU-SAM, LightM-UNet, and EfficientSAM, our model achieved competitive Dice (0.79) with low memory (~1.6 GB) and modest runtime (~1 h), demonstrating a balanced trade-off between accuracy and resource efficiency for real-world constrained environments.

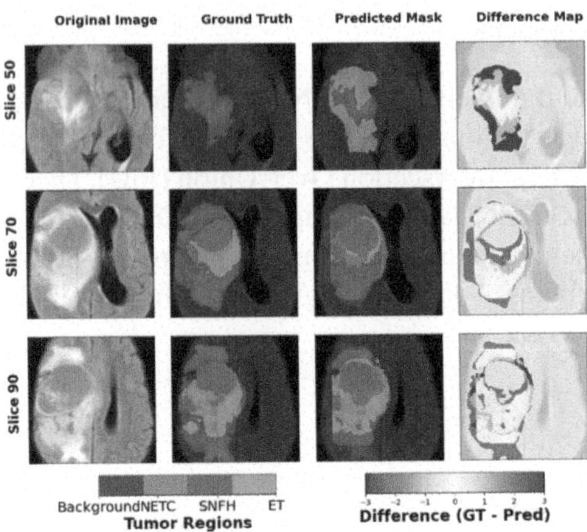

Fig. 3. Representative slices from a test case showing (left to right): the MRI image, ground truth mask, model prediction, and overlay of prediction on the image, highlighting segmentation performance.

Table 2. Benchmarking Lightweight 3D Segmentation Models

Model	Key Contribution	Dataset(s)	Runtime (CPU)	Dice	Memory Requirement
Our Model	Lightweight 3D segmentation for RCS; tested in SPARK workshop	Brain Tumor MRI	~1 h (Ryzen 5), 314 s/epoch	0.79	**~1.6 GB**
CVPR 2024 CPU-SAM [18]	Optimized SAM for laptops (CPU)	Multi-modal MRI	~4.6 s/case	0.75	**Not reported**
LightM-UNet [19]	Mamba-based lightweight UNet	Multi-organ CT/MRI	<10 s/case (GPU + CPU hybrid)	0.96	**Very low**
EfficientSAM [20]	Distilled SAM with EfficientFormer backbone	Med Seg datasets	9 × faster CPU inference	+ 0.7–1.1% mIoU vs MobileSAM	**Not reported**

4.3 Real-World Validation: SPARK Academy 2025 Workshop

The participants from the workshop who contributed to the pipeline's implementation represent a diverse global team, with 48 members from Africa, 4 from South Asia, and 1 from Europe. This international group comprised of young researchers, graduate students, and radiology residents and junior consultants. All 53 participants completed the implementation of the pipeline within 1 week. Most participants used Windows (40%) and Mac (40%), while the rest used Linux (20%). The RAM varied widely and ranged from 4 GB to 64 GB. Nearly all (95.5%) had over 5 GB of storage. Processor architectures were diverse, from Core i3 and Ryzen 5 to Core i7 and Apple M1/M2 Pro, highlighting real-world computational diversity. The participants recorded an average training time of 1-hour, average Dice scores of 0.5 (training set) and 0.7 (test set), and average HD95 scores of 21 (training set) and 17.75 (test set). It was observed that participants with lesser computing resource had a longer training time, but the resources had no substantial effect on the model performances. Participants expressed enthusiasm about the accessibility of 3D tumor segmentation on their personal computers with low computing resource and how it helped improve their knowledge of AI.

Table 3. Model Training Time on Participants' Systems.

Training Time Range	Device processor	Number of Participants	Percentage
<1 h	Core i9	20	22.7%
1 h–1.5 h	Core i7	24	27.3%
1.5 h–2 h	Core i5	21	23.9%
2 h–3 h	Core i3	10	11.4%
>3 h	Core i3	13	14.8%

The Time Taken to complete the training as seen in Table 3, for most of the participants was consistent with this study's original result, and the variation in the training tri.

5 Discussion

We designed and tested a lightweight 3D U-Net that can train and run on CPUs only, making brain tumor segmentation more accessible in low-resource settings. Training took less than three hours on a standard workstation, with minimal memory use. Careful optimization of data loading, cropping, and CPU-friendly augmentations kept the process stable and efficient. A step-by-step pipeline provided a practical and easy to follow workflow for assessment of model performance on many types of CPU configuration and for capacity building in low-resource settings. Our model produced good Dice scores across tumor sub-regions (validation: from 0.75 to 0.84 and test data: 0.73 to 0.85), showing good model generalizability. The SNFH was harder to segment, probably due to its lower contrast. HD95 errors were higher on test data but still acceptable for clinical use. While GPU-based models still give better accuracy, our results prove that 3D brain tumor segmentation can work on everyday hardware. Future improvements like quantization, knowledge distillation, and better augmentations could help.

6 Conclusion

This study demonstrates that a carefully optimized 3D U-Net can be successfully trained and deployed on CPU-only systems, reducing reliance on specialized GPU hardware. Using the BraTS-Africa 2024 dataset, our lightweight architecture achieved competitive segmentation performance while maintaining low memory (~1.6 GB) and modest runtime (~1 h). Comparative analysis with recent CPU-efficient models, such as CPU-SAM, LightM-UNet, and EfficientSAM, highlights that our method provides a balanced trade-off between accuracy and resource efficiency, making it particularly suitable for constrained environments. Importantly, the entire pipeline, including dataset, preprocessing, training scripts, and evaluation routines, is openly available on protocols.io, enabling reproducibility and adaptation on standard desktops and laptops. By lowering the hardware barrier, this approach empowers researchers, clinicians, and students in low-resource settings to leverage 3D deep learning for neuroimaging, fostering sustainable, community-led innovation where access to GPUs is limited or cost-prohibitive.

7 Disclosure of Interests.

The authors have no competing interests to declare that are relevant to the content of this article.

Acknowledgments. This study was supported by the MAI Lab. The authors also thank all participants of the SPARK Academy 2025 Workshop for their contributions to the real-world validation of the pipeline.

References

1. Uwishema, O., et al.: Epidemiology and etiology of brain cancer in Africa: a systematic review. Brain Behav. **13**, e3112 (2023). https://doi.org/10.1002/brb3.3112
2. Daimary, D., Bora, M.B., Amitab, K., Kandar, D.: Brain tumor segmentation from MRI images using hybrid convolutional neural networks. Procedia Comput. Sci. **167**, 2419–2428 (2020). https://doi.org/10.1016/j.procs.2020.03.295
3. Ronneberger, O., Fischer, P., Brox, T.: U-Net: convolutional networks for biomedical image segmentation (2015). http://arxiv.org/abs/1505.04597. https://doi.org/10.48550/arXiv.1505.04597
4. Çiçek, Ö., Abdulkadir, A., Lienkamp, S.S., Brox, T., Ronneberger, O.: 3D U-Net: learning dense volumetric segmentation from sparse annotation. In: Ourselin, S., Joskowicz, L., Sabuncu, M.R., Unal, G., Wells, W. (eds.) Medical Image Computing and Computer-Assisted Intervention – MICCAI 2016, pp. 424–432. Springer, Cham (2016). https://doi.org/10.1007/978-3-319-46723-8_49
5. Menze, B.H., et al.: The multimodal brain tumor image segmentation benchmark (BRATS). IEEE Trans. Med. Imaging **34**, 1993–2024 (2015). https://doi.org/10.1109/TMI.2014.2377694
6. Prastawa, M., Bullitt, E., Ho, S., Gerig, G.: A brain tumor segmentation framework based on outlier detection. Med. Image Anal. **8**, 275–283 (2004). https://doi.org/10.1016/j.media.2004.06.007

7. Karimi, D., Salcudean, S.E.: Reducing the hausdorff distance in medical image segmentation with convolutional neural networks. IEEE Trans. Med. Imaging **39**, 499–513 (2020). https://doi.org/10.1109/TMI.2019.2930068
8. Eelbode, T., et al.: Optimization for medical image segmentation: theory and practice when evaluating with dice score or jaccard index. IEEE Trans. Med. Imaging **39**, 3679–3690 (2020). https://doi.org/10.1109/TMI.2020.3002417
9. Quantization Friendly MobileNet (QF-MobileNet) Architecture for Vision Based Applications on Embedded Platforms | Request PDF. https://www.researchgate.net/publication/348046747_Quantization_Friendly_MobileNet_QF-MobileNet_Architecture_for_Vision_Based_Applications_on_Embedded_Platforms. Accessed 24 June 2025
10. Reddy, K.R., Dhuli, R.: A novel lightweight CNN architecture for the diagnosis of brain tumors using MR images. Diagnostics (Basel) **13**, 312 (2023). https://doi.org/10.3390/diagnostics13020312
11. (PDF) Optimized Brain Tumor Segmentation for Resource Constrained Settings: VGG-Infused U-Net Approach. https://www.researchgate.net/publication/388822406_Optimized_Brain_Tumor_Segmentation_for_Resource_Constrained_Settings_VGG-Infused_U-Net_Approach. Accessed 24 June 2025
12. A lightweight deep learning model for automatic segmentation and analysis of ophthalmic images | Scientific Reports. https://www.nature.com/articles/s41598-022-12486-w. Accessed 24 June 2025
13. Kanmounye, U.S., Karekezi, C., Nyalundja, A.D., Awad, A.K., Laeke, T., Balogun, J.A.: Adult brain tumors in Sub-Saharan Africa: a scoping review. Neuro Oncol. **24**, 1799–1806 (2022). https://doi.org/10.1093/neuonc/noac098
14. Stroke lesion segmentation from low-quality and few-shot MRIs via similarity-weighted self-ensembling framework. https://conferences.miccai.org/481-Paper0252. Accessed 24 June 2025
15. Adewole, M., et al.: Expanding the Brain Tumor Segmentation (BraTS) data to include African Populations (BraTS-Africa) (version 1). Cancer Imaging Arch. (2024). 10.7937/v8h6-8×67
16. Adewole, M., et al.: The BraTS-Africa dataset: expanding the brain tumor segmentation data to capture african populations. Radiol. Artif. Intell. **7**, e240528 (2025). https://doi.org/10.1148/ryai.240528
17. 2D and 3D U-Net Architectures. https://codefinity.com/blog/2D-and-3D-U-Net-Architectures. Accessed 24 June 2025
18. Dao, T.T., Ye, X., Scarsbrook, J., Balarupan, G., Ribeiro, F.L., Bollmann, S.: Modality-specific strategies for medical image segmentation using lightweight SAM architectures. In: Presented at the CVPR 2024: Segment Anything in Medical Images on Laptop, 24 June 2024 (2024)
19. Liao, W., Zhu, Y., Wang, X., Pan, C., Wang, Y., Ma, L.: LightM-UNet: mamba assists in lightweight UNet for medical image segmentation (2024). http://arxiv.org/abs/2403.05246. https://doi.org/10.48550/arXiv.2403.05246
20. (PDF) Efficient SAM for Medical Image Analysis, https://www.researchgate.net/publication/375895620_Efficient_SAM_for_Medical_Image_Analysis. Accessed 22 Aug 2025

Fetal Abdomen-Guided Ultrasound Reconstruction via Autoencoder For Optimal Frame Selection and Segmentation

Ahmed Hazem Youssef[1](✉), Halim Benhabiles[1], Tanya Akumu[2], Marawan Elbatel[3], and Karim Lekadir[2]

[1] IMT Nord Europe, Institut Mines-Télécom, Université de Lille, Center for Digital Systems, 59000 Lille, France
ahmedhazem10119@gmail.com

[2] Department of Mathematics and Computer Science, Universitat de Barcelona, Gran Via de les Corts Catalanes 585, 08007 Barcelona, L'Eixample, Spain

[3] Department of Electronic and Computer Engineering, The Hong Kong University of Science and Technology, Hong Kong, China

Abstract. Regular pregnancy monitoring is critical to ensure the health and safety of both infants and mothers. However, in low-resource settings, access to regular and accurate monitoring remains limited due to the lack of sophisticated medical equipment and shortage of trained sonographers. Efforts in the development of affordable portable ultrasound probes, has permitted to propose alternative solutions. Nevertheless, these devices often produce images with degraded quality making their analysis challenging. In this work, we propose a methodology for selecting the most informative frame (optimal frame) obtained from such devices and the segmentation of its region of interest (ROI) namely the fetal abdomen. Our method is mainly based on ROI-guided image reconstruction approach, potentially allowing better reconstruction from an optimal frame than a non-optimal one. A frame proximity-based ranking algorithm is applied to refine the selection process of the optimal frame. Evaluations on a public dataset released within an international challenge (ACOUSLIC-AI 2024) show that our method achieves promising performance surpassing leading methods.

Keywords: Fetal ultrasound images · Optimal frame selection · Autoencoder · ROI segmentation · Low-resource settings

1 Introduction

Fetal growth restriction (FGR) is one of the main causes of neonatal mortality and adverse neonatal outcomes, especially in low-resource countries [1]. As stated by the WHO (World Health Organization), neonatal mortality rates in Sub-Saharan Africa and southern Asia are the highest ones in the world, namely 27

U. Anazodo et al. (Eds.): MIRASOL 2025, LNCS 16398, pp. 62–71, 2026.
https://doi.org/10.1007/978-3-032-13654-1_7

and 21 deaths per thousand respectively [2]. This is largely due to the lack of continuous and accurate pregnancy monitoring, which stems from the scarcity of trained sonographers and the high cost of sonographic equipment [3].

To address these challenges, alternative solutions have been developed particularly based on affordable and easy to use ultrasound devices [4]. As illustrated in Fig. 1, these devices utilize obstetric sweep protocol (OSP), allowing novice operators to acquire a sequence of ultrasound images without extensive training [5]. However, a significant limitation is the degraded quality of the images produced. Indeed, most of the captured frames are non-optimal, often exhibiting high levels of noise that hinder accurate abdominal circumference (AC) estimation obtained from the fetal abdomen region. It is worth noting that AC is a key biometric measurement commonly used to detect early signs of fetal growth restriction (FGR) [6,7].

In this article, we propose a methodology designed to automatically select the optimal frame for each case and perform segmentation of the fetal abdomen on that frame. The main contributions of our work are:

- We propose an autoencoder-based model that learns to reconstruct the fetal abdomen region from optimal frames, while producing blank or near-zero pixel outputs for non-optimal frames. This region-aware reconstruction strategy implicitly encourages the model to differentiate between informative and non-informative inputs, thereby enhancing its ability to prioritize optimal frames. Furthermore, this reconstruction strategy permits the automatic derivation of binary masks that delineate the reconstructed region (i.e., fetal abdomen segmentation).
- We design a customized optimal frame selection strategy that combines geometric criteria with a proximity-based ranking mechanism. Each frame in the sequence is initially ranked based on the area and circularity of the reconstructed fetal abdomen region. This ranking is then optimized by evaluating each frame's proximity to its neighboring frames within the sequence. This two-step approach promotes the selection of the most representative frame.
- We benchmark our method on both optimal frame selection and fetal abdomen segmentation tasks, comparing its performance with submitted methods from the ACOUSLIC-AI challenge (Abdominal Circumference Operator-agnostic UltraSound measurement)[1] at MICCAI 2024.

2 Related Work

Several studies have been proposed on the analysis of ultrasound images, often focusing on various downstream tasks, such as measuring fetal anatomical structures and estimating characteristics like gestational age (GA) and fetal weight [8–10]. In most cases, the proposed methods rely on manually selected optimal frames curated by domain experts. In the following, we will present two works that proposed both tasks of optimal frame selection and segmentation.

[1] https://acouslic-ai.grand-challenge.org.

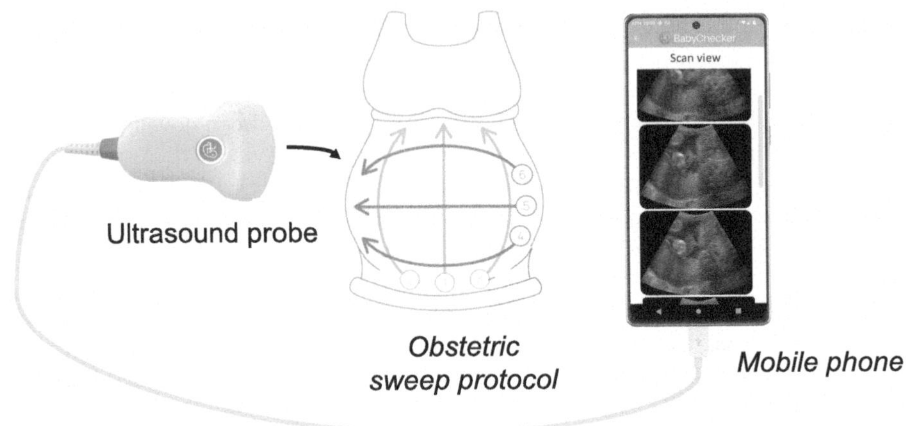

Fig. 1. Obstetric sweep protocol.

Plotka et al. [11] introduced a multi-task deep learning framework that localizes the optimal frame, classifies fetal body parts, and performs biometric measurements. Their method used a multi-task encoder-decoder neural network with ConvLSTM cells to analyze fetal ultrasound video sequences, enabling simultaneous frame classification (to determine if a frame is optimal or not), segmentation, and measurement. Their method demonstrated fetal biometric measurements comparable to those performed by five experienced sonographers. More specifically, the differences in measurement values within the range of intra- and inter-observer variability were about 1.04 cm for Head circumference (HC), and 1.06 cm for AC. However, as stated by the authors, the model was trained and evaluated on data acquired by costly sonographic equipment and therefore does not specifically address the challenges posed by low-quality images, an issue the authors cite as a direction for future work.

Van den Heuvel et al. [12] utilized images obtained from low-cost ultrasound probes to identify the optimal frame and calculate the head circumference (HC), from which they estimated GA. The authors used two neural networks: a VGG-inspired network trained to detect frames containing the fetal head, and a U-Net architecture to automatically compute the HC for the frames where the head was detected. Their evaluation compared the automatically estimated HC and GA against reference measurements obtained manually by experienced sonographers. This selection of the optimal frame achieved an accuracy of 0.97. For the HC (segmentation of the fetal head) they obtained a mean absolute difference (MAD) of 10.3 mm compared to the estimated one by experts.

In contrast to the aforementioned approaches, our method does not rely on a separate classifier to identify the optimal frame. Instead, it evaluates the reconstructed regions of interest to determine the most representative frame. This design simplifies the pipeline and enhances its efficiency, as the reconstruction of the selected frame is also directly used for segmentation, eliminating the need for additional processing stages.

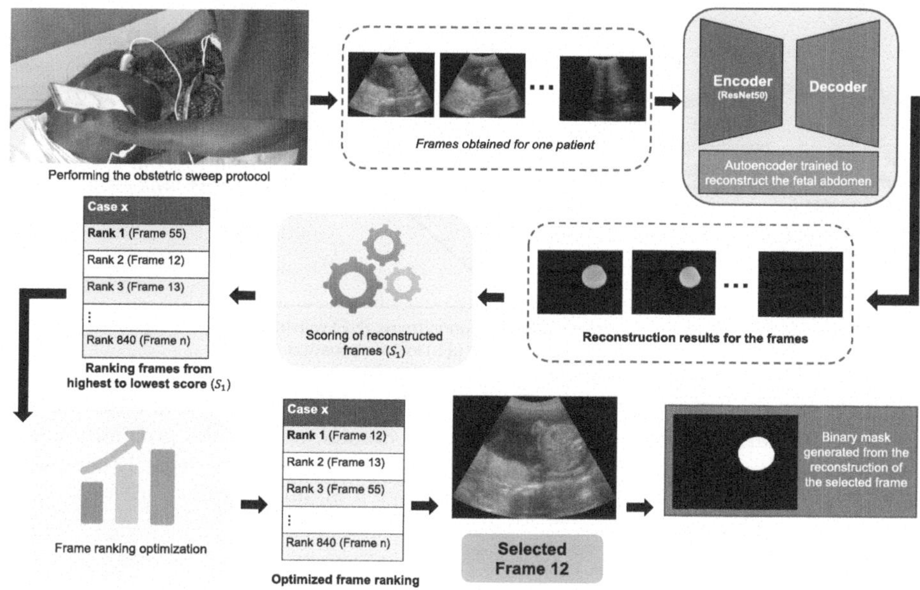

Fig. 2. Overview of our method: Each case's ultrasound frames are passed through an AE trained to reconstruct the fetal abdomen. A score is assigned to each frame based on the reconstructed region's area and circularity. Frames are initially ranked by this score, then refined using a proximity-based adjustment. The top-ranked frame is selected as optimal, and its reconstruction is used to assess segmentation performance.

3 Methodology

As illustrated in Fig. 2, our method for optimal frame selection and fetal abdomen segmentation is mainly based on a ROI-guided image reconstruction approach. In the following, we detail the different steps of the method.

3.1 Image Reconstruction

For the image reconstruction, we employed an Auto Encoder (AE) architecture. In our case, the encoder backbone is based on a ResNet50 pretrained on ImageNet. Both the encoder and decoder parameters are fine-tuned during training to optimize reconstruction performance. The AE takes as input an ultrasound image and learns to reconstruct either its ROI, namely the fetal abdomen, if it is an optimal frame, or a black image otherwise. This setup pushes the model to reconstruct the ROI only when relevant anatomical structures are present. During inference, the resulting reconstructions are used to score the frames and identify the optimal one.

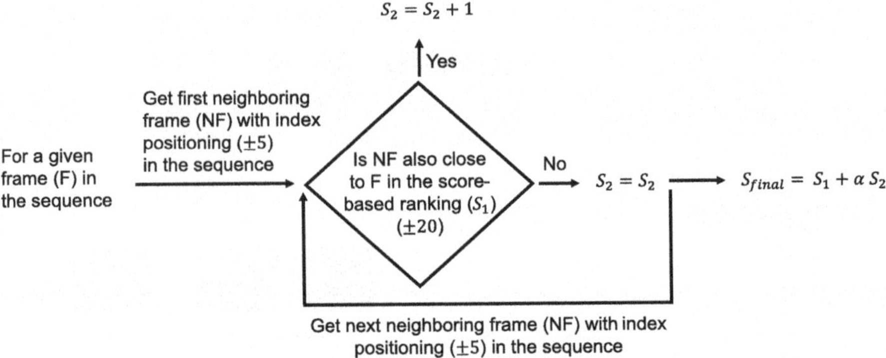

Fig. 3. Proximity calculation: Let S_1 denote the score of the frame based on the area and circularity of the reconstructed fetal abdomen, and S_2 the proximity count set initially to zero. Both S_1 and S_2 are normalized, and the final score is computed as $S_1 + \alpha S_2$, where α is a weighting factor (set to 0.3 in our experiments).

3.2 Scoring of Frames

The reconstructed frames are scored based on the area and circularity of their corresponding ROI, denoted as S_1:

$$S_1 = \alpha * area + \beta * circularity \tag{1}$$

where α and β are coefficients, empirically determined. Circularity is computed using the following equation:

$$circularity = \frac{4\pi \cdot area}{perimeter^2} \tag{2}$$

After scoring, the frames are ranked in descending order, under the assumption that those with the largest area and most circular ROI correspond to the optimal cases, as guided by the model's training strategy. The scores are then further refined through an optimization step based on the frames proximity, as described in the following section.

3.3 Frame Ranking Optimization

Based on the experimented dataset, we observed that the optimal frames are adjacent in terms of index positioning in the sequence. For this reason, we proposed to optimize the initial scores (obtained in the previous step) by favoring the frames that are close to each other in terms of index positioning as well as score ranking. The algorithm used for this score optimization is presented in Fig. 3. After this step, the frames are re-ranked accordingly.

4 Experiments and Results

To evaluate the performance of our method we tested it on the dataset proposed in the frame of the ACOUSLIC-AI challenge 2024[2]. This international challenge was dedicated to optimal frame selection and fetal abdomen segmentation tasks from low cost ultrasound imaging devices. It is worth mentioning that we contacted the dataset owners to request access to the hidden test set used for evaluating the submitted algorithms by the participants. However, our request was declined, as the test set is intended to remain private for future benchmarking purposes. For this reason we exploited only the public part of the dataset for the evaluation of our method. Additionally, we asked the teams that achieved the highest scores in the competition (Top 3 teams) to get their data split on the public part for fair comparison. Only one team, Tanya et al. [13], shared with us the data split and associated results. The authors proposed two independent models for the two tasks namely a ResNet based classifier for optimal frame selection and an nnUNet based fetal abdomen segmentation model. Evaluating on additional public fetal ultrasound datasets, is an important direction for future work to further assess generalizability.

4.1 Dataset

The public part of the dataset used in this study consists of 300 annotated cases collected from three public health units in Sierra Leone. Each case yielded approximately 840 ultrasound frames, of which only 0.85% were labeled as optimal and 1.78% as suboptimal (lower quality than optimal ones but still accepted for segmentation tasks).

4.2 Preprocessing

Normalization. The images were normalized to the [0, 1] range and resized from (744, 562) to (376, 288) to reduce computational load while preserving image characteristics and aspect ratio. In addition, black frames in the produced sequences were excluded from the dataset prior to training.

Data Augmentation. Eight types of data augmentation were applied: rotation, zoom, translation, Gaussian noise, sharpening, intensity variation, elastic deformation, and speckle noise. The last two were included to simulate ultrasound-specific distortions [14]. For each batch, two augmentation types were randomly selected, and each image had a 60% chance of being augmented.

Data Balancing. Due to the strong imbalance between optimal/suboptimal and non-optimal frames, a custom batch balancer was implemented. Each training batch (size 16) included 8 optimal/suboptimal and 8 non-optimal frames. To prevent underrepresentation of certain cases, the balancer ensured that all cases contributed approximately equally to the set of non-optimal frames.

[2] https://acouslic-ai.grand-challenge.org.

4.3 Training Setup

We employed 5-fold cross-validation, training each fold for 50 epochs. The Adam optimizer was used with an initial learning rate of 1e-5. A learning rate warmup was applied during the first 10 epochs to reach 1e-4, followed by exponential decay down to a minimum of 1e-7. Training each epoch took 5 min on a GPU (NVIDIA RTX 4000), and inference for one patient takes 18 s.

4.4 Evaluation Metrics

To evaluate performance, we used the proposed metrics by the ACOUSLIC-AI challenge:

Weighted Frame Selection Score (WFSS). Measures the appropriateness of the selected frame. A score of 1 is given for optimal, 0.6 for suboptimal, and 0.0 for non-optimal selections.

Soft Dice Similarity Coefficient (DSC_{soft}). Measures spatial overlap between the predicted and ground truth masks. If the ground truth is not available for the predicted frame, it is compared to the nearest annotated frame in the sequence. The score is adjusted by a penalty term proportional to the temporal distance between frames:

$$DSC_{soft} = DSC_{nearest} \left(1 - \frac{|predicted\ frame - nearest|}{max\_frame\_tolerance}\right) \tag{3}$$

where $max\_frame\_tolerance$ is set by the challenge to 15 frames.

Normalized Absolute Error (NAE). Quantifies the relative error in estimated abdominal circumference (AC):

$$\mathrm{NAE}_{AC} = \frac{|AC_{gt} - AC_{pred}|}{\max(AC_{gt}, AC_{pred}, \epsilon)} \tag{4}$$

Total Score. A weighted sum combining segmentation quality and measurement accuracy:

$$\mathrm{Score} = 0.5\,(1 - \mathrm{NAE}_{AC}) + 0.25\,DSC_{soft} + 0.25\,WFSS \tag{5}$$

4.5 Results

Table 1 summarizes the average performance across the five folds of our method compared to the method from Tanya et al. [13]. The table shows that our method, using ResNet50-based backbone, achieved the highest performance, underscoring the effectiveness of our ultrasound images analysis methodology.

Table 1. Mean performance comparison (± standard deviation) across 5 folds between our method and that of Tanya et al. [13]. ↑ indicates higher is better, ↓ indicates lower is better.

Method	WFSS ↑	DSC soft ↑	NAE ↓	Total Score ↑
Tanya et al. [13]	0.64 ± 0.08	0.67 ± 0.09	0.18 ± 0.07	0.73 ± 0.07
Our method (ResNet18-AE)	0.75 ± 0.03	0.78 ± 0.02	0.17 ± 0.01	0.79 ± 0.01
Our method (ResNet50-AE)	**0.82 ± 0.06**	**0.82 ± 0.05**	**0.15 ± 0.04**	**0.83 ± 0.05**

Table 2. Performance without frame ranking optimization (mean ± standard deviation).

Our method	WFSS	DSC soft	NAE	Total Score
(ResNet50-AE)	0.77 ± 0.05	0.80 ± 0.04	0.16 ± 0.03	0.80 ± 0.04

Ablation Study. Table 2 shows the average results across the five folds without frame ranking optimization. A noticeable drop in WFSS is observed, suggesting that ranking optimization contributes significantly to selecting optimal frames.

To determine the optimal ranges of index positioning and score-based ranking, we conducted an ablation study on these two hyperparameters. Figure 4a, shows that the best range for index positioning is ±5, while Fig. 4b, shows that the best range for score-based ranking is ±20.

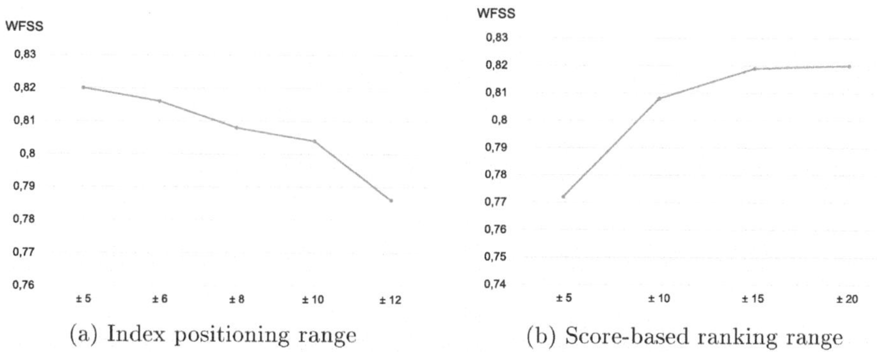

(a) Index positioning range (b) Score-based ranking range

Fig. 4. Comparison of ablation studies for index and rank window sizes.

5 Conclusion

In this study, we proposed a methodology for selecting optimal frames and segmenting the fetal abdomen from low quality ultrasound images. Experiments on the ACOUSLIC-AI dataset demonstrate promising performance across several evaluation metrics. These results underscore the value of combining reconstruction analysis with rank proximity information to efficiently distinguish between optimal and non-optimal frames. A direction for future work is optimizing the analysis pipeline for deployment on mobile phones.

Acknowledgments. This work has been supported by the research center of IMT Nord Europe and AIMIX project (ERC Grant Agreement ID 101044779), Universitat de Barcelona.

Disclosure of Interests. The authors have no competing interests to declare that are relevant to the content of this article.

References

1. Accrombessi, M., Zeitlin, J., Massougbodji, et al.: What do we know about risk factors for fetal growth restriction in Africa at the time of sustainable development goals? A scoping review. Paediatr. Perinat. Epidemiol. **32**(2), 184–196 (2018)
2. WHO. Newborn mortality. https://www.who.int/news-room/fact-sheets/detail/levels-and-trends-in-child-mortality-report-2021
3. Sendra-Balcells, C., Campello, V.M., Torrents-Barrena, J., Ahmed, et al.: Maternal fetal ultrasound planes from low-resource imaging settings in five African countries (2023)
4. Ranger, B.J., Bradburn, E., Chen, et al.: Portable ultrasound devices for obstetric care in resource-constrained environments: mapping the landscape. Gates Open Res. **7**, 133 (2023)
5. Gomes, R.G., Vwalika, B., Lee, C., Willis, A., Sieniek, et al.: AI system for fetal ultrasound in low-resource settings (2022)
6. Long, M., Nakahara, A., Elmayan, et al.: Fetal growth restriction defined by abdominal circumference alone predicts perinatal mortality. Am. J. Obstet. Gynecol. **226**(1), S179 (2022)
7. Marchand, C., Köppe, J., Köster, H.A., Oelmeier, et al.: Fetal growth restriction: comparison of biometric parameters. J. Personalized Med. **12**(7) (2022)
8. Giaxi, P., Vivilaki, V., Sarella, A., Harizopoulou, V., Gourounti, K.: Artificial intelligence and machine learning: an updated systematic review of their role in obstetrics and midwifery. Cureus, **17**(3) (2025)
9. Naz, S., et al.: Use of artificial intelligence for gestational age estimation: a systematic review and meta-analysis. Front. Global Women's Health, **6** (2025)
10. Umans, E., Dewilde, K., Williams, H., Deprest, J., Van den Bosch, T.: Artificial intelligence in imaging in the first trimester of pregnancy: a systematic review. Fetal Diagn. Ther. **51**(4), 343–356 (2024)
11. Płotka, S., Klasa, A., Lisowska, et al.: Deep learning fetal ultrasound video model match human observers in biometric measurements. Phys. Med. Biol. **67**(4), 045013 (2022)
12. van den Heuvel, T.L.A., Petros, H., Santini, et al.: Automated fetal head detection and circumference estimation from free-hand ultrasound sweeps using deep learning in resource-limited countries. Ultrasound Med. Biol. **45**(3), 773–785 (2019)

13. Akumu, T., Martin-Isla, C., Campello, V.M., Fabila, J.: Fetal abdominal detection and segmentation in 3D blind sweep ultrasound. ACOUSLIC-AI Challenge submission report (2024)
14. Choi, H., Jeong, J.: Despeckling algorithm for removing speckle noise from ultrasound images. Symmetry **12**(6), 938 (2020)

Recovering Diagnostic Value: Super-Resolution–Aided Echocardiographic Classification in Resource-Constrained Imaging

Krishan Agyakari Raja Babu[1](✉), Om Prabhu[2], Annu[3], and Mohanasankar Sivaprakasam[1]

[1] Indian Institute of Technology Madras, Chennai, India
rajababu@alumni.iitm.ac.in
[2] All India Institute of Medical Sciences, New Delhi, India
[3] Indian Institute of Technology Hyderabad, Sangareddy, India

Abstract. Automated cardiac interpretation in resource-constrained settings (RCS) is often hindered by poor-quality echocardiographic imaging, limiting the effectiveness of downstream diagnostic models. While super-resolution (SR) techniques have shown promise in enhancing magnetic resonance imaging (MRI) and computed tomography (CT) scans, their application to echocardiography—a widely accessible but noise-prone modality—remains underexplored. In this work, we investigate the potential of deep learning-based SR to improve classification accuracy on low-quality 2D echocardiograms. Using the publicly available CAMUS dataset, we stratify samples by image quality and evaluate two clinically relevant tasks of varying complexity: a relatively simple Two-Chamber vs. Four-Chamber (2CH vs. 4CH) view classification and a more complex End-Diastole vs. End-Systole (ED vs. ES) phase classification. We apply two widely used SR models—Super-Resolution Generative Adversarial Network (SRGAN) and Super-Resolution Residual Network (SRResNet), to enhance poor-quality images and observe significant gains in performance metric—particularly with SRResNet, which also offers computational efficiency. Our findings demonstrate that SR can effectively recover diagnostic value in degraded echo scans, making it a viable tool for AI-assisted care in RCS, achieving more with less.

Keywords: Super-Resolution · Cardiac Classification · Image Enhancement · Resource-Constrained Settings

1 Introduction

Echocardiography is one of the most widely used cardiac imaging modalities, valued for its real-time capability, portability, and affordability. These attributes make it particularly critical in resource-constrained settings (RCS), including rural clinics and low- and middle-income countries (LMICs) [10,12,19]. However, the diagnostic utility of echocardiography is often undermined by poor

U. Anazodo et al. (Eds.): MIRASOL 2025, LNCS 16398, pp. 72–82, 2026.
https://doi.org/10.1007/978-3-032-13654-1_8

image quality. Studies have reported that echocardiographic scans performed with handheld or low-end devices in LMICs are frequently suboptimal for clinical interpretation [6,8]. This is primarily due to factors such as limited imaging hardware, variability in operator expertise, and difficult acquisition conditions (e.g., in emergency or bedside scenarios) [20,22,23].

Poor-quality echo scans not only hinder human interpretation but also significantly degrade the performance of automated tools for view classification, chamber quantification, and disease prediction— technologies that are increasingly being deployed to address the shortage of trained specialists in such regions [4,14,24]. Despite the growing reliance on AI-assisted diagnostic pipelines, relatively few studies address the pre-processing bottleneck of enhancing low-fidelity echocardiographic inputs to improve downstream performance [5,9].

Super-resolution (SR) has emerged as a promising solution for enhancing the quality of medical images, particularly in scenarios where high-resolution acquisition is limited by hardware constraints. While SR techniques have achieved notable success in high-contrast modalities such as magnetic resonance imaging (MRI) and computed tomography (CT) [15,21], their application to echocardiography—arguably one of the most noise-prone and variable modalities—remains limited. The combination of speckle noise, inconsistent probe positioning, and patient-dependent acoustic windows makes SR in echocardiography a uniquely challenging problem [1,11]. Moreover, most downstream AI models assume access to high-quality input images, overlooking a critical bottleneck in RCS, where degraded image quality is often the norm rather than the exception [12,24].

Our contributions are summarized as follows:

- We investigate the underexplored application of super-resolution in 2D echocardiography and position it as a lightweight, model-agnostic preprocessing step for enhancing AI-based clinical interpretation in RCS.
- We demonstrate that super-resolution significantly improves classification performance on degraded scans, with SRResNet offering a favorable trade-off between diagnostic accuracy and computational efficiency.

2 Related Work

SR techniques in echocardiography have evolved from traditional signal processing approaches to recent deep learning-based models. Early efforts predominantly explored temporal super-resolution, aiming to improve the frame rate of echocardiographic videos. Gifani et al. [11] introduced a sparse representation-based method that reconstructed intermediate frames using learned dictionaries. Similarly, Afrakhteh et al. [2] proposed a high-precision interpolation technique leveraging non-polynomial functions to enhance temporal continuity in ultrasound sequences. While effective in improving temporal resolution, these methods offered limited enhancement in spatial fidelity—an equally critical factor for diagnostic interpretation.

The development of deep learning methods has opened new avenues for spatial SR in ultrasound imaging. Abdel-Nasser and Omer [1] applied a CNN-based architecture for general ultrasound image enhancement, reporting improvements in structural details and contrast. Cammarasana et al. [7] proposed a patch-based SR method for 2D ultrasound images and videos, demonstrating gains in spatial resolution across generic anatomical scenes. Similarly, Li et al. [18] integrated speckle reduction with deep learning-based SR to enhance segmentation performance in ultrasonic echo images. Unlike prior efforts that primarily focus on perceptual or technical image quality, our study assesses how SR translates into clinically meaningful improvements—especially under RCS.

3 Methodology

We adopt a two-stage pipeline, as illustrated in Fig. 1, to assess the diagnostic utility of SR in enhancing low-quality echocardiographic images. Let $\mathcal{D} = \{(x_i, y_i, q_i)\}_{i=1}^{N}$ denote the dataset, where $x_i \in \mathbb{R}^{H \times W}$ is a 2D echocardiographic frame, $y_i \in \mathcal{Y}$ is the corresponding diagnostic label (e.g., view type or cardiac phase), and $q_i \in \{\text{good}, \text{medium}, \text{poor}\}$ is the image quality metadata provided by clinical experts. Rather than using pixel resolution as a proxy for image quality—which is often misleading—we utilize this clinically validated q_i to stratify the dataset into three disjoint subsets: $\mathcal{D}_{\text{good}}$, $\mathcal{D}_{\text{medium}}$, and $\mathcal{D}_{\text{poor}}$.

We treat $\mathcal{D}_{\text{poor}}$ as a representative of images acquired in RCS. To assess diagnostic performance across varying image quality, we define a classifier $g_\phi : \mathbb{R}^{H \times W} \rightarrow \mathcal{Y}$ trained independently on each subset and tested across all quality levels. We consider two clinically relevant classification tasks: (i) a simple view classification of two-chamber vs. four-chamber (2CH vs. 4CH) and (ii) a complex cardiac phase classification of End-Diastole vs. End-Systole (ED vs. ES). The goal is to observe how image quality affects g_ϕ's ability to extract diagnostic features without spurious biases.

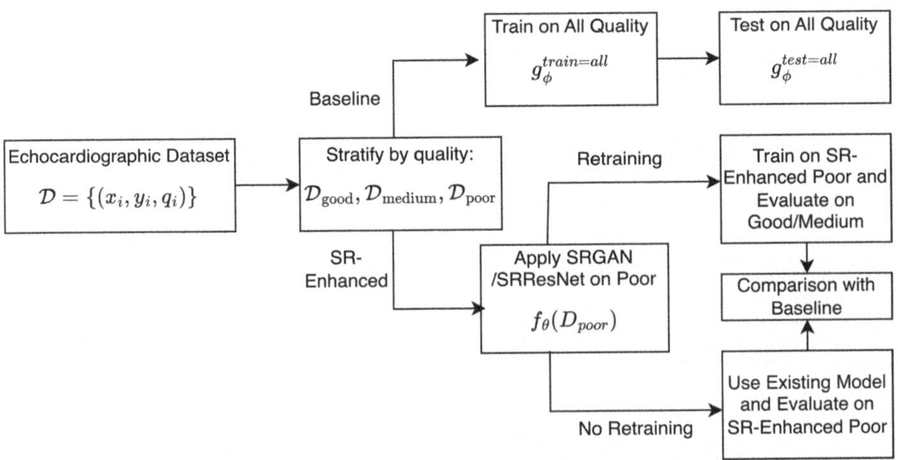

Fig. 1. Proposed workflow for super-resolution–aided echo-classification.

For enhancement, we introduce a super-resolution module $f_\theta : \mathbb{R}^{H \times W} \rightarrow \mathbb{R}^{rH \times rW}$, where r is the upsampling factor. We investigate two SR architectures: SRGAN and SRResNet, pretrained on natural images and fine-tuned on echocardiographic images. The SR-enhanced poor quality subset is then denoted as $\mathcal{D}_{\text{poor}}^{\text{SR}} = \{(f_\theta(x_i), y_i) \mid (x_i, y_i) \in \mathcal{D}_{\text{poor}}\}$. We re-evaluate g_ϕ using $\mathcal{D}_{\text{poor}}^{\text{SR}}$ to quantify the gains in classification metric post enhancement. This approach allows us to systematically evaluate the extent to which super-resolution can recover diagnostic information from clinically degraded echocardiograms and improve the utility of downstream AI models in RCS.

4 Experiments

4.1 Dataset and Quality Stratification

We conduct our experiments on the publicly available "Cardiac Acquisitions for Multi-structure Ultrasound Segmentation" (CAMUS) dataset [16], which consists of 2D echocardiographic sequences from 500 patients. For each patient, ED and ES frames are annotated in both 2CH and 4CH views, yielding a total of 2,000 labeled frames. Importantly, the dataset includes expert-provided image quality annotations—categorized as good, medium, or poor—which we leverage for clinically grounded stratification. Unlike resolution-based proxies, this metadata reflects real-world diagnostic usability, making it more relevant for evaluating model performance in practical settings.

Table 1 summarizes the distribution of echo images across views, cardiac phases and image quality in the dataset. In the 2CH view, 43.4% of frames are rated as good, 42.8% as medium, and 13.8% as poor. Similarly, in the 4CH view, 57.6% of frames are rated as good, 33.0% as medium, and 9.4% as poor. While most frames fall into the good or medium categories, a clinically meaningful subset—232 poor-quality frames—reflects imaging conditions typical of RCS, and thus serves as our primary focus for enhancement and evaluation.

Table 1. Image distribution by view, phase, and quality in the CAMUS dataset.

View	Phase	Good	Medium	Poor	Total
2CH	ED	217	214	69	500
	ES	217	214	69	500
4CH	ED	288	165	47	500
	ES	288	165	47	500
Total		**1010**	**758**	**232**	**2000**

Figure 2 presents representative echocardiographic frames across the three quality tiers, spanning both views and cardiac phases. As evident from the visual examples, poor-quality frames exhibit low contrast, speckle noise, and structural ambiguity, posing substantial challenges for both human readers and AI systems. These visual limitations motivate the application of super-resolution as a pre-processing step to enhance diagnostic value under degraded conditions.

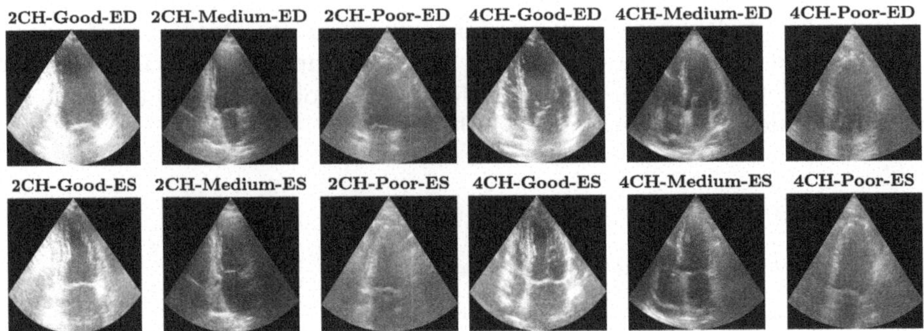

Fig. 2. Representative echocardiographic frames from the CAMUS dataset across quality levels (Good, Medium, Poor), views (2CH, 4CH), and phases (ED, ES).

4.2 Classification Tasks and Setup

To assess the downstream utility of SR, we define two clinically meaningful classification tasks: (i) a simpler view classification task distinguishing between 2CH and 4CH echocardiographic views, and (ii) a more challenging cardiac phase classification task differentiating between ED and ES frames. These tasks serve to evaluate both coarse and fine-grained diagnostic distinctions.

We adopt a ResNet-18 [13] model pretrained on ImageNet as the classification backbone. The model is fine-tuned on echocardiographic images using a batch size of 16 for 10 epochs, with cross-entropy loss as the objective function and the Adam optimizer (learning rate 1×10^{-4}).

To ensure class balance and control for potential sampling bias, we construct uniformly stratified datasets by taking the number of poor-quality samples as the reference. Specifically, for the 2CH vs 4CH classification task, we use 138 2CH (69 ED + 69 ES) and 94 4CH (47 ED + 47 ES) images. For the ED vs ES classification task, we use 116 ED (69 2CH + 47 4CH) and 116 ES (69 2CH + 47 4CH) images. In both cases, 80% of the data is used for training and 20% for testing. This setup allows us to rigorously evaluate the diagnostic discriminative power of images under different quality conditions, and later assess whether super-resolution improves this performance.

4.3 Super-Resolution Integration

To train the SR models on domain-specific data, we construct synthetic low-resolution and high-resolution image pairs using the 1,010 good-quality echocardiographic frames from the CAMUS dataset. Each high-quality image is degraded via bicubic downsampling by a factor of 4 to simulate low-resolution inputs, forming paired data for supervised SR training.

We leverage two widely adopted SR architectures—SRResNet and SRGAN [17]—both pretrained on the DIV2K dataset [3]. SRResNet is a lightweight model trained with pixel-wise mean squared error (MSE) loss, favoring accurate structural reconstruction and fast inference. In contrast, SRGAN employs a more complex generative adversarial framework, combining a perceptual loss with an adversarial loss to generate visually sharper and more realistic outputs. We fine-tune SRResNet for 4,000 epochs and SRGAN for 8,000 epochs on echocardiographic data using a scaling factor of 4.

Following training, we apply both models to enhance the 232 poor-quality frames, generating SR-enhanced counterparts. These enhanced images are subsequently used as input for downstream classification tasks to assess whether diagnostic value can be recovered under degraded imaging conditions. A visual comparison of the super-resolved outputs is shown in Fig. 3, focusing on a representative image patch. Although SRGAN, a perceptual-driven generative model, yields a Peak Signal-to-Noise Ratio (PSNR) of 32.70 dB and a Structural Similarity Index Measure (SSIM) of 0.7164, it fails to recover finer structural details. In contrast, SRResNet achieves notably higher reconstruction fidelity (PSNR = 38.98 dB, SSIM = 0.9214) despite being less computationally demanding—making it a more practical choice for deployment in RCS.

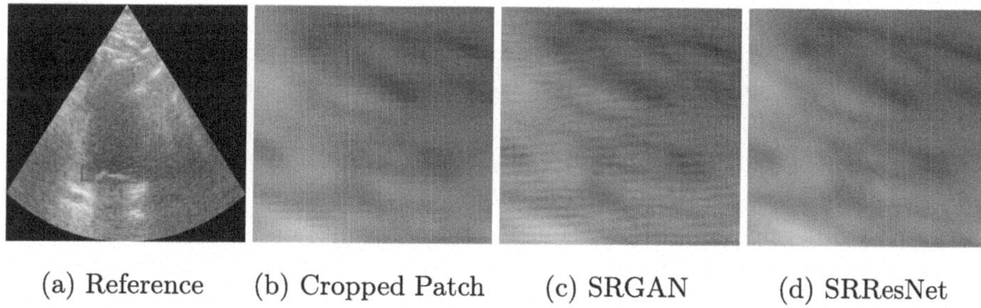

(a) Reference (b) Cropped Patch (c) SRGAN (d) SRResNet

Fig. 3. SR-enhanced outputs for a poor-quality 2CH ED image: (c) SRGAN—32.70 dB/0.7164; (d) SRResNet — 38.98 dB/0.9214 (PSNR/SSIM).

5 Results and Discussion

Table 2 presents the baseline classification accuracy for both view (2CH vs. 4CH) and phase (ED vs. ES) tasks across all combinations of training and testing image quality, prior to applying SR. Figure 4 illustrates the percentage improvement achieved when SR-enhanced poor-quality images are used for training or testing.

Table 2. View and phase classification accuracy(%) across image quality levels.

Train ↓/Test →	Quality	View Accuracy	Phase Accuracy
Good	Good	100	87
	Medium	90	83
	Poor	82	77
Medium	Good	92	85
	Medium	100	81
	Poor	91	81
Poor	Good	77	74
	Medium	90	77
	Poor	100	79

Based on these results, following observations can be noted:

1. **Diagnostic Collapse: The Cost of Image Degradation Across Tasks and Quality Levels:** As shown in Table 2, in the case of view classification, the model trained on good-quality images performs perfectly on similar data (100% accuracy), but its accuracy drops sharply—by 10% on medium and 18% on poor-quality images. Conversely, the model trained on poor-quality data, while achieving 100% accuracy on its own domain, experiences a 23% drop when tested on good images and 10% on medium. In comparison, phase classification, although it does not reach perfect accuracy in any setting, exhibits lower sensitivity to quality shifts—with an 11.5% drop from Good→Poor and only 6.3% from Poor→Good. These findings reveal that image degradation leads to greater diagnostic collapse in the simpler 2CH vs. 4CH view classification—which relies on anatomical structures—compared to the more abstract ED vs. ES phase classification that leverages functional cues. Notably, medium-quality images consistently yield robust performance across test conditions, suggesting they strike an effective balance between noise and structural fidelity—making them a potential "sweet spot" for SR optimization in RCS, which may offer the best trade-off between performance and resource demands.
2. **Recovery Power of Super-Resolution Across Architectures:** As illustrated in Fig. 4, integrating SR-enhanced poor-quality images into the classification pipeline yields notable gains in both view and phase tasks. As expected, the simpler 2CH vs. 4CH classification benefits more than the functionally complex ED vs. ES task. Interestingly, when model is trained on SR-enhanced poor images and tested on good-quality data, we observe an average accuracy improvement of 14.9% in the view task and 11.5% in the phase task—underscoring both the cross-quality generalizability and the restorative potential of SR. Notably, SRResNet—despite being architecturally simpler than the adversarially-trained SRGAN—achieves higher average improvement across all test cases (7.83% vs. 6.6%). This may be attributed to SRGAN's reliance

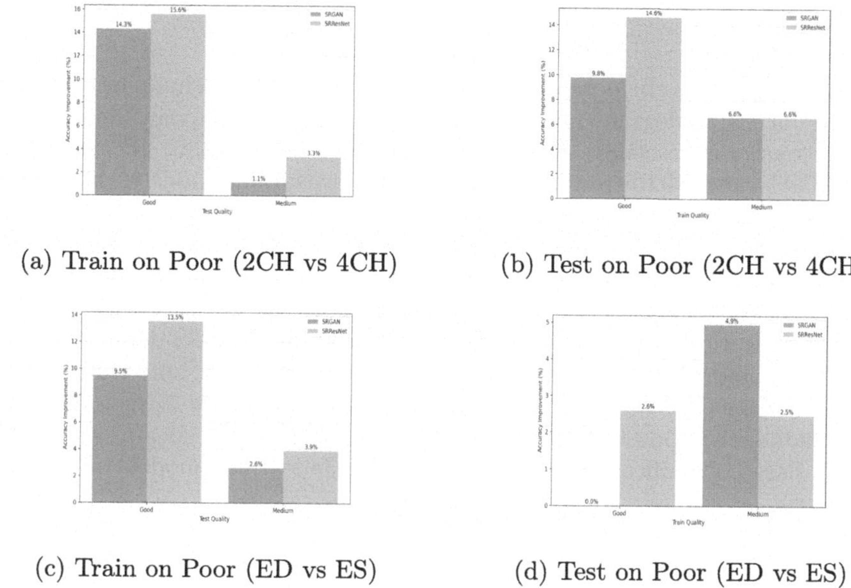

(a) Train on Poor (2CH vs 4CH)

(b) Test on Poor (2CH vs 4CH)

(c) Train on Poor (ED vs ES)

(d) Test on Poor (ED vs ES)

Fig. 4. Percentage improvement in classification performance after integrating super-resolution across two classification tasks (2CH vs 4CH and ED vs ES).

on perceptual and adversarial losses, which prioritize visual realism but can introduce subtle artifacts or distortions that compromise the anatomical precision essential in medical imaging. In contrast, pixel-accurate SR methods like SRResNet may be better suited for echocardiographic recovery in RCS, where both structural fidelity is critical and computational overhead must be minimized.

3. **Silent Gains with SR-Resolved Evaluation:** Figures 4(a) and (c) represent scenarios where SR-enhanced poor-quality images are used for retraining, while Figs. 4(b) and (d) show the case where these enhanced images are used solely during evaluation, without modifying the trained models. Remarkably, even without retraining, we observe measurable gains: in the 2CH vs. 4CH task, accuracy improves by an average of 12.2% when models are trained on good-quality data and by 6.6% when trained on medium. In the more complex ED vs. ES classification, the improvements are more modest—1.3% and 3.7% for models trained on good and medium-quality data, respectively. While these gains may appear subtle, they underscore the practical value of SR as a lightweight, test-time enhancement strategy—particularly beneficial in RCS where retraining may not be feasible and even small performance boosts can have meaningful clinical impact.

6 Conclusion

This study highlights the potential of super-resolution (SR) techniques to restore diagnostic utility in degraded echocardiographic images, a frequent limitation in resource-constrained settings (RCS). By enhancing poor-quality 2D echo scans with SRGAN and SRResNet, we demonstrate measurable gains in both view and phase classification accuracy—particularly with SRResNet, which consistently outperforms while remaining computationally efficient. Notably, performance improvements are observed even when SR is applied only at inference, underscoring its value as a lightweight preprocessing tool.

A key limitation of this study is its reliance on the CAMUS dataset. Future work should address this by validating the generalizability and clinical utility of SR-enhanced echocardiography across larger and more diverse datasets, and extending the approach to more complex clinical tasks—such as segmentation and disease classification—using advanced deep learning models tailored for RCS.

Disclosure of Interests. The authors have no competing interests to declare that are relevant to the content of this article.

References

1. Abdel-Nasser, M., Omer, O.A.: Ultrasound image enhancement using a deep learning architecture. In: Proceedings of the International Conference on Advanced Intelligent Systems and Informatics 2016, pp. 639–649. Springer, Cham (2017). https://doi.org/10.1007/978-3-319-48308-5_61
2. Afrakhteh, S., Jalilian, H., Iacca, G., Demi, L.: Temporal super-resolution of echocardiography using a novel high-precision non-polynomial interpolation. Biomed. Sig. Process. Control **78**, 104003 (2022). https://doi.org/10.1016/j.bspc.2022.104003
3. Agustsson, E., Timofte, R.: Ntire 2017 challenge on single image super-resolution: dataset and study. In: 2017 IEEE Conference on Computer Vision and Pattern Recognition Workshops (CVPRW), pp. 1122–1131 (2017). https://doi.org/10.1109/CVPRW.2017.150
4. Akkus, Z., Aly, Y.H., Attia, I.Z., et al.: Artificial intelligence (ai)-empowered echocardiography interpretation: a state-of-the-art review. J. Clin. Med. **10**(7) (2021). https://doi.org/10.3390/jcm10071391
5. Ashrafian, P., Yazdani, M., Heidari, M., Shahriari, D., Hacihaliloglu, I.: Vision-language synthetic data enhances echocardiography downstream tasks (2024). https://arxiv.org/abs/2403.19880
6. Becker, D.M., Tafoya, C.A., Becker, S.L., et al.: The use of portable ultrasound devices in low- and middle-income countries: a systematic review of the literature. Trop. Med. Int. Health **21**(3), 294–311 (2015). https://doi.org/10.1111/tmi.12657

7. Cammarasana, S., Nicolardi, P., Patanè, G.: Super-resolution of 2d ultrasound images and videos. Med. Biol. Eng. Comput. **61**(10), 2511–2526 (2023). https://doi.org/10.1007/s11517-023-02818-x
8. Chamsi-Pasha, M.A., Sengupta, P.P., Zoghbi, W.A.: Handheld echocardiography: current state and future perspectives. Circulation **136**(22), 2178–2188 (2017). https://doi.org/10.1161/CIRCULATIONAHA.117.026622
9. Fernández-Rodríguez, A., López-Rubio, E., Torres-Salomón, P., et al.: Enhancing echocardiography quality with diffusion neural models. In: Bioinformatics and Biomedical Engineering, pp. 169–181. Springer, Cham (2024). https://doi.org/10.1007/978-3-031-64636-2_13
10. Frija, G., Blažić, I., Frush, D.P., et al.: How to improve access to medical imaging in low- and middle-income countries ? eClinicalMedicine **38**, 101034 (2021). https://doi.org/10.1016/j.eclinm.2021.101034
11. Gifani, P., Behnam, H., Haddadi, F., et al.: Temporal super resolution enhancement of echocardiographic images based on sparse representation. IEEE Trans. Ultrason. Ferroelectr. Freq. Control **63**(1), 6–19 (2016). https://doi.org/10.1109/TUFFC.2015.2493881
12. Hamza, I., Pellikka, P.A., Abdulla, A., Ahmad, M.: Artificial intelligence echocardiography in resource-limited regions: applications and challenges. Echocardiography **41**(10), e15939 (2024). https://doi.org/10.1111/echo.15939. pMID: 39367770
13. He, K., Zhang, X., Ren, S., Sun, J.: Deep residual learning for image recognition. In: 2016 IEEE Conference on Computer Vision and Pattern Recognition (CVPR), pp. 770–778 (2016). https://doi.org/10.1109/CVPR.2016.90
14. Hirata, Y., Kusunose, K.: AI in echocardiography: state-of-the-art automated measurement techniques and clinical applications
15. Isaac, J.S., Kulkarni, R.: Super resolution techniques for medical image processing. In: 2015 International Conference on Technologies for Sustainable Development (ICTSD), pp. 1–6 (2015). https://doi.org/10.1109/ICTSD.2015.7095900
16. Leclerc, S., Smistad, E., Pedrosa, J., Ostvik, A., et al.: Deep learning for segmentation using an open large-scale dataset in 2D echocardiography. IEEE Trans. Med. Imaging **38**(9), 2198–2210 (2019). https://doi.org/10.1109/TMI.2019.2900516
17. Ledig, C., Theis, L., Huszar, F., et al.: Photo-realistic single image super-resolution using a generative adversarial network (2017). https://arxiv.org/abs/1609.04802
18. Li, S., Zhang, M., Li, Y., Sumi, C.: Segmentation with speckle reduction and super-resolution by deep leaning for human ultrasonic echo image. In: 2021 43rd Annual International Conference of the IEEE Engineering in Medicine & Biology Society (EMBC), pp. 3663–3667 (2021). https://doi.org/10.1109/EMBC46164.2021.9630440
19. Marangou, J., Beaton, A., Aliku, T.O., et al.: Echocardiography in indigenous populations and resource poor settings. Heart Lung Circ. **28**(9), 1427–1435 (2019). https://doi.org/10.1016/j.hlc.2019.05.176
20. Michelis, K.C., Narotsky, D.L., Choi, B.G.: Cardiovascular Imaging in Global Health Radiology, pp. 207–224. Springer, Cham (2019). https://doi.org/10.1007/978-1-4614-0604-4_18
21. Shin, M., Seo, M., Lee, K., Yoon, K.: Super-resolution techniques for biomedical applications and challenges

22. Vedanthan, R., Choi, B.G., Baber, U., et al.: Bioimaging and subclinical cardiovascular disease in low- and middle-income countries. J. Cardiovasc. Transl. Res. **7**(8), 701–710 (2014). https://doi.org/10.1007/s12265-014-9588-y
23. Wierda, E., Bouma, B.J., van den Brink, R.B.A.: Handheld echocardiography in patients with cardiovascular disease: to use or not to use, that is the question
24. Xochicale, M., Thwaites, L., Yacoub, S., et al.: A machine learning case study for ai-empowered echocardiography of intensive care unit patients in low- and middle-income countries (2023). https://arxiv.org/abs/2212.14510

SharpXR: Structure-Aware Denoising for Pediatric Chest X-Rays

Ilerioluwakiiye Abolade[1(✉)], Emmanuel Idoko[2], Solomon Odelola[3], Promise Omoigui[4], Adetola Adebanwo[5], Aondana Iorumbur[6], Udunna Anazodo[7], Alessandro Crimi[8], and Raymond Confidence[7]

[1] Federal University of Agriculture Abeokuta, Abeokuta, Nigeria
aboladeilerioluwakiiye@gmail.com
[2] University of Lagos, Lagos, Nigeria
[3] University of Nigeria, Nsukka, Nigeria
[4] University of Benin, Benin, Nigeria
[5] Olabisi Onabanjo University, Ago-Iwoye, Nigeria
[6] Federal University of Technology, Akure, Nigeria
[7] McGill University, Montreal, Canada
[8] AGH University of Krakow, Krakow, Poland

Abstract. Pediatric chest X-ray imaging is essential for early diagnosis, particularly in low-resource settings where advanced imaging modalities are often inaccessible. Low-dose protocols reduce radiation exposure in children but introduce substantial noise that can obscure critical anatomical details. Conventional denoising methods often degrade fine details, compromising diagnostic accuracy. In this paper, we present **SharpXR**, a structure-aware dual-decoder U-Net designed to denoise low-dose pediatric X-rays while preserving diagnostically relevant features. SharpXR combines a Laplacian-guided edge-preserving decoder with a learnable fusion module that adaptively balances noise suppression and structural detail retention. To address the scarcity of paired training data, we simulate realistic Poisson-Gaussian noise on the Pediatric Pneumonia Chest X-ray dataset. SharpXR outperforms state-of-the-art baselines across all evaluation metrics while maintaining computational efficiency suitable for resource-constrained settings. SharpXR-denoised images improved downstream pneumonia classification accuracy from 88.8% to 92.5%, underscoring its diagnostic value in low-resource pediatric care. Code is available at https://github.com/ileri-oluwa-kiiye/SharpXR.

Keywords: Structure-Aware Denoising · Pediatric X-rays · Low-Dose Imaging · Dual-Decoder Networks

1 Introduction

X-ray imaging is a cornerstone of pediatric diagnostics, especially in resource-constrained settings where access to advanced modalities like CT or MRI is limited [1]. It is commonly used to detect conditions such as pneumonia and

U. Anazodo et al. (Eds.): MIRASOL 2025, LNCS 16398, pp. 83–92, 2026.
https://doi.org/10.1007/978-3-032-13654-1_9

bone fractures in children. Due to heightened sensitivity to ionizing radiation, pediatric imaging relies on low-dose protocols [2,3]. However, these protocols significantly degrade image quality, introducing noise that can obscure subtle anatomical features critical for diagnosis [4,5]. This is particularly concerning in children, whose smaller and less ossified structures are more easily masked by noise [6,7]. The resulting diagnostic uncertainty often leads to misdiagnoses or repeat scans, exacerbating risks in low-resource environments [8].

Conventional denoising methods, including traditional filters and deep learning models, often struggle with the extreme noise levels and structural sensitivity of pediatric scans [9,10]. Most models are trained on adult or non-medical images and generalize poorly to pediatric anatomy [7]. Even established architectures like REDCNN have been shown to underperform in pediatric contexts [4]. Moreover, many methods over-smooth critical structures or fail under very low signal-to-noise ratios, impairing diagnostic utility [11,12].

To address these challenges, we propose **SharpXR**, a structure-aware dual-decoder U-Net designed for denoising pediatric low-dose chest X-rays in resource-limited settings. The architecture features two decoding branches: one optimized for smooth anatomical reconstruction and another enhanced with Laplacian-based skip connections to recover fine structures. A learnable attention-based fusion module combines these outputs, allowing the model to balance structural fidelity and noise suppression adaptively. Our main contributions are:

- We introduce **SharpXR**, a dual-decoder U-Net with structure-aware attention fusion for pediatric low-dose X-ray denoising.
- We propose a multi-scale Laplacian enhancement module to preserve anatomical boundaries without over-sharpening.
- We simulate Poisson-Gaussian noise on a real pediatric chest X-ray dataset to reflect realistic low-dose conditions.
- We demonstrate state-of-the-art denoising performance and improved pneumonia classification, underscoring clinical utility.

2 Related Work

2.1 Low-Dose Medical Image Denoising

Low-dose imaging reduces radiation exposure but introduces quantum noise that can obscure anatomical structures and reduce diagnostic accuracy [1,4]. Traditional denoising methods like BM3D [13] and non-local means [14] are effective against Gaussian noise but often oversmooth clinically important features [5].

Deep learning models have since outperformed these methods. DnCNN [15] introduced residual learning to estimate and subtract noise directly, while RED-CNN [4] applied an encoder-decoder structure with skip connections for low-dose CT denoising. ResUNet++ [16] incorporated dense skip paths and attention to preserve fine structures, and Attention U-Net [17] used gating mechanisms to suppress irrelevant features and enhance localization.

More recent transformer-based models offer improved context modeling. Hformer [18] uses windowed attention to recover high-frequency content in noisy CT images, and SwinIR [19] blends convolution and attention for better local-global integration. However, these methods often struggle under extreme noise levels typical of pediatric scans.

2.2 Pediatric-Specific Challenges and Solutions

Pediatric scans are especially sensitive to noise due to smaller, evolving anatomy [1,6]. Adult-trained models often generalize poorly, missing pediatric-specific structures [7]. The AAPM Task Group 273 [3] highlights this issue, calling for pediatric-specific model validation.

Public datasets like the Kermany collection [20] support pediatric research but lack paired clean-noisy images, limiting supervised denoising. To address this, synthetic noise is modeled using Poisson-Gaussian distributions [21,22] to simulate realistic degradation.

Few models target pediatric denoising directly. Sharp U-Net [23] added sharpening kernels to enhance edge recovery, and Luo et al. [12] applied Laplacian-based enhancements in fluoroscopy. However, these models lack adaptive fusion to balance detail preservation and noise suppression across diverse anatomical regions.

Yoon et al. [24] improved classification using structure-aware representations via masked autoencoders, showing the importance of preserving fine detail, though their work did not address denoising directly.

Our proposed SharpXR fills this gap with a dual-decoder U-Net and learnable fusion designed for structure-preserving denoising in pediatric chest X-rays.

3 Method

Denoising pediatric low-dose chest X-rays poses two major challenges: (1) the absence of paired clean-noisy datasets for supervised training and (2) the need to preserve subtle anatomical details that are critical for diagnosis. Our method addresses these challenges through realistic noise modeling and a dual-decoder network tailored for structural fidelity.

3.1 Noise Simulation

Because publicly available pediatric datasets lack paired low-dose and standard-dose images, we simulate realistic radiographic degradation to enable supervised training. We adopt a hybrid Poisson-Gaussian noise model [4,15,21]. Poisson noise accounts for photon-counting variability, while Gaussian noise simulates electronic and system-related disturbances. The noisy image

$$\tilde{X} = \frac{1}{\eta} \cdot \text{Poisson}(\eta X) + \mathcal{N}(0, \sigma^2),$$

where X denotes the clean image normalized to $[0,1]$, $\eta \in [50,300]$ controls the Poisson rate, and $\sigma \in [5,30]$ is the standard deviation of the additive Gaussian noise. Each parameter is sampled independently per image to reflect variability in exposure and hardware characteristics across scans.

3.2 Proposed Architecture: Dual-Decoder U-Net with Learnable Fusion

We propose a dual-decoder U-Net architecture designed to jointly achieve effective noise reduction and preservation of anatomical detail in pediatric chest X-rays. The model is composed of a shared encoder, two task-specific decoders, and a lightweight fusion module that adaptively blends their outputs.

Encoder. The encoder follows a standard U-Net backbone using double-convolution blocks with feature channels $\{64, 128, 256, 512\}$, culminating in a bottleneck of 1024 channels. It extracts multi-scale features $\{f_1, f_2, f_3, f_4\}$ from the input image.

Decoder for Noise Reduction. The first decoder, D_{denoise}, uses conventional skip connections to reconstruct a smoothed output image:

$$D_{\text{denoise}} : \{f_4, f_3, f_2, f_1\} \rightarrow \hat{X}_{\text{denoise}}.$$

This pathway is designed to suppress noise while preserving large-scale structural information (Fig. 1).

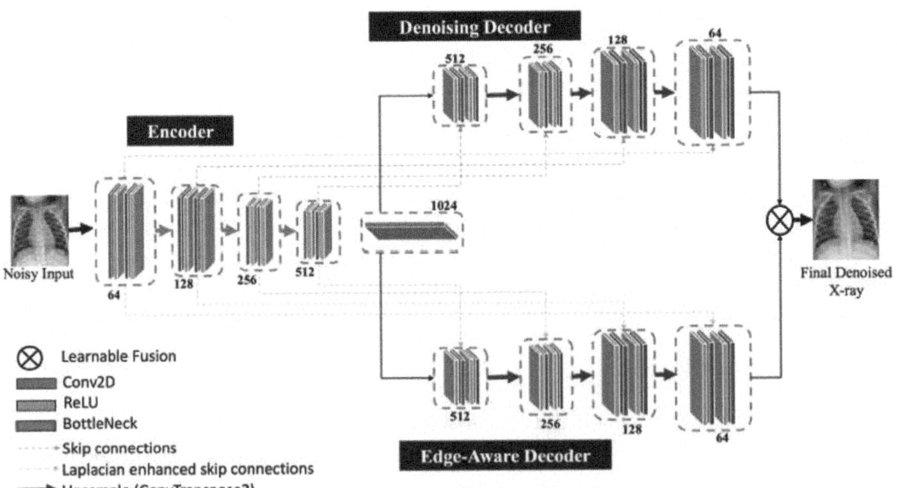

Fig. 1. SharpXR architecture. A dual-decoder U-Net with Laplacian-enhanced skip connections and attention-based fusion for denoising pediatric chest X-rays.

Decoder for Edge Preservation. The second decoder, D_{edge}, enhances the skip connections by adding Laplacian-filtered residuals, which amplify anatomical boundaries:

$$f'_i = f_i + \mathcal{L}(f_i), \quad D_{\text{edge}} : \{f'_4, f'_3, f'_2, f'_1\} \rightarrow \hat{X}_{\text{edge}}.$$

Here, $\mathcal{L}$ denotes the 2D Laplacian operator:

$$\mathcal{L} = \begin{bmatrix} -1 & -1 & -1 \\ -1 & 8 & -1 \\ -1 & -1 & -1 \end{bmatrix}.$$

This decoder is tailored to preserve fine anatomical features that are often degraded in low-dose acquisitions.

Learnable Fusion Module. The two decoder outputs are fused using a small convolutional network that generates pixel-wise attention weights:

$$\hat{X} = \alpha_1 \odot \hat{X}_{\text{denoise}} + \alpha_2 \odot \hat{X}_{\text{edge}},$$

where α_1 and α_2 are softmax-normalized weights computed from the concatenated decoder outputs, and $\odot$ denotes element-wise multiplication. This mechanism allows the model to locally balance smoothness and edge fidelity based on contextual information.

3.3 Loss Function and Training Strategy

The network is optimized using the Root Mean Square Error (RMSE) loss:

$$\mathcal{L}_{\text{RMSE}} = \sqrt{\frac{1}{N}\sum_{i=1}^{N}(\hat{X}_i - X_i)^2},$$

which penalizes intensity deviations between the prediction $\hat{X}$ and the clean target X, promoting globally consistent reconstructions.

Training was performed for up to 50 epochs using the Adam optimizer (learning rate 1×10^{-4}, batch size of 4) on an NVIDIA T4 GPU. All random seeds were fixed for reproducibility.

4 Experiments and Results

4.1 Dataset Description

We utilized the Pediatric Pneumonia Chest X-ray dataset [20], comprising 5,856 frontal-view chest radiographs from pediatric patients aged 1–5 years. This dataset has been extensively validated in pediatric diagnostic modelling applications [24–27]. All images were resized to 256×256 pixels and normalized to $[0, 1]$. Data augmentation techniques included random horizontal flipping ($p = 0.5$), brightness and contrast jittering ($\pm 10\%$), and random rotation ($\pm 15^\circ$). The dataset was split using stratified sampling into 75% training (4,391 images), 10% validation (586 images), and 15% testing (879 images).

4.2 Evaluation Metrics

We evaluate SharpXR against several state-of-the-art denoising baselines, including BM3D [13], DnCNN [15], REDCNN [4], ResUNet++ [16], Attention U-Net [17], Sharp U-Net [23], and Hformer [18], on the pediatric chest X-ray dataset with simulated Poisson-Gaussian noise. Performance is assessed using standard quantitative metrics: Root Mean Square Error (RMSE), Peak Signal-to-Noise Ratio (PSNR), Structural Similarity Index (SSIM), and Signal-to-Noise Ratio (SNR). Lower RMSE and higher PSNR, SSIM, and SNR values indicate better denoising performance. For completeness, the metrics are defined as follows:

$$\text{RMSE} = \sqrt{\frac{1}{MN}\sum_{i,j}(X_{ij} - \hat{X}_{ij})^2}, \quad \text{PSNR} = 10\log_{10}\left(\frac{MAX^2}{\text{MSE}}\right),$$

$$\text{SSIM}(X,\hat{X}) = \frac{(2\mu_X\mu_{\hat{X}} + c_1)(2\sigma_{X\hat{X}} + c_2)}{(\mu_X^2 + \mu_{\hat{X}}^2 + c_1)(\sigma_X^2 + \sigma_{\hat{X}}^2 + c_2)}, \quad \text{SNR} = 10\log_{10}\left(\frac{\sum_{i,j} X_{ij}^2}{\sum_{i,j}(X_{ij} - \hat{X}_{ij})^2}\right).$$

4.3 Overall Benchmark Performance

Table 1 summarizes test results. **SharpXR** achieves the best performance across all metrics.

Table 1. Quantitative comparison of denoising models on the pediatric chest X-ray dataset. Metrics include RMSE (lower is better), PSNR, SSIM, and SNR (higher is better). The best value in each column is highlighted in bold.

Model	RMSE	PSNR	SSIM	SNR
Attention U-Net	0.0193 ± 0.001	34.44 ± 0.12	0.8876 ± 0.004	28.76 ± 0.15
Sharp U-Net	0.0172 ± 0.001	35.40 ± 0.11	0.9261 ± 0.002	29.72 ± 0.14
HFormer	0.0271 ± 0.002	31.46 ± 0.21	0.8464 ± 0.005	25.78 ± 0.24
ResUNet++	0.0317 ± 0.003	31.89 ± 0.20	0.8569 ± 0.004	26.17 ± 0.23
REDCNN	0.0181 ± 0.001	34.95 ± 0.13	0.9140 ± 0.003	29.20 ± 0.16
DnCNN	0.0203 ± 0.001	33.97 ± 0.14	0.8945 ± 0.003	28.29 ± 0.18
BM3D	0.0346 ± 0.003	29.45 ± 0.25	0.7003 ± 0.006	22.85 ± 0.30
SharpXR	**0.0170 ± 0.001**	**35.52 ± 0.12**	**0.9263 ± 0.002**	**29.84 ± 0.13**

4.4 Noise-Level Specific Performance

Tables 2 and 3 present PSNR and SSIM values under various combinations of Poisson (η) and Gaussian (σ) noise. SharpXR consistently outperforms all baselines across varying noise conditions, demonstrating robustness to low and high photon counts and electronic noise.

Table 2. Denoised PSNR values per noise level.

σ	η	Attention U-Net	DnCNN	HFormer	REDCNN	ResUNet++	Sharp U-Net	SharpXR
5	300	36.367	36.166	33.151	37.218	33.777	37.760	37.819
10	200	35.092	34.915	32.295	35.856	32.696	36.475	36.583
15	150	33.981	33.812	31.476	34.736	31.826	35.445	35.569
20	100	32.897	32.726	30.648	33.698	30.803	34.491	34.641
25	50	31.517	31.163	29.535	32.385	29.357	33.332	33.514
30	100	31.428	31.125	29.522	32.405	29.378	33.343	33.534

Table 3. Denoised SSIM values per noise level.

σ	η	Attention U-Net	DnCNN	HFormer	REDCNN	ResUNet++	Sharp U-Net	SharpXR
5	300	0.921	0.932	0.861	0.946	0.879	0.951	0.950
10	200	0.904	0.916	0.832	0.929	0.852	0.938	0.938
15	150	0.888	0.897	0.809	0.910	0.833	0.927	0.927
20	100	0.872	0.874	0.779	0.891	0.809	0.914	0.914
25	50	0.848	0.833	0.744	0.864	0.778	0.896	0.898
30	100	0.846	0.828	0.742	0.864	0.776	0.892	0.897

4.5 Qualitative Results

Figure 2 shows qualitative results on pediatric chest X-rays. For each case (a–d), the red box in the top row indicates a region of interest, which is enlarged directly below. SharpXR removes noise while preserving diagnostic structures such as ribs and lung contours.

4.6 Downstream Classification Performance

We evaluated SharpXR's clinical utility using identical classifiers trained on noisy, denoised, and clean X-rays, achieving 88.8%, 92.5%, and 93.7% accuracy, respectively. This confirms that SharpXR enhances diagnostic performance nearly to clean-image levels.

4.7 Ablation Study

We evaluate the individual contributions of the dual decoder, Laplacian enhancement, and fusion module. As shown in Table 4, the dual decoder alone improves performance over the single-decoder baseline. Adding Laplacian filtering without fusion slightly reduces SSIM, suggesting over-sharpening. Full SharpXR, combining both with learnable fusion, achieves the best overall performance.

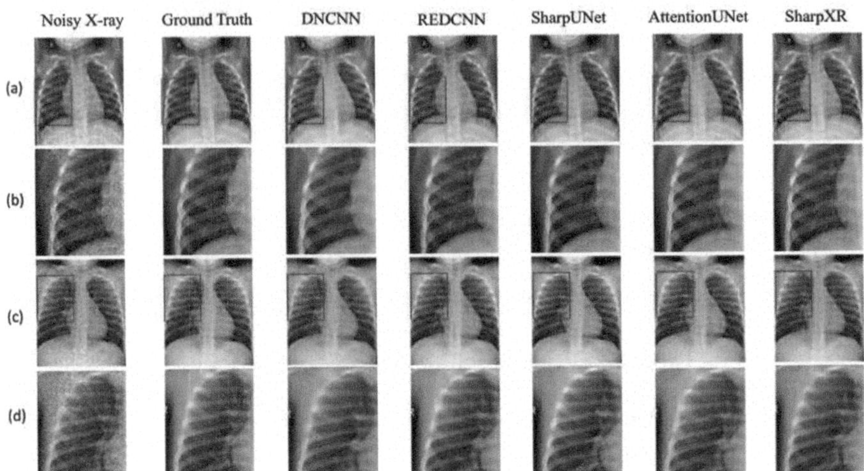

Fig. 2. Qualitative comparison of denoised pediatric chest X-rays. For each case (a–d), the red box marks a region of interest that is enlarged in the row below. SharpXR preserves structural detail while effectively reducing noise. (Color figure online)

Table 4. Ablation study on SharpXR components. Best values are in bold.

Model Variant	RMSE	PSNR	SSIM	SNR
Single Decoder	0.0175	35.23	0.9257	29.52
Dual Decoder Only	0.0173	35.43	0.9257	29.64
Dual Decoder + Laplacian (no fusion)	0.0175	35.28	0.9212	29.56
SharpXR (Full)	**0.0170**	**35.52**	**0.9263**	**29.84**

4.8 Discussion

SharpXR shows strong denoising performance across metrics and noise levels. The dual-decoder with learnable fusion maintains diagnostic structures while suppressing noise. These qualities make SharpXR promising for real-world pediatric imaging, especially in low-dose and low-resource contexts. Future studies include the collection and use of real low-dose ground truth datasets with realistic noise, which are unavailable at the current stage. Our dataset covers children aged 1–5 years; generalization to neonates and older children requires further validation due to anatomical differences.

5 Conclusion

We proposed SharpXR, a structure-aware dual-decoder U-Net tailored for denoising pediatric chest X-rays while preserving anatomical detail. By combining Laplacian-enhanced skip connections with a learnable fusion mechanism, SharpXR balances noise suppression with structural fidelity. Experimental results on simulated low-dose data demonstrate consistent improvements over traditional and learning-based baselines. However, the absence of real clinical low-dose scans in our evaluation remains a limitation. In future work, we aim to validate SharpXR on real pediatric acquisitions, explore self-supervised training to address limited labeled data, and investigate generalization to other modalities such as CT or ultrasound in resource-constrained settings.

Acknowledgements. We thank the ML Collective community for their generous compute resources and insightful weekly feedback, which helped shape the direction of this work. We are also grateful to AFRICAI for their mentorship. This work was partly supported by the Italian Ministry of University and Research (MUR) under project PE0000013 – Future of Artificial Intelligence Research (FAIR).

References

1. Power, S.P., et al.: Computed tomography and patient risk: facts, perceptions and uncertainties. World J. Radiol. **8**(12), 902–915 (2016)
2. Nagy, E., et al.: Pediatric CT made easy. Pediatr. Radiol. **53**, 581–588 (2023)
3. AAPM Task Group Report 273: recommendations on best practices for AI and machine learning for computer-aided diagnosis in medical imaging. https://www.aapm.org/pubs/reports/detail.asp?docid=235 (2023)
4. Chen, H., et al.: Low-dose CT with a residual encoder-decoder convolutional neural network. IEEE Trans. Med. Imaging **36**(12), 2524–2535 (2017)
5. Zhao, L., et al.: Nuclear Medicine Pediatric Imaging. In: StatPearls. StatPearls Publishing (2024)
6. Arthur, R.: Interpretation of the paediatric chest X-ray. Paediatr. Respir. Rev. **1**(1), 41–50 (2000)
7. Rajaraman, S., et al.: Effects of model initialization on deep model generalization: a study with adult and pediatric chest X-ray images. PLOS Digit. Health **3**(1), e0000286 (2024)
8. Rauf, S., et al.: Noise-induced systematic errors in ratio imaging: serious artefacts and correction with multi-resolution denoising. J. Microsc. **228**(2), 123–131 (2007)
9. Alzubaidi, L., et al.: Review of deep learning: concepts, CNN architectures, challenges, applications. J. Big Data **8**(1), 1–74 (2021)
10. Suzuki, K.: Overview of deep learning in medical imaging. Radiol. Phys. Technol. **10**(3), 257–273 (2017)
11. Rajaraman, S., et al.: Pre-trained CNNs for improved malaria detection. PeerJ **6**, e4568 (2018)
12. Luo, Y., et al.: Edge-enhancement DenseNet for fluoroscopy denoising. Int. J. Comput. Assist. Radiol. Surg. **17**(10), 1897–1906 (2022)
13. Dabov, K., et al.: Image denoising by sparse 3-D transform-domain collaborative filtering. IEEE Trans. Image Process. **16**(8), 2080–2095 (2007)
14. Buades, A., et al.: A non-local algorithm for image denoising. In: CVPR, vol. 2, pp. 60–65. IEEE (2005)
15. Zhang, K., et al.: Beyond a Gaussian denoiser: residual learning for image denoising. IEEE Trans. Image Process. **26**(7), 3142–3155 (2017)
16. Jha, D., et al.: ResUNet++: an advanced architecture for medical image segmentation. arXiv:1911.07067 (2019)

17. Oktay, O., et al.: Attention U-net: learning where to look for the pancreas. arXiv:1804.03999 (2018)
18. Zhang, S.Y., et al.: Hformer: highly efficient vision transformer for low-dose CT denoising. Nucl. Sci. Tech. **34**, 61 (2023)
19. Liang, J., et al.: SwinIR: image restoration using swin transformer. arXiv:2108.10257 (2021)
20. Kermany, D., et al.: Labeled OCT and chest X-ray images for classification. Mendeley Data, v2. https://doi.org/10.17632/rscbjbr9sj.2
21. Foi, A., et al.: Practical Poissonian-Gaussian noise modeling for raw-data. IEEE Trans. Image Process. **17**(10), 1737–1754 (2008)
22. Wang, J., et al.: X-ray image blind denoising in hybrid noise based on CNNs. In: WI-IAT '21 Companion (2021). https://doi.org/10.1145/3498851.3498952
23. Zunair, H., Hamza, A.B.: Sharp U-Net: depthwise ConvNet for biomedical segmentation. arXiv:2004.10342 (2020)
24. Yoon, T., Kang, D.: Enhancing pediatric pneumonia diagnosis via masked autoencoders. Sci. Rep. **14**, 3292 (2024)
25. Kundu, R., et al.: Pneumonia detection using ensemble deep learning. PLoS ONE **16**(9), e0256630 (2021)
26. Singh, S., et al.: Efficient pneumonia detection using vision transformers. Sci. Rep. **14**, 1722 (2024)
27. Al Reshan, M.S., et al.: Pneumonia detection from chest X-rays using MobileNet. Healthc. (Basel) **11**(11), 1561 (2023)

Uncertainty-Aware Evaluation of Deep Learning Object Detectors Under Scarce and Evolving Test Datasets

Esla Timothy Anzaku[1,2](✉), Mohammed Aliy Mohammed[3,4], Stefan Magez[5], Sofie Van Hoecke[3], Arnout Van Messem[6], and Wesley De Neve[1,2,3]

[1] Department of Electronics and Information Systems, Ghent University, Ghent, Belgium

[2] Center for Biosystems and Biotech Data Science, Ghent University Global Campus, Incheon, South Korea

eslatimothy.anzaku@ugent.be

[3] IDLab, Department of Electronics and Information Systems, Ghent University, Ghent, Belgium

[4] School of Biomedical Engineering, Jimma Institute of Technology, Jimma University, Jimma, Oromia, Ethiopia

[5] Department of Bio-engineering Sciences, Vrije Universiteit Brussel, Brussel, Belgium

[6] Department of Mathematics, University of Liège, Liège, Belgium

Abstract. In data-scarce domains, building reliable deep learning models often requires the relabeling, merging, or expansion of existing datasets. While these steps improve dataset quality and diversity, they complicate model evaluation: changes in evaluation outcomes may arise from dataset changes rather than real model improvement, making standard evaluation protocols difficult to interpret. We demonstrate this challenge in the context of trypanosome parasite detection, where model comparisons across three dataset versions yield inconsistent conclusions. Conventional metrics such as mean average precision (mAP) fail to reveal critical reliability issues, particularly when test data evolves. To address this, we propose a complementary evaluation approach based on predictive uncertainty. By evaluating how effectively model confidence discriminates correct from false detections, including those arising on out-of-distribution samples, we obtain a more stable and informative signal of model reliability across dataset versions. Our findings show that uncertainty-aware evaluation exposes overlooked failure modes, enables more meaningful comparisons across evolving datasets, and highlights models that maintain reliable confidence estimates under distribution shift.

Keywords: DNN Reliability · Neglected Tropical Diseases · Object Detection · Trypanosome Parasite Detection

U. Anazodo et al. (Eds.): MIRASOL 2025, LNCS 16398, pp. 93–102, 2026.
https://doi.org/10.1007/978-3-032-13654-1_10

1 Introduction

In data-scarce domains like parasite detection for neglected tropical diseases (NTDs), reliable deep neural networks (DNNs) often require iterative dataset development through relabeling, merging fragmented datasets, and expanding diversity. These necessary steps, however, complicate evaluation: as test datasets evolve, standard protocols that assume a fixed test set may yield ambiguous signals of progress, where observed changes reflect modifications in dataset composition or annotation rather than genuine model improvements.

We demonstrate these challenges in trypanosome parasite detection with DNNs, a key element of AI-based microscopy diagnostics for trypanosomiasis screening. We constructed three dataset versions: V1, comprising microscopy images of unstained thick blood smears from a public dataset; V2, a refinement of V1 with annotation corrections from model feedback and manual inspection; and V3, which extends V2 with additional samples of a different species and staining protocol from another public source. This progression reflects a plausible development path in data-scarce domains, where annotation quality and biological diversity improve over time, but also complicates evaluation by making it unclear whether observed changes in model effectiveness (using conventional object detection metrics) stem from better models or from evolving datasets. Full details are in Sect. 3.

Ambiguity in evaluation directly affects the deployment of DNNs for safety-critical applications. In point-of-care diagnostics, these models must combine high predictive accuracy with reliable uncertainty estimates, yet conventional evaluation often misses failure modes such as overconfidence on non-targets or underconfidence on visible targets. Without consistent methods to assess reliability across evolving datasets, it is difficult to measure progress or ensure safe deployment.

To address this challenge, we propose an uncertainty-aware evaluation approach that complements conventional object detection evaluation methods, which rarely assess uncertainty quality or leverage out-of-distribution samples. Our approach incorporates near out-of-distribution (near-OOD) samples. We define near-OOD samples as plausible microscopy images—such as other parasite vectors stained with Giemsa or parasite-free slides from the same acquisition process—that are likely in trypanosomiasis screening but do not contain the parasite. Such samples provide a stable reference for evaluating model behavior across dataset versions. Since declines in predictive uncertainty quality often stem from shortcut learning or overfitting, our approach quantifies how well uncertainty separates true from false positives and in-distribution from near-OOD samples. Crucially, it targets a core aspect of trustworthiness: the ability of a model to assign higher confidence to correct predictions and lower confidence to errors, a property that remains meaningful across dataset versions. Anchoring evaluation on this property provides a robust complementary path for assessing reliability as datasets and models co-evolve.

1.1 Contributions

We make the following contributions to the evaluation of DNNs in data-scarce NTD settings:

- **Problem Formalization: Evaluation Instability Under Dataset Evolution.** We identify and formalize the issue of evaluation instability when test datasets evolve over time, and demonstrate its impact on the reliability of DNN assessment in domains where datasets are scarce and fragmented.
- **Dataset Contributions: Enhanced Annotations and Biological Diversity.** We release two new dataset versions: V2, with corrected and refined annotations, and V3, which introduces additional species and microscopy capture conditions. These improvements support richer species diversity and better reflect real-world diagnostic variability.
- **Uncertainty-Aware Evaluation approach.** We propose a complementary approach that assesses the quality of the predictive uncertainty over near-OOD samples, in addition to a selected number of standard metrics, to assess model reliability. This approach provides richer model effectiveness information across dataset revisions, especially when standard evaluation approaches fail to reflect meaningful evaluation signals.
- **Reliability Analysis: Experiments and Insights Across Evolving Datasets.** We perform comprehensive experiments by training 8 object detection models and evaluating them across 3 dataset versions. Our analysis reveals critical failure modes and limitations of standard metrics under dataset evolution.

2 Related Work

DNN object detectors have the potential to transform microscopy-based diagnostics by enabling faster and more accessible disease detection in low-resource settings. Several initiatives have explored the integration of AI into global health diagnostics [5,13,17], but their success hinges on access to diverse, well-annotated datasets that support the development of reliable models.

Such datasets are essential because DNNs can generalize from minimal task specifications (e.g., microscopy images and parasite bounding boxes) when provided with a suitable objective function. However, this same capacity raises reliability concerns, as DNNs are prone to overfitting to dataset-specific biases or spurious correlations (e.g., hospital-specific artifacts or imaging markers) that undermine generalization under distribution shifts [4,7,18]—a particularly serious issue in clinical contexts where failures are hard to detect. Moreover, practices like label refinement, dataset merging, and incremental expansion, while critical for improving dataset quality, can introduce instability into evaluation [14,16]. As datasets evolve alongside model development, there is a growing need for evaluation strategies that can track model reliability and effectiveness consistently across dataset versions.

3 Methodology and Experimental Setup

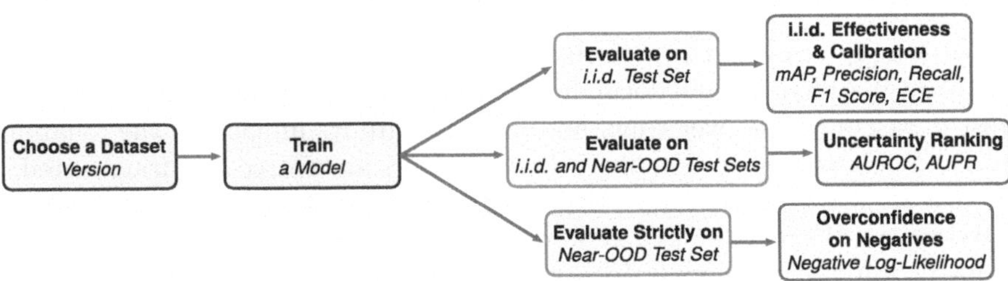

Fig. 1. Overview of our experimental pipeline. For each dataset version, models are trained and evaluated under three conditions: i.i.d., mixed, and near-OOD. Each yields distinct metrics capturing detection accuracy, uncertainty ranking, and overconfidence on negative samples.

This section describes the experimental setup for evaluating an object detector's reliability across evolving datasets, including dataset preparation, training, and evaluation, to capture changes in the model's effectiveness and uncertainty across dataset versions. Each model is evaluated in three test settings—under identically and independently distributed (i.i.d.) assumptions, mixed i.i.d. and near-OOD, and strictly OOD—using metrics for accuracy, uncertainty ranking, and overconfidence on near-OOD (Fig. 1). This process is repeated for each model–dataset pair, and results are visualized through performance trends and confidence histograms. While our primary aim is to establish a baseline of the inherent uncertainty produced by detectors, the framework is general and can also evaluate post-training uncertainty quantification methods. The following paragraphs describe each pipeline component in detail.

Datasets. The Tryp dataset [2] comprises microscopy images of *T. brucei brucei* from unstained thick smears, but we found annotation issues in it. Other datasets offer point labels [13], cropped images [19,20], or motion analysis [12], limiting their utility for DNN-based object detection.

Therefore, we curated three dataset versions. Version 1 (V1) comprises of Tryp as-is, that is, the train, validation and test partitions are the same. Version V2 (V2) comprises of the same images and dataset partitions in V1 (that is, the train, validation and test partitions are the same as those of V1), but improves the annotation quality via model-assisted and manual relabeling. Version V3 (V3) extends V2 with re-annotated giemsa-stained images of *T. cruzi* from [13], increasing domain and species diversity. To evaluate robustness and false positive suppression, we assemble a near-OOD set from two sources: (i) microscopy images from non-infected blood samples [2], and (ii) microscopy images of other infections (e.g., *Plasmodium, M. tuberculosis*) [1] captured under a similar imaging protocol as in [13].

Dataset statistics for validation splits of V1–V3 are shown in Table 1, with sample images in Figs. 2 and 3. Full dataset details are in the cited source publications.

Models. We selected object detectors representative of major architectural paradigms: two-stage (Faster R-CNN [15]) and one-stage (RetinaNet [8], RTMDet [11], RT-DETRv2 [10]) object detectors. Faster-RCNN and RetinaNet are classic deep learning object detectors, while RTMDet and RT-DETRv2 are more recent detectors. Faster-RCNN, RetinaNet, and RTMDet employ a CNN-based architecture design; RT-DETRv2 adopts a transformer-based decoder with end-to-end detection via global attention. Each model is instantiated with two backbone capacities: lightweight variants (Faster-RCNN-r18, RetinaNet-r18, RTMDet-small, RT-DETRv2-r18) and larger variants (Faster-RCNN-r50, RTMDet-tiny, RT-DETRv2-r50). ResNet backbones (*r18*, *r50*) follow [6], while *small* and *tiny* refer to the RTMDet variants introduced in [11].

Training Procedure. All models were initialized with MS-COCO [9] pre-trained weights and fine-tuned separately on each dataset version (V1–V3). Faster-RCNN, RetinaNet, and RTMDet were trained using MMDetection [3], while RT-DETRv2 followed the authors' official implementation [10]. For a given model, the same training procedure was applied across all dataset versions, but training settings differed between models according to their respective default implementations. The final checkpoint was chosen as the epoch with the highest validation mAP. Detailed training settings are documented in the accompanying code repository.[1]

Evaluation Methodology. Evaluation spans both accuracy and uncertainty-based metrics to capture model behavior under distributional shift. Two test sets are used: (i) a labeled i.i.d. set representing the same distribution as the training data, and (ii) a curated near-OOD set, composed of microscopy images collected under similar conditions but confirmed to lack trypanosome parasites. While i.i.d. predictions may include true and false positives, near-OOD predictions are expected to be sparse or have low confidence.

We report standard detection metrics (precision, recall, mAP, F1 Score) on the i.i.d. test set. To assess uncertainty quality, we compute Expected Calibration Error (ECE), Area Under the ROC Curve (AUROC), Area Under Precision/Recall Curve (AUPR-in, AUPR-out), and negative log-likelihood (NLL). AUPR-in treats i.i.d. predictions as positives, while AUPR-out treats near-OOD predictions as positives to quantify separability. NLL penalizes overconfident incorrect predictions, particularly useful under distribution shifts where there are no objects of interest, and we do not expect any predictions. Together, these metrics evaluate both detection quality and the structure of predictive uncertainty. Refer to Fig. 1 for an illustration of the evaluation methodology.

[1] https://github.com/esla/ObjectDetectionURE.

4 Results and Analyses

Lightweight and larger backbone variants show similar trends, with lightweight models slightly outperforming. We therefore report results mainly for lightweight models.

Table 1. Summary of test datasets: V1, V2, V3 (progressively refined trypanosome detection); Near-OOD (similar-condition microscopy images without parasites).

Dataset Version	Total #Images	Total #Annotations	Parasite Species	Blood Smear	Staining Protocol
V1	610	8,697	*T. brucei brucei*	Thick	Unstained
V2	610	9,914	*T. brucei brucei*	Thick	Unstained
V3	724	10,474	*T. brucei brucei, T. cruzi*	Thin, Thick	Stained, Unstained
Near-OOD	2,467	0	None (*Plasmodium, M. tuberculosis*)	Thin, Thick	Stained, Unstained

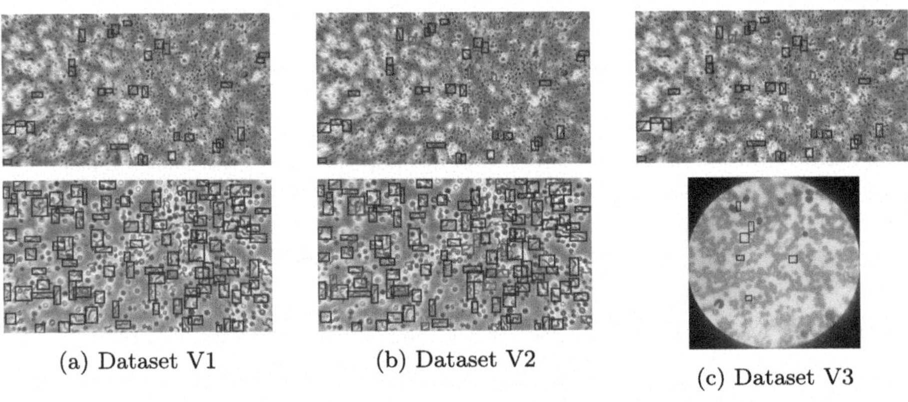

(a) Dataset V1 (b) Dataset V2 (c) Dataset V3

Fig. 2. Sample images from the three dataset versions. All blue and green boxes represent valid annotations. Green boxes highlight annotation differences between V1 and V2, which share the same underlying images but differ in annotation quality. (Color figure online)

Diverging Trends Without a Clear Winner. Figure 4 shows that no model consistently excels across all metrics. Standard detection metrics (mAP, Precision, Recall, F1) reveal performance trade-offs, while uncertainty metrics (AUROC, AUPR, ECE) capture differences in calibration and ranking quality, underscoring the need for multidimensional evaluation to assess reliability across i.i.d. and near-OOD scenarios. Top-row metrics generally improve across dataset versions, reflecting better in-distribution detection, but fail to adequately capture changes in confidence behavior under near-OOD shifts. This limitation is evident in the bottom-row metrics, where several models show a noticeable decline in AUPR-out in V3, likely due to increased visual similarity between near-OOD samples and *T. cruzi* images, which hinders confident rejection of near-OOD samples. These results highlight that accuracy metrics alone

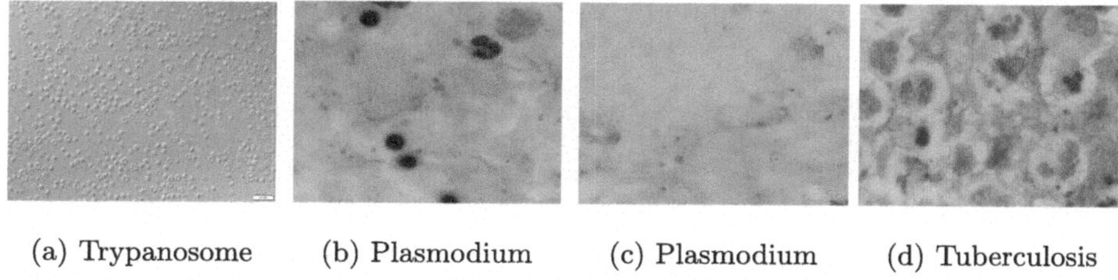

Fig. 3. A sample image from each of the dataset sources that constitutes the near-OOD test dataset.

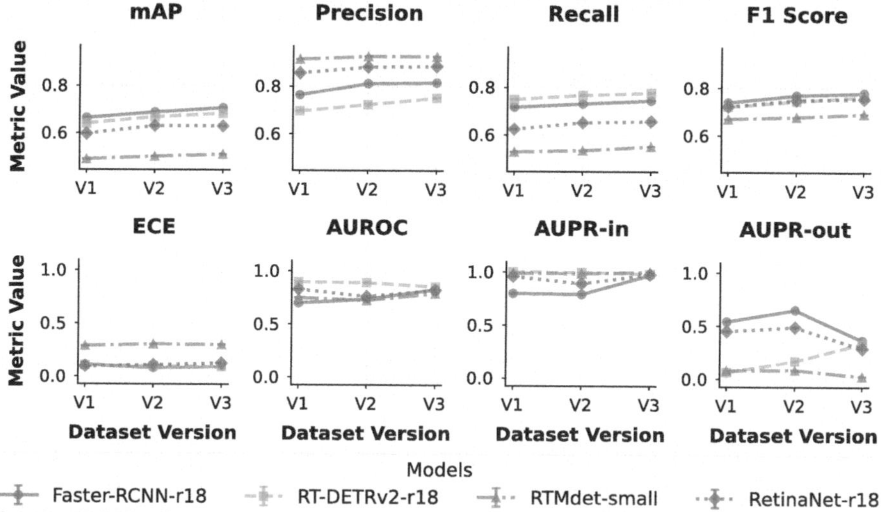

Fig. 4. Top row: standard i.i.d. evaluation metrics for detection performance, including mAP, Precision, Recall, and F1 Score. Bottom row: reliability-oriented metrics, with ECE measuring calibration on the i.i.d. set, and AUROC, AUPR-in, and AUPR-out assessing OOD detection performance. For all metrics, higher values indicate better performance, except for ECE, where lower is better.

are insufficient for diagnosing reliability issues. While uncertainty-aware metrics help quantify confidence–correctness alignment and in-/out-distribution separability, they remain insensitive to the severity of miscalibrations and may overlook rare but critical overconfidence. This shortcoming is exemplified in Fig. 5, where Faster-RCNN-r18 produces highly confident false positives on near-OOD samples despite favorable scores on mAP, F1, ECE, and AUPR-out.

The Benefit of NLL as a Complementary Metric. None of the earlier reported metrics adequately penalize overconfident incorrect predictions. As shown in Fig. 5, models like Faster-RCNN-r18 assign high confidence to false positives and near-OOD inputs while still achieving reasonable scores on mAP, F1 Score, AUPR-out, and ECE. To better capture this failure mode, we report

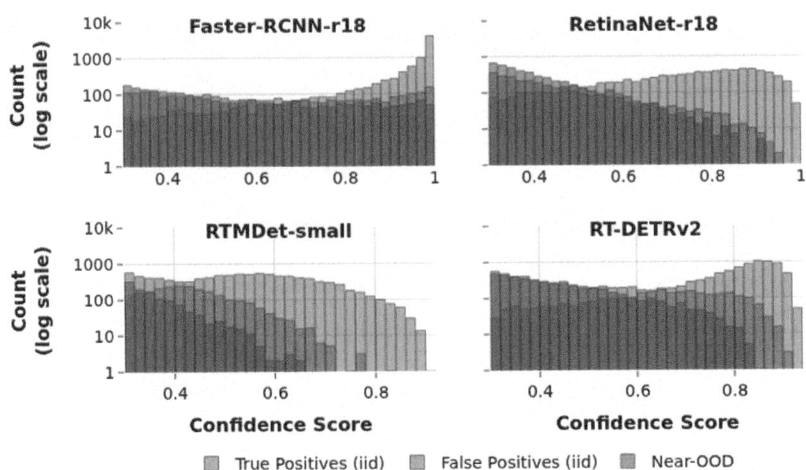

Fig. 5. Distribution of prediction confidence scores for i.i.d. true positives, i.i.d. false positives, and near-OOD predictions.

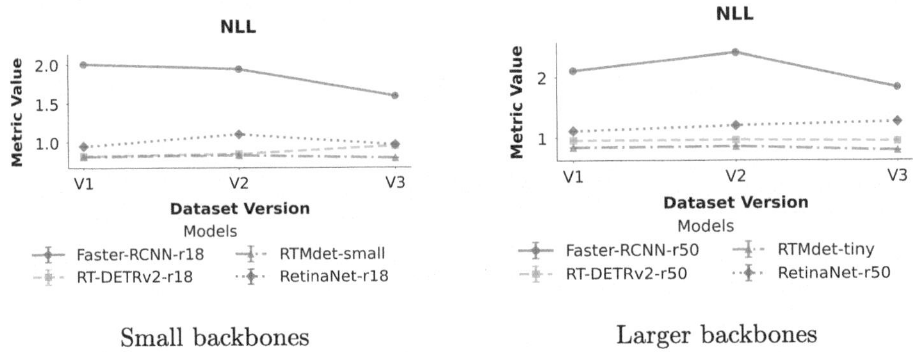

Fig. 6. NLL for all models across dataset versions. Lower values indicate better uncertainty calibration on near-OOD samples.

the NLL on near-OOD datapoints, where no detections are expected. As shown in Fig. 6, Faster-RCNN models have the highest NLL values, as NLL penalizes over-confident predictions. Based on Fig. 4, Fig. 5, and Fig. 6, RTMDet and RT-DETRv2 models demonstrate more desirable uncertainty behavior than the other evaluated models, even though they do not consistently rank best in the other metrics.

5 Conclusions

In data-scarce domains, iterative dataset improvements are often necessary but complicate model evaluation. We show that standard object detection and OOD metrics miss key reliability issues, including high-confidence in-distribution errors, poor in-/out-distribution separation, and overconfidence on near-OOD samples. By augmenting these metrics with NLL and confidence histograms, we reveal overlooked failure modes and more fully characterize model behavior. Our results underscore the need for broader evaluation as datasets evolve. RTMDet and RT-DETRv2 exhibit more desirable uncertainty behavior alongside strong accuracy, making them promising for real-world diagnostics. RetinaNet also balances accuracy and uncertainty effectively. Though designed for data-scarce settings, our evaluation approach could be applied to richer domains to enable more comprehensive assessments.

Disclosure of Interests. The authors have no competing interests to declare that are relevant to the content of this article.

References

1. Automated Blood Smear Analysis for Mobile Malaria Diagnosis. In: Karlen, W. (ed.) Mobile Point-of-Care Monitors and Diagnostic Device Design, pp. 132–149. CRC Press, 0 edn. (2018)
2. Anzaku, E.T., et al.: Tryp: a dataset of microscopy images of unstained thick blood smears for trypanosome detection. Scientific Data **10**(1), 716 (2023). https://doi.org/10.1038/s41597-023-02608-y
3. Chen, K., et al.: MMDetection: Open MMLab Detection Toolbox and Benchmark (2019). https://doi.org/10.48550/arXiv.1906.07155
4. Compton, R., Zhang, L., Puli, A., Ranganath, R.: When more is less: incorporating additional datasets can hurt performance by introducing spurious correlations. In: Proceedings of the 8th Machine Learning for Healthcare Conference, pp. 110–127 (2023)
5. Ewnetu, Y., et al.: A digital microscope for the diagnosis of Plasmodium falciparum and Plasmodium vivax, including P. falciparum with hrp2/hrp3 deletion. PLOS Global Public Health **4**(5), e0003091 (2024). https://doi.org/10.1371/journal.pgph.0003091
6. He, K., Zhang, X., Ren, S., Sun, J.: Deep residual learning for image recognition. In: IEEE Conference on Computer Vision and Pattern Recognition, pp. 770–778 (2016). https://doi.org/10.1109/CVPR.2016.90
7. Jabbour, S., Fouhey, D., Kazerooni, E., Sjoding, M.W., Wiens, J.: Deep learning applied to chest X-Rays: exploiting and preventing shortcuts. In: Proceedings of the 5th Machine Learning for Healthcare Conference, pp. 750–782 (2020)
8. Lin, T.Y., Goyal, P., Girshick, R.B., He, K., Doll r, P.: Focal loss for dense object detection. In: 2017 IEEE International Conference on Computer Vision (ICCV), pp. 2999–3007 (2017)
9. Lin, T.Y., et al.: Microsoft COCO: common objects in context. In: Fleet, D., Pajdla, T., Schiele, B., Tuytelaars, T. (eds.) European Conference on Computer Vision, pp. 740–755 (2014). https://doi.org/10.1007/978-3-319-10602-1_48
10. Lv, W., Zhao, Y., Chang, Q., Huang, K., Wang, G., Liu, Y.: RT-DETRv2: Improved Baseline with Bag-of-Freebies for Real-Time Detection Transformer (2024). https://doi.org/10.48550/arXiv.2407.17140
11. Lyu, C., et al.: RTMDet: An Empirical Study of Designing Real-Time Object Detectors (2022). https://doi.org/10.48550/arXiv.2212.07784

12. Martins, G.L., Ferreira, D.S., Ramalho, G.L.B.: Collateral motion saliency-based model for *Trypanosoma cruzi* detection in dye-free blood microscopy. Comput. Biol. Med. **132**, 104220 (2021). https://doi.org/10.1016/j.compbiomed.2021.104220
13. Morais, M.C.C., et al.: Automatic detection of the parasite Trypanosoma cruzi in blood smears using a machine learning approach applied to mobile phone images. PeerJ **10**, e13470 (2022). https://doi.org/10.7717/peerj.13470
14. Recht, B., Roelofs, R., Schmidt, L., Shankar, V.: Do ImageNet classifiers generalize to ImageNet? In: Chaudhuri, K., Salakhutdinov, R. (eds.) Proceedings of the 36th International Conference on Machine Learning, vol. 97, pp. 5389–5400 (2019)
15. Ren, S., He, K., Girshick, R., Sun, J.: Faster R-CNN: towards real-time object detection with region proposal networks. In: Advances in Neural Information Processing Systems, vol. 28 (2015)
16. Ruckert, J., et al.: ROCOv2: Radiology Objects in COntext Version 2, An Updated Multimodal Image Dataset (2023). https://zenodo.org/records/8333645
17. Ward, P., et al.: Affordable artificial intelligence-based digital pathology for neglected tropical diseases: a proof-of-concept for the detection of soil-transmitted helminths and Schistosoma mansoni eggs in Kato-Katz stool thick smears. PLoS Negl. Trop. Dis. **16**(6), e0010500 (2022). https://doi.org/10.1371/journal.pntd.0010500
18. Zech, J.R., Badgeley, M.A., Liu, M., Costa, A.B., Titano, J.J., Oermann, E.K.: Variable generalization performance of a deep learning model to detect pneumonia in chest radiographs: a cross-sectional study. PLoS Med. **15**(11), e1002683 (2018). https://doi.org/10.1371/journal.pmed.1002683
19. Zhang, C., et al.: Deep learning for microscopic examination of protozoan parasites. Comput. Struct. Biotechnol. J. **20**, 1036–1043 (2022). https://doi.org/10.1016/j.csbj.2022.02.005
20. Zhang, Y., Jiang, H., Ye, T., Juhas, M.: Deep learning for imaging and detection of microorganisms. Trends Microbiol. **29**(7), 569–572 (2021). https://doi.org/10.1016/j.tim.2021.01.006

Advancing Fetal Ultrasound Image Quality Assessment in Low-Resource Settings

Dongli He, Hu Wang, and Mohammad Yaqub(✉)

Mohamed bin Zayed, University of Artificial Intelligence, Abu Dhabi, United Arab Emirates
{dongli.he,hu.wang,mohammad.yaqub}@mbzuai.ac.ae

Abstract. Accurate fetal biometric measurements, such as abdominal circumference, play a vital role in prenatal care. However, obtaining high-quality ultrasound images for these measurements heavily depends on the expertise of sonographers, posing a significant challenge in low-income countries due to the scarcity of trained personnel. To address this issue, we leverage FetalCLIP, a vision-language model pretrained on a curated dataset of over 210,000 fetal ultrasound image-caption pairs, to perform automated fetal ultrasound image quality assessment (IQA) on blind-sweep ultrasound data. We introduce FetalCLIP$_{CLS}$, an IQA model adapted from FetalCLIP using Low-Rank Adaptation (LoRA), and evaluate it on the ACOUSLIC-AI dataset against six CNN and Transformer baselines. FetalCLIP$_{CLS}$ achieves the highest F1 score of 0.757. Moreover, we show that an adapted segmentation model, when repurposed for classification, further improves performance, achieving an F1 score of 0.771. Our work demonstrates how parameter-efficient fine-tuning of fetal ultrasound foundation models can enable task-specific adaptations, advancing prenatal care in resource-limited settings. The experimental code is available at: https://github.com/donglihe-hub/FetalCLIP-IQA.

Keywords: Fetal ultrasound · Image quality assessment · Parameter-efficient fine-tuning

1 Introduction

Ultrasound has long played a vital role in routine prenatal care [26]. Fetal biometric measurements, such as abdominal circumference, biparietal diameter, and femur length, are essential for assessing fetal growth and monitoring high-risk pregnancies. However, acquiring accurate measurements requires considerable expertise from sonographers to identify appropriate standard planes. In regions where experienced personnel are scarce, blind-sweep ultrasound is often employed [19]. During the acquisition of blind-sweep scans, novice operators use low-cost portable probes to perform free-hand sweeps without viewing the ultrasound images. As a result, these scans usually yield large volumes of low-quality data that may lack the precise anatomical planes conventionally required

U. Anazodo et al. (Eds.): MIRASOL 2025, LNCS 16398, pp. 103–113, 2026.
https://doi.org/10.1007/978-3-032-13654-1_11

for accurate biometric assessment. Automating the identification of high-quality frames within these sweeps is therefore crucial for enhancing workflow efficiency and enabling scalable prenatal screening in resource-limited settings.

Recent advances in machine learning have demonstrated remarkable capabilities in medical domains. One important application is automated assessment of fetal ultrasound image quality [27], which is particularly useful for free-hand ultrasound sequences acquired through blind-sweep protocols as it enables accurate biometric measurements even in the absence of skilled sonographers.

To develop robust machine learning models, a common strategy is to train them on data collected from the same regions where they will be deployed. However, real-world deployment of medical models in resource-constrained settings is often limited by scarce data and computational resources. An effective alternative is to leverage large pretrained models, commonly referred to as foundation models, which are trained on extensive datasets and subsequently adapted to specific downstream tasks through transfer learning.

In this work, we build our models upon FetalCLIP [13], a vision-language foundation model specifically trained for fetal ultrasound, to address the task of fetal ultrasound image quality assessment (IQA). Our objective is to assist less-experienced sonographers in identifying high-quality frames for abdominal circumference measurement. To enable efficient model adaptation, we employ parameter-efficient fine-tuning using Low-Rank Adaptation (LoRA) [7], which allows efficient task-specific tuning with minimal trainable parameters. This approach is well-suited for deployment in settings with limited computational resources. Our main contributions are as follows:

- We propose FetalCLIP$_{CLS}$, a model adapted from the fetal ultrasound foundation model FetalCLIP using LoRA. FetalCLIP$_{CLS}$ consistently outperforms strong CNN and Transformer baselines on the fetal ultrasound IQA task, while requiring only a small number of trainable parameters.
- We demonstrate the feasibility of repurposing a segmentation model for IQA tasks. This adapted segmentation model, named FetalCLIP$_{SEG}$, further improves classification performance, achieving a higher F1 score and recall compared to classification approaches.

2 Related Work

Methods for image quality assessment (IQA) [12] can be broadly classified into two categories: statistical approaches and machine learning-based techniques. Statistical methods include human visual system-based metrics such as the Structural Similarity Index (SSIM) [25], transform domain-based techniques like BLIINDS-II [17], and natural scene statistics-based approaches such as the Information Fidelity Criterion (IFC) [21].

Machine learning-based IQA methods can be categorized into traditional machine learning, convolutional neural network (CNN)-based, and Transformer-based approaches. Traditional machine learning methods such as BRISQUE [15] rely on hand-crafted features derived from natural scene statistics to estimate image quality. CNN-based methods such as IQA-CNN [9] extract deep features from image data to improve prediction accuracy. More recently, Transformer-based methods like TRIQ [28] have been proposed to address the locality bias inherent in CNNs by capturing long-range dependencies, leading to improved performance on IQA tasks.

State-of-the-art fetal ultrasound IQA methods predominantly rely on machine learning. Wu et al. [27] employ two CNNs to assess ultrasound scans of the fetal abdominal region, demonstrating performance comparable to subjective ratings from medical experts. Cengiz et al. [2] propose an automatic method to evaluate the quality of predicted segmentation masks. Boumeridja et al. [1] introduce a super-resolution technique that enhances ultrasound image resolution to improve downstream classification performance in low-resource settings.

The majority of models for fetal ultrasound IQA have adopted pretrained models due to their superior performance [20]. However, most publicly available foundation models are pretrained on natural images, which may not be optimal for domain-specific tasks. FetalCLIP is a foundation model pretrained on fetal ultrasound image-caption pairs using Contrastive Language-Image Pretraining (CLIP) [16]. We are interested in evaluating its transferability in low-resource settings.

3 Methodology

3.1 Task Formulation

We formulate fetal ultrasound image quality assessment (IQA) as a binary classification problem to identify frames containing clear anatomical structures suitable for fetal biometric measurement. Formally, given a 2D ultrasound frame X, the model predicts a binary label $y \in \{0, 1\}$, where $y = 1$ indicates the presence of relevant anatomical structures and $y = 0$ indicates their absence.

3.2 Model Architecture

We utilize the image encoder from FetalCLIP [13], which was trained on over 210,000 fetal ultrasound image-caption pairs using CLIP [16]. As illustrated in Fig. 1(a), this encoder captures rich semantic and spatial features by aligning ultrasound images with their corresponding textual descriptions during pretraining.

In our approach, we keep the pretrained image encoder frozen and insert LoRA modules within the attention and feed-forward blocks to enable parameter-efficient fine-tuning.

Ultrasound scans are passed through the encoder to produce a one-dimensional embedding that encapsulates semantic and structural information about the images. A classification head, consisting of a single fully connected layer, is appended to the image encoder to map the embedding to a binary prediction indicating whether the frame is suitable for fetal biometric measurement. We refer to this classification pipeline as FetalCLIP_{CLS}, as illustrated in Fig. 1(b).

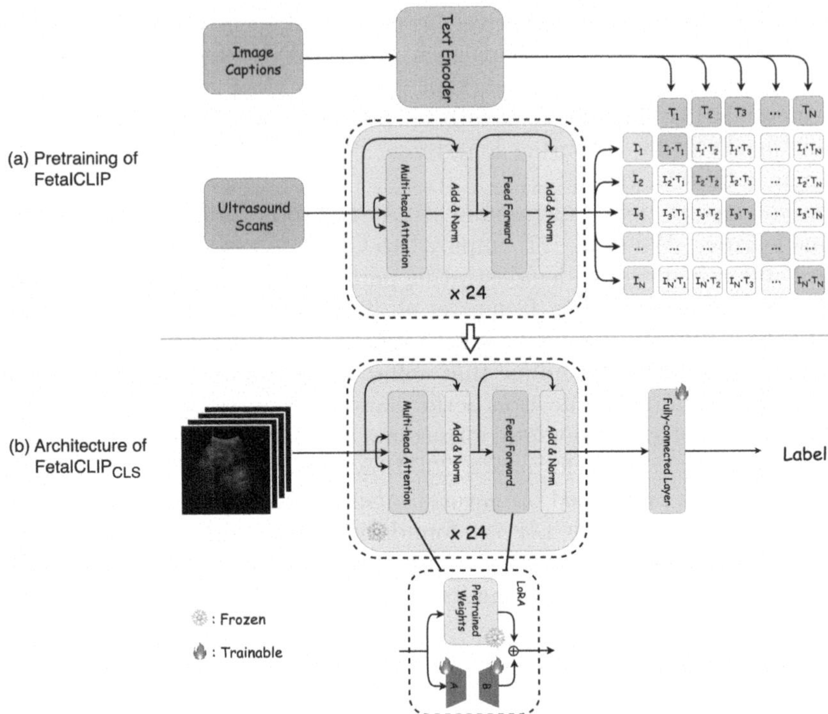

Fig. 1. (a) Contrastive pretraining of FetalCLIP, where the model learns to align ultrasound scans with their corresponding text descriptions. (b) FetalCLIP$_{CLS}$ utilizes the pretrained image encoder from FetalCLIP followed by a linear head to predict whether a frame is optimal for fetal biometric measurement. The image encoder remains frozen during FetalCLIP$_{CLS}$ training, with only the LoRA modules and linear head being trainable.

3.3 Leveraging Segmentation Model for Classification Task

We propose that the ability to segment key anatomical structures serves as an effective proxy for image quality in standard plane classification. Based on the hypothesis that a model capable of perfect segmentation can also classify an image without error, we propose FetalCLIP$_{SEG}$, a segmentation model whose predictions are converted via a thresholding strategy into binary classification labels.

This model leverages the same frozen image encoder from FetalCLIP with embedded LoRA modules for parameter-efficient fine-tuning, followed by a lightweight U-shaped network inspired by UNETR [6]. Ground truth masks, as detailed in Sect. 4.1, are used during training to guide the segmentation process.

4 Experiments

4.1 Dataset

We use a 2D B-mode fetal ultrasound dataset from the MICCAI 2024 ACOUSLIC-AI Challenge [18]. The dataset was originally developed for operator-agnostic abdominal circumference measurement in low-income countries. It comprises fetal abdominal ultrasound scans collected from pregnant women between 20 and 32 weeks of gestation in Sierra Leone and Tanzania. The scans were acquired by novice users with only one hour of training using a low-cost portable ultrasound probe, and each scan is accompanied by an expert-annotated mask. Since not all frames contain abdominal structures, some masks may be empty. The blind-sweep acquisition protocol involves six sweeps over the gravid abdomen: three transverse sweeps in a caudocranial direction and three sagittal sweeps from the patient's left to right, with each sweep containing 140 frames.

In the challenge, a total of 300 cases are provided for training. On average, only 2.6 out of every 100 frames contain abdominal structures, resulting in a highly imbalanced label distribution. To mitigate this imbalance, we exclude sweeps without annotated masks. This filtering process yields a more balanced dataset, increasing the average number of frames with clear abdominal planes to 8.6 per 100 scans.

Since the official validation and test sets are not publicly available, we randomly split the 300 cases into 210 for training, 30 for validation, and 60 for testing. To prevent data leakage, all sweeps from the same patient are assigned to the same split. This results in 52,500 training, 8,540 validation, and 16,380 test ultrasound images.

The dataset provides annotation masks labeled as either optimal or suboptimal for abdominal circumference measurement. As both types contain sufficient anatomical information for biometric measurement, we assign $y = 1$ to both, although optimal frames are preferred when available.

4.2 Evaluation Metrics

We evaluate model performance using widely adopted classification metrics that focus on different aspects: overall correctness (accuracy), the ability to identify optimal frames (precision and recall), and the harmonic mean of precision and recall (F1 score).

4.3 Implementation Details

During preprocessing, each image is first padded to a square shape. Data augmentation is then applied to the training set, including color jittering, contrast-limited adaptive histogram equalization (CLAHE) [29], and affine transformations. We generate two additional augmented versions for each training frame. All images are subsequently resized to 224×224 pixels.

We train all models using the AdamW optimizer [11] with a fixed learning rate of 3×10^{-4} for 5 epochs. All experiments are conducted on a single NVIDIA RTX 4090 GPU. Each experiment is repeated five times, and the checkpoint with the lowest validation loss is selected for evaluation on the test set.

All classification models are optimized using binary cross-entropy loss, while FetalCLIP$_{SEG}$ is trained with the Dice loss [14]. To convert segmentation outputs into binary labels, we apply a thresholding strategy: if the number of foreground pixels in the predicted mask exceeds 1% of the image area, the frame is labeled as $y = 1$; otherwise, it is labeled as $y = 0$. This 1% threshold was chosen empirically. Given an average mask ratio of 9.5% and a minimum of 1.8%, a 1% threshold provides a reasonable criterion for distinguishing valid anatomical structures from spurious noise.

4.4 Results

We compare FetalCLIP$_{CLS}$ against six baseline models: three CNN-based models—DenseNet [8], EfficientNet [23], and VGG [22]—and three Transformer-based models—Swin Transformer [10], DeiT [24], and Vision Transformer (ViT) [5].

Table 1. Model performance on fetal ultrasound IQA. We compare FetalCLIP$_{CLS}$ against six baseline models. Metrics are reported as mean ± standard deviation across five independent runs. The last column indicates the number of trainable parameters for each model. The best average scores for each metric are **bolded**, while second-best scores are <u>underlined</u>.

Architecture	Models	Accuracy↑	F1 Score↑	Precision↑	Recall↑	# Trainable
CNN	DenseNet	0.9516 ± 0.002	0.7024 ± 0.028	<u>0.7805</u> ± 0.026	0.6420 ± 0.059	7.0 M
	EfficientNet	0.9537 ± 0.004	0.7253 ± 0.030	0.7725 ± 0.025	0.6855 ± 0.053	4.0 M
	VGG	0.9510 ± 0.002	0.7084 ± 0.021	0.7580 ± 0.023	0.6671 ± 0.048	134 M
Transformer	Swin	0.9565 ± 0.003	0.7429 ± 0.039	**0.7864** ± 0.032	0.7113 ± 0.087	1.7 M
	DEIT	0.9554 ± 0.001	0.7466 ± 0.014	0.7619 ± 0.035	0.7363 ± 0.059	2.4 M
	ViT$_{400M}$	<u>0.9560</u> ± 0.003	<u>0.7506</u> ± 0.019	0.7657 ± 0.042	**0.7417** ± 0.067	2.4 M
	FetalCLIP$_{CLS}$	**0.9575** ± 0.001	**0.7570** ± 0.007	0.7782 ± 0.034	<u>0.7397</u> ± 0.041	2.4 M

The CNN baselines are pretrained on ImageNet-1K [4], and we perform full-parameter fine-tuning for these models. For the Transformer baselines, we freeze the encoder and apply LoRA, following an architecture similar to FetalCLIP$_{CLS}$. In fact, the only difference lies in the choice of encoder for Transformer-based models. We choose Transformer baselines with a comparable number of parameters to FetalCLIP$_{CLS}$ to ensure a fair comparison. Swin Transformer and DeiT are pretrained on ImageNet-22K, while the ViT baseline is pretrained on WIT-400M image-text pairs using CLIP [3].

As shown in Table 1, FetalCLIP$_{CLS}$ achieves superior performance in accuracy and F1 score compared to all baselines while maintaining competitive precision and recall with a small number of trainable parameters. Note that FetalCLIP is pretrained on over 210,000 fetal ultrasound image-caption pairs. Although the ViT baseline is pretrained on 400M image-text pairs—a dataset more than 1,900 times larger than that used for FetalCLIP—FetalCLIP$_{CLS}$ still outperforms ViT$_{400M}$ in accuracy, F1 score, and precision. This demonstrates the effectiveness of foundation models' domain-specific pretraining and highlights the advantages of FetalCLIP$_{CLS}$ for tasks requiring high-quality fetal ultrasound representations.

We observe that Transformer-based models consistently outperform CNN-based models in accuracy, F1 score, and recall. Although CNNs were once the preferred choice for many machine learning practitioners, they have been surpassed by Transformers for fetal ultrasound IQA tasks, according to our experimental results. As a general recommendation, practitioners seeking superior model performance should favor Transformer-based models over CNNs when developing fetal ultrasound IQA systems.

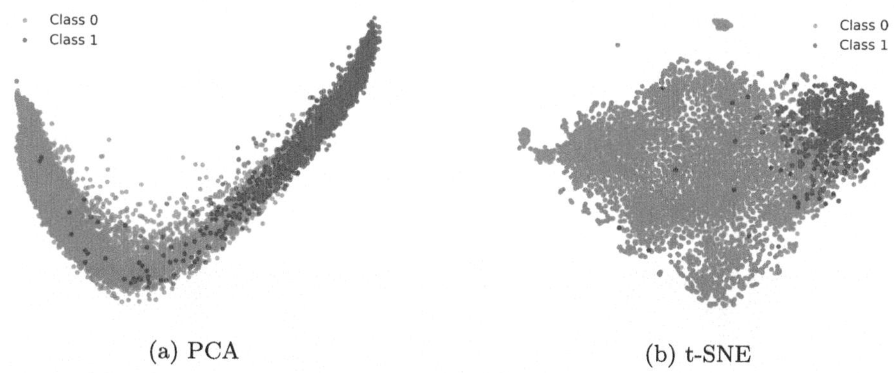

Fig. 2. Visualizations of the feature embeddings extracted from the penultimate layer of FetalCLIP$_{CLS}$, using PCA and t-SNE.

We present PCA and t-SNE visualizations of the feature embeddings extracted from FetalCLIP$_{CLS}$ in Fig. 2. Both plots exhibit relatively low intra-class variance, indicating discriminative embeddings. However, noticeable inter-class overlap is observed, which may be attributed to the similarity of suboptimal frames adjacent to the optimal ones, reflecting the inherent difficulty of the dataset.

4.5 Repurposing Segmentation Model for IQA

As shown in Table 2, FetalCLIP$_{SEG}$ achieves a higher F1 score and recall compared to FetalCLIP$_{CLS}$. Moreover, the strong Dice score (0.7244) evaluated on the test dataset confirms that the segmentation decoder produces accurate masks. The results demonstrate the feasibility of repurposing a segmentation model for classification tasks through mask thresholding and show that an adapted segmentation model, supervised with pixel-wise annotations, can achieve remarkable classification performance, even surpassing a dedicated classification model on critical metrics.

Table 2. Model performance of FetalCLIP$_{CLS}$ and FetalCLIP$_{SEG}$. The best average scores for each metric are **bolded**.

Models	DICE↑	Accuracy↑	F1 Score↑	Precision↑	Recall↑	# Trainable
FetalCLIP$_{CLS}$	/	**0.9575** $\pm$ 0.001	0.7570 $\pm$ 0.007	**0.7782** $\pm$ 0.034	0.7397 $\pm$ 0.041	2.4 M
FetalCLIP$_{SEG}$	0.7244 $\pm$ 0.007	0.9543 $\pm$ 0.001	**0.7708** $\pm$ 0.005	0.6988 $\pm$ 0.010	**0.8599** $\pm$ 0.017	4.0 M

The primary drawback of FetalCLIP$_{SEG}$ is its reduced precision (0.6988), approximately 8% lower than that of FetalCLIP$_{CLS}$. These results suggest that while the segmentation-based approach is effective in identifying relevant anatomical structures, it may also produce more false positives by potentially classifying suboptimal or ambiguous frames as positive. This can in turn raise the risk of inaccurate biometric measurements derived from misclassified frames.

On the other hand, since the encoder remains frozen and the introduced decoder is lightweight, FetalCLIP$_{SEG}$ maintains computational efficiency and remains well-suited for deployment in environments with limited computational resources.

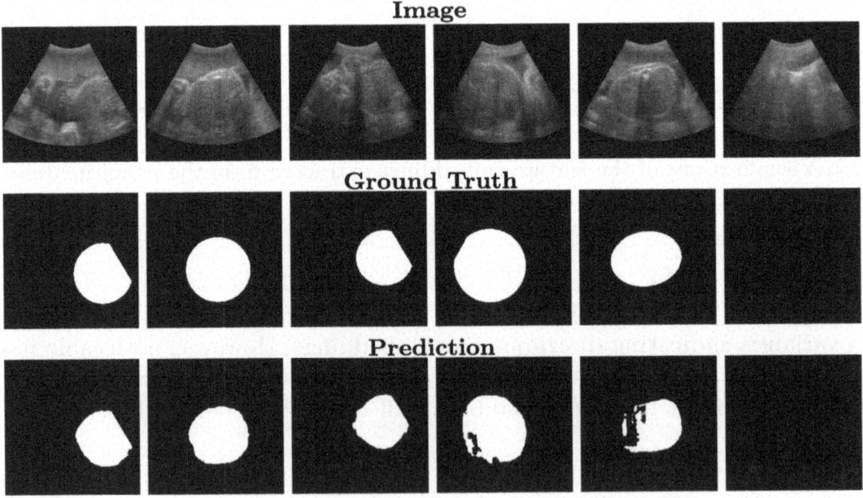

Fig. 3. Each vertical set displays the ultrasound image (top), ground truth annotation (middle) and its corresponding predicted segmentation mask (bottom). In the rightmost example, both the ground truth and prediction contain no mask, indicating a correctly segmented frame.

Figure 3 presents six sets of ultrasound image, ground truth annotations, along with their predicted segmentation masks, illustrating FetalCLIP$_{SEG}$'s ability to accurately localize the abdominal region in fetal ultrasound frames.

5 Conclusion

In this work, we investigated the application of FetalCLIP for fetal ultrasound IQA, aiming to assist less-experienced sonographers in identifying suitable frames for abdominal circumference measurement from blind sweeps.

Our results demonstrate that FetalCLIP produces more effective ultrasound-specific embeddings than several of the most commonly used foundation models. FetalCLIP$_{CLS}$ consistently outperforms CNN and Transformer baselines in both accuracy and F1 score while maintaining competitive precision and recall with low computational overhead. Additionally, incorporating mask supervision in FetalCLIP$_{SEG}$ yields an improved F1 score and recall for optimal frame selection, albeit at the cost of precision.

Our study highlights that ultrasound-specific foundation models can improve diagnostic accuracy while being deployable in a resource-efficient way. Future work may extend this framework to additional fetal biometry tasks and incorporate datasets spanning a broader range of difficulty levels, facilitating more diverse and robust analyses aimed at improving prenatal care in low-resource settings.

References

1. Boumeridja, H., et al.: Enhancing fetal ultrasound image quality and anatomical plane recognition in low-resource settings using super-resolution models. Sci. Rep. **15**, 8376 (2025). https://doi.org/10.1038/s41598-025-91808-0
2. Cengiz, S., Almakky, I., Yaqub, M.: Fusqa: fetal ultrasound segmentation quality assessment (2023). https://arxiv.org/abs/2303.04418
3. Cherti, M., et al.: Reproducible scaling laws for contrastive language-image learning. In: 2023 IEEE/CVF Conference on Computer Vision and Pattern Recognition (CVPR), pp. 2818–2829. IEEE (2023). https://doi.org/10.1109/cvpr52729.2023.00276
4. Deng, J., Dong, W., Socher, R., Li, L.J., Li, K., Fei-Fei, L.: Imagenet: a large-scale hierarchical image database. In: 2009 IEEE Conference on Computer Vision and Pattern Recognition, pp. 248–255 (2009). https://doi.org/10.1109/CVPR.2009.5206848
5. Dosovitskiy, A., et al.: An image is worth 16x16 words: transformers for image recognition at scale (2021). https://arxiv.org/abs/2010.11929
6. Hatamizadeh, A., et al.: Unetr: transformers for 3d medical image segmentation (2021). https://arxiv.org/abs/2103.10504
7. Hu, E.J., et al.: Lora: low-rank adaptation of large language models (2021). https://arxiv.org/abs/2106.09685

8. Huang, G., Liu, Z., van der Maaten, L., Weinberger, K.Q.: Densely connected convolutional networks (2018). https://arxiv.org/abs/1608.06993
9. Kang, L., Ye, P., Li, Y., Doermann, D.: Convolutional neural networks for no-reference image quality assessment. In: 2014 IEEE Conference on Computer Vision and Pattern Recognition, pp. 1733–1740 (2014). https://doi.org/10.1109/CVPR.2014.224
10. Liu, Z., et al.: Swin transformer: hierarchical vision transformer using shifted windows (2021). https://arxiv.org/abs/2103.14030
11. Loshchilov, I., Hutter, F.: Decoupled weight decay regularization (2019). https://arxiv.org/abs/1711.05101
12. Ma, C., Shi, Z., Lu, Z., Xie, S., Chao, F., Sui, Y.: A survey on image quality assessment: insights, analysis, and future outlook (2025). https://arxiv.org/abs/2502.08540
13. Maani, F., et al.: Fetalclip: a visual-language foundation model for fetal ultrasound image analysis (2025). https://arxiv.org/abs/2502.14807
14. Milletari, F., Navab, N., Ahmadi, S.A.: V-net: fully convolutional neural networks for volumetric medical image segmentation (2016). https://arxiv.org/abs/1606.04797
15. Mittal, A., Moorthy, A.K., Bovik, A.C.: No-reference image quality assessment in the spatial domain. IEEE Trans. Image Process. **21**(12), 4695–4708 (2012). https://doi.org/10.1109/TIP.2012.2214050
16. Radford, A., et al.: Learning transferable visual models from natural language supervision (2021). https://arxiv.org/abs/2103.00020
17. Saad, M.A., Bovik, A.C., Charrier, C.: Blind image quality assessment: a natural scene statistics approach in the dct domain. IEEE Trans. Image Process. **21**(8), 3339–3352 (2012). https://doi.org/10.1109/TIP.2012.2191563
18. Sappia, M.S., et al.: Acouslic-ai challenge report: Fetal abdominal circumference measurement on blind-sweep ultrasound data from low-income countries. Med. Image Anal. **105**, 103640 (2025). https://doi.org/10.1016/j.media.2025.103640
19. Self, A., et al.: Developing clinical artificial intelligence for obstetric ultrasound to improve access in underserved regions: protocol for a computer-assisted low-cost point-of-care ultrasound (calopus) study. JMIR Res. Protoc. **11**(9), e37374 (2022). https://doi.org/10.2196/37374
20. Sendra-Balcells, C., et al.: Generalisability of fetal ultrasound deep learning models to low-resource imaging settings in five African countries (2023). https://arxiv.org/abs/2209.09610
21. Sheikh, H., Bovik, A.: Image information and visual quality. IEEE Trans. Image Process. **15**(2), 430–444 (2006). https://doi.org/10.1109/TIP.2005.859378
22. Simonyan, K., Zisserman, A.: Very deep convolutional networks for large-scale image recognition (2015). https://arxiv.org/abs/1409.1556
23. Tan, M., Le, Q.V.: Efficientnet: rethinking model scaling for convolutional neural networks (2020). https://arxiv.org/abs/1905.11946
24. Touvron, H., Cord, M., Jégou, H.: Deit iii: revenge of the vit (2022). https://arxiv.org/abs/2204.07118
25. Wang, Z., Bovik, A., Sheikh, H., Simoncelli, E.: Image quality assessment: from error visibility to structural similarity. IEEE Trans. Image Process. **13**(4), 600–612 (2004). https://doi.org/10.1109/TIP.2003.819861
26. Whitworth, M., Bricker, L., Neilson, J., Dowswell, T.: Ultrasound for fetal assessment in early pregnancy. Cochrane Database Syst. Rev. (2010). https://doi.org/10.1002/14651858.CD007058.pub3

27. Wu, L., Cheng, J.Z., Li, S., Lei, B., Wang, T., Ni, D.: Fuiqa: fetal ultrasound image quality assessment with deep convolutional networks. IEEE Trans. Cybern. **47**(5), 1336–1349 (2017). https://doi.org/10.1109/TCYB.2017.2671898
28. Yang, S., et al.: Maniqa: multi-dimension attention network for no-reference image quality assessment (2022). https://arxiv.org/abs/2204.08958
29. Zuiderveld, K.: Contrast Limited Adaptive Histogram Equalization, pp. 474–485. Academic Press Professional, Inc., USA (1994)

Enhancing and Accelerating Vessel Annotations in Medical Imaging

Mehran Azimbagirad[1(✉)], Shahab Aslani[1], Shanshan Wang[1], Pardeep Vasudev[1], Alfred So[1,2], John McCabe[1], and Joseph Jacob[1]

[1] Department of Computer Science, Satsuma Lab, Hawkes's Institute, University College London, London, UK
m.azimbagirad@ucl.ac.uk

[2] Guy's Cancer Centre, Guy's and St Thomas' NHS Foundation Trust, London, UK

Abstract. Generating high-quality ground truth data for deep learning segmentation models is typically labor-intensive and for large scale data such as 3D medical scans, can be extremely challenging to achieve. This challenge is particularly evident for vessel segmentation in organs such as the lungs, liver, and brain, where comprehensive manual annotations can take weeks for a single case and still be incomplete. Factors such as partial volume effects, imaging artifacts, and anatomical abnormalities frequently lead to missing vascular labels, further complicating annotations. In this work, we present a novel tool designed to generate near-complete vessel labels with minimal manual effort. Our approach begins with an automatic thresholding step to detect lung vessels coarsely, followed by connecting isolated voxels/components that are disconnected from proximal branches. To enhance connectivity, we employ a modified Dijkstra algorithm with targeted function optimization to link fragmented structures, ensuring a more anatomically consistent segmentation. For efficiency, red voxels are connecting on Torch Tensor-based operators saving calculation time. We evaluated our method on 55 lung CT scans from the publicly available CARVE challenge and their manually-labeled vessel ground truth. Results demonstrate that our tool significantly improves connectivity (Largest Connected Component Ratio >15%) and increases volume vessel labels 1.54% (±0.51%) compared to standard manual annotations. This approach has the potential to streamline vessel segmentation across multiple organs, improving both research efficiency and clinical applications. The tools will be available on GitHub for assessment and future use.

Keywords: Ground Truth Generation · Vessel Segmentation · Dijkstra Algorithm · CT Image Processing

1 Introduction

Computed Tomography (CT) imaging enables visualization and analysis of the vascular system, which is essential for tasks including disease diagnosis and

U. Anazodo et al. (Eds.): MIRASOL 2025, LNCS 16398, pp. 114–123, 2026.
https://doi.org/10.1007/978-3-032-13654-1_12

prognosis, surgical planning, and tumor evaluation [1]. CT criteria such as vessel occlusion, narrowing, and wall irregularity can be used to predict vascular involvement in disease states such as malignant tumours [2]. Alongside advanced segmentation techniques such as graph cuts, graph-based analysis [3], active contour and optimization techniques [4], Deep Learning (DL) models [5] can efficiently extract and separate vessel networks from CT images. However, the accuracy of such methods can suffer due to issues such as partial volume effects, imaging artifacts, and anatomical abnormalities. Specifically, for DL models, accurate and fully segmented vessels are often necessary for model training as ground truth.

Regarding the lung, the pulmonary vessels are intricate structures of varying sizes and density, making manual annotation a challenging and time-consuming task. However, missing or inaccurately labeled vessels can cause segmentation models to learn incorrect features, reducing their ability to generalize across different datasets and clinical scenarios. Incomplete/wrong segmentations can lead to vessel discontinuities which are crucial for down-stream tasks such as vessel skeletonization. Ensuring precise and comprehensive ground truth data is essential when developing DL models that can reliably assist in clinical applications, such as surgical planning and disease prognostication.

Instead of exhaustive manual annotations, a more efficient approach involves using an initial threshold to roughly detect vessels, followed by automated methods to refine the segmentation (see also [6]). Disconnected components due to partial volume effects, abnormalities in vessel branches, or imaging artifacts can be reconnected using techniques such as graph-based algorithms or topology-aware corrections. Additionally, missed vessel regions can be identified and corrected with minimal manual intervention, ensuring a more complete and accurate ground truth. This semi-automated approach significantly reduces annotation time while maintaining the accuracy required for DL models to learn robust vessel structures.

1.1 Previous Works

Several deep and non-deep learning approaches have been proposed for lung vessel segmentation in CT images. As an example for a non-DL model, in [7] a supervised recursive segmentation method using geometric active contours was used to extract lung tissues and reconstruct 3D models of lungs and vessel trees. Their method was implemented and tested on fifteen CT datasets of low resolution (inter slice distance 2mm). Limitations of their work was a limited number of subjects, a relatively time consuming methodology that was sensitive to partial volume effects and potential mislabeling of the bronchus wall as vessel.

In another work [8], a framework for labeling and classifying tree-like structures was proposed using local information and a specifically designed graph cut method. In their pipeline, an airway segmentation was used to refine the vessels labels. The performance of the algorithm was tested on a set of 20 independent clinical cases showing good overall results particularly when the quality of the airway segmentation was good. The proposed method was also applied to data from the CARVE challenge [9]. However, depending on the quality of the data, proper airway segmentation may not be possible due to the partial volume effect, mucus plugging in the airways, or thick airway walls.

Deep learning methods are capable of attaining high quality vessel segmentation [10–12]. As an example, a 3D U-Net-based algorithm developed by Nam et al. [10], showed promising results (AUC=0.97) in pulmonary vessel segmentation, particularly in assessing vascular remodeling in COPD patients. They trained their model on the VESSEL 12 dataset which contained contrast-enhanced CT images, while their model was designed for noncontrast CT analysis. They also excluded a considerable portion of the initial consecutively collected cohort (92 out of 373 patients).

1.2 Our Suggestion

In the reviewed studies on lung vessel segmentation, several limitations have been identified such as a reliance on 2D analysis rather than 3D, limitations in spatial continuity (connectivity) and accuracy of segmented vessel structures. Many studies focus on high-resolution and high-dose CT scans, making them less applicable to low-dose CT scans (e.g., lung cancer screening datasets) which are forming an increasingly large proportion of chest CT imaging. A further major challenge is the lack of publicly available code, making it difficult to evaluate and compare different segmentation methods objectively. Lastly, the structural similarity between bronchial walls and small vessels means that bronchial walls may not uncommonly be labelled as vessels. This potential cause of segmentation inaccuracies has not been explicitly investigated in past studies, and may be a problem in locations where vessel and airway structures are closely intertwined. Our method addresses these gaps by leveraging a more comprehensive and accessible approach to lung vessel segmentation.

2 Methodology

In this work, we propose a novel pipeline to generate high-quality ground truth data for lung vessel segmentation with minimal manual effort, suitable for both deep learning and traditional vessel segmentation methods. Our approach begins with an automatic preprocessing step to roughly threshold vessels in the CT image. Next, we refine the vessel labels by removing airway walls using a pre-trained nnU-Net model [13] designed for airway segmentation. To address vessel discontinuities, we employ a tensor-based optimization strategy that utilizes a modified Dijkstra algorithm to identify the most optimal connections between disconnected components. Finally, we apply a region-growing method to ensure complete and accurate vessel connectivity (see Fig. 1).

2.1 Dataset

We used a dataset from a previous MICCAI challenge [9] comprising 10 ground truth labelled CTs and 45 automatic and semi-automatic generated labels for

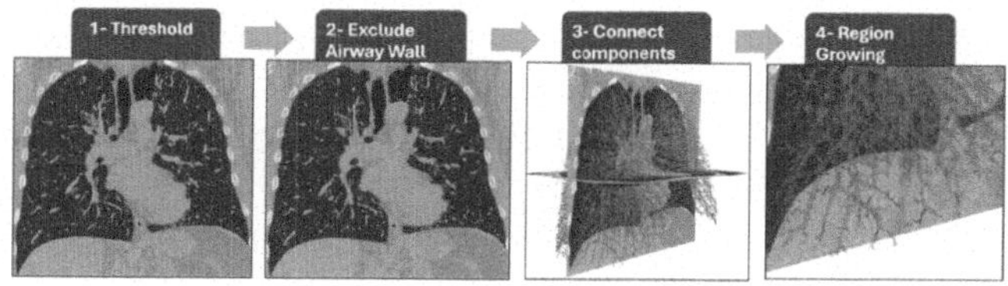

Fig. 1. Suggested pipeline to generate lung vessel segmentation with minimal supervision. In the second step, red labels are related to airway walls which should be distinguished from vessel labels. In the third and last step, yellow and red voxels are connecting those discontinuous vessels. (Color figure online)

vessels. These 55 non-contrast volumetric low dose (30 mAs at 120–140 kV) thoracic CT scans were taken from the Dutch-Belgium lung cancer screening study (NELSON). The CTs contained abnormalities including emphysema and interstitial lung disease of variable severity. Scans were acquired on a 16 or 64-slice CT scanner (Philips Medical systems, Cleveland, OH) at full inspiration. Axial images were reconstructed to 512×512 matrices with a 1.0 mm thickness at 0.7 mm increment, by means of a soft reconstruction kernel (Philips B). More information on CT scan acquisition can be found in [14]. The 55 CT scans and their labels are publicly available through the ANODE challenge [15].

2.2 Thresholding

Although thresholding can be a manual step, previous studies suggest that Alveolar septa and small pulmonary vessels occupy 10% of the total lung parenchyma volume [16]. Based on this, we designed an automated loop to determine the optimal threshold for detecting approximately 7–10% of the lung volume as vessels to be sure all detected voxels belong to vessels. For this purpose, we first applied a lung mask and rescaled the density voxel values to a range of (0–1). We then incrementally increased the threshold from 0.1 to 1 (any voxel over the threshold is vessel), stopping when the detected vessel volume fell within the desired range. Our results show that a high percentage of the identified voxels correspond to actual vessels. However, some voxels along the lobar fissures and bronchial walls were incorrectly classified as vessels, highlighting potential areas for further refinement.

2.3 Airway Exclusion

While refining the threshold vessel label, it became apparent that airway walls could be labeled as vessels using thresholding. We therefore employed an airway segmentation model [13] to exclude airways and their walls from the field of view. This airway segmentation model was trained on 25 volumetric ground truths by combination of pulmonary toolkit [13] and a 2D dilated U-NET. As post processing a morphological closing procedure was then applied to re-connect airway segments that were disconnected. The final airway tree mask was obtained by performing a largest connected components procedure (for more details see [13]). Whilst even a rough segmentation of the airway tree can effectively eliminate most of the airway walls, any remaining voxels can be easily removed using Slicer tools such as 'Remove Island' [17]. This step for each subject roughly took 15 min.

2.4 Connecting Components

When attempting to connect disconnected vessels, one approach is to identify high-density voxels between two nearby vessel components and connect them. The Dijkstra algorithm [18] which has been efficiently implemented in *heapq* Python module is helpful for this task. However, due to artifacts, disease, or partial volume effects, high-density voxels are not always available (otherwise they would have been detected during the thresholding phase). To address this, we limited the search region by requiring a direct connection between two close components with 2 voxel freedom movement. The Dijkstra algorithm can then operate within this constrained region, either moving through high-density areas or with limited voxel movement. Since we connect each isolated component to its closest neighbor, some components may still remain isolated after the first iteration. Adding more loops ensure almost all components are connected.

2.5 Region Growing

Though disjointed vessels may be successfully connected using the constrained Dijkstra algorithm, the connection may be a single voxel in size and therefore smaller in diameter than the size of the vessel either side of the connection (see Fig. 3, (I) (1)). For example, two components with a diameter of 5 voxels might be linked by a connection only 1 voxel wide, making it a weak connection. To strengthen these connections, we employ a region-growing technique with two key conditions: (1) the growth is restricted to voxels with similar density values, ensuring anatomical consistency, and (2) the expansion does not exceed the diameter of the original disjoint components, preventing over-segmentation.

2.6 Model Evaluation

We evaluated our model against an established manually-acquired ground truth to assess the improvement in vessel label connectivity. For this purpose, we used 10 subjects that had been manually segmented, despite containing disconnected components. After connecting all the vessels, a radiologist reviewed the connections to accept or reject new labels.

We evaluated our vessel label generation pipeline using two approaches. First, we directly generated vessel labels from the original scans. Second, we used a state-of-the-art model, Total Segmentator [19,20], to generate vessel labels on the CARVE dataset. We then compared the performance of both methods. We used several similarity and connectivity metrics to compare and evaluate both models such as DICE, AUC, Sensitivity, Specificity, Accuracy [21], Number of Connected Components (NC), Largest Connected Component Ratio (LCCR), and Shortest Path Length (SPL). Time consumption for all steps in pipeline (Fig. 1) including 3 loops for connectivity for a normal scan is between 15 to 20 min which considerably faster than the reviewed approaches.

3 Results

In the initial phase of our model implementation, we applied the connectivity step to the 10 ground truth segmentations. Figure 2 illustrates the results for these 10 subjects after applying step 3 of our pipeline, which focuses on connecting disjoint components to enhance segmentation. The volume of the labeled vessel using our pipeline increased averagely 1.54% (±0.51%).

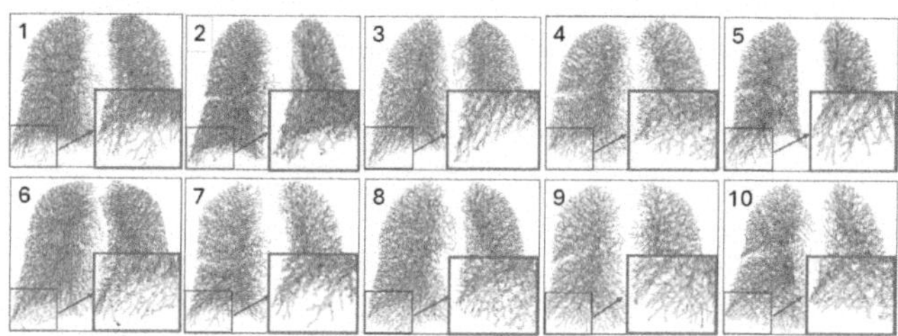

Fig. 2. Connecting components using tensor based optimization and modified Dijkstra algorithm. Green label shows the original ground truth, Yellow label shows the connection in the first loop and red label shows the connections in second loop. (Color figure online)

Using our proposed pipeline (Fig. 1) to generate vessel segmentations from the original scan, our results demonstrate a DICE score of 0.76, closely matching the state-of-the-art model, Total Segmentator, which achieved a DICE score of 0.77. However, our approach outperforms Total Segmentator in key metrics such as AUC, Hausdorff Distance, and Sensitivity and connectivity metrics. Additionally, both models yield identical results in terms of Specificity and Accuracy (see Table 1).

Although for the remaining 45 subjects we do not have ground truth, still the connectivity metrics estimated by our pipeline were 17.84 (±10.87), 0.53 (±0.07), and 92.84 (±135.74), while those for Total Segmentator were 3043.91 (±934.39), 0.31 (±0.08), and 31.04 (±66.88) for NC, LCCR, and SPL, respectively. Our pipeline outperformed Total Segmentator across all three metrics.

Table 1. Comparison results of two models, i.e., Our pipeline versus Total Segmentator model to generate vessel labels on 10 CT scans with respect to the ground truth label.

Metric type	Metric Name	Our pipeline	Total Segmentator
Volume similarity	DICE ↑	0.76 (± 0.02)	**0.77**(± 0.02)
	AUC ↑	**0.92** (± 0.03)	0.86 (± 0.03)
	Hausdorff Distance ↓	**19.4** (± 4.41)	113.5 (± 39.05)
	Sensitivity ↑	**0.84** (± 0.06)	0.73 (± 0.05)
	Specificity ↑	0.99 (± 0.00)	0.99 (± 0.00)
	Acuracy ↑	0.99 (± 0.00)	0.99 (± 0.00)
	REFVOL*	855.53 (± 0.02)	855.53 (± 0.02)
	SEGVOL*	1028.42(± 193.56)	758.97 (± 151.93)
Connectivity	NC ↓	**19.00** (± 18.27)	3182.10 (± 1082.96)
	LCCR ↑	**0.51** (± 0.03)	0.31 (± 0.06)
	SPL* ↑	**73.90** (± 129.60)	15.62 (± 103.95)

* unit is 10^3 voxel, AUC = Area Under Curve, REFVOL = Reference Volume, SEGVOL = Segmentation Volume, NC = number of components, LCCR = Largest Connected Component Ratio, SPL = Shortest Path Length, ↓: lower value is better, ↑: higher value is better

4 Discussion

We propose a pipeline to generate and refine high-quality ground truth vessel segmentations in the lung, enabling more reliable evaluation of segmentation models. While our pipeline requires partial supervision to ensure accuracy, only a limited number of subjects may be required to train an effective classification model, using a deep learning approach, for application across the entire dataset. Since our pipeline needs a partial supervision step, we recommend using it on limited data for preparing the ground truth. Following a DL model such as nnUNet would be robust enough for segmentation task.

Dijkstra's algorithm has been widely used for finding optimal paths [22]. However, it does not always guarantee the best connection between two vessel components, as it primarily relies on optimizing a single criterion, such as minimizing or maximizing density. In our modified version of Dijkstra's algorithm, by constraining the search to a region of interest (ROI) and incorporating both high-density voxels and anatomical directionality, the accuracy of vessel connectivity can be significantly improved.

Some connections in our pipeline remain incomplete such as when considering the direction of two disconnected components as a third criterion (see Fig. 3, (I) (2)). In addition, we should extend only the vessels which are not ended naturally. To address this, we aim to introduce an additional step that balances three key factors: optimal directionality, highest-density voxels, and a restricted search region to determine the most accurate path. While this approach may increase computational cost, it has the potential to enhance segmentation accuracy and better align with human decision-making.

Although the improvement in segmented vessel volumes may seem minor (1.54% (±0.51%, and see Table 1), the improvement can be crucial for downstream tasks such as skeletonization of the segmentation. Discontinuity of vessels can lead to the exclusion of large vessel segments during skeletonization, resulting in large errors in measurement and interpretation. Therefore, even small enhancements in connectivity play a vital role in ensuring the accuracy and reliability for downstream applications.

The labels generated by our pipeline exhibit a slightly larger volume compared to the original ground truth. Upon examining the false positive voxels in the scans, we found that most corresponded to actual vessels, some being clearly identifiable, while others remained ambiguous and which were not included in the original ground truth. As a result, the accuracy reported in Table 1 likely underestimates our model's true performance, as the original ground truth itself contains missed vessels (see Fig. 3, (II)).

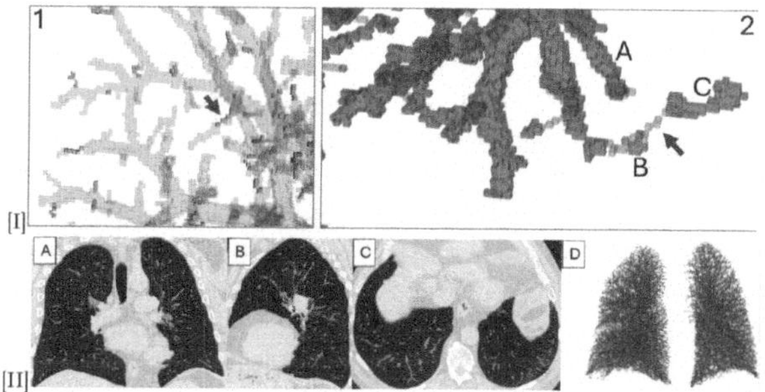

Fig. 3. [I] Two drawbacks of our pipeline for connecting components. 1)Blue arrow shows a narrow connection between two components. 2) Blue arrow shows a wrong connection since component C should be connected to component A instead of B. [II] Extra voxels detected by our method which were missed in the ground truth data and/or Total Segmentator. A-D, coronal, sagittal, axial and 3D rendered of vessels by ground truth (green) and our tool (red). (Color figure online)

5 Conclusion

In this study, we introduced a fast pipeline for generating high-quality ground truth for lung vessel segmentation with minimal manual effort. Our approach integrates thresholding, airway wall exclusion, tensor-based path optimization, and region growing to enhance vessel connectivity and improve segmentation accuracy. Compared to state-of-the-art models like TotalSegmentator, our pipeline achieved competitive results, demonstrating strong performance in key metrics such as AUC, Hausdorff Distance, Sensitivity, and connectivity metrics while maintaining comparable Specificity and Accuracy. Despite these advances, some vessel connections remain incomplete. To further enhance segmentation quality, we plan to incorporate vessel direction as an additional criterion alongside voxel density and regional constraints.

References

1. Selle, D., Preim, B., Schenk, A., Peitgen, H.O.: Analysis of vasculature for liver surgical planning. IEEE Trans. Med. Imaging **21**(11), 1344–57 (2002). https://doi.org/10.1109/TMI.2002.801166
2. Franken, L.C., et al.: Multidetector computed tomography assessment of vascular involvement in perihilar cholangiocarcinoma. Quant Imaging Med Surg. **11**(11), 4514–4521 (2021). https://doi.org/10.21037/qims-20-1303
3. Homann, H., Vesom, G. , Noble, J.A.: Vasculature segmentation of CT liver images using graph cuts and graph-based analysis. In 5th IEEE International Symposium on Biomedical Imaging, pp. 53–56, Paris (2008). https://doi.org/10.1109/ISBI.2008.4540930
4. Kaftan, J., Tek, H., Aach, T.: A two-stage approach for fully automatic segmentation of venous vascular structures in liver CT images. **7259** (2009). https://doi.org/10.1117/12.812407
5. Chierici, A., Lareyre, F., Salucki, B., Iannelli, A., Delingette, H., Raffort, J.: Vascular liver segmentation: a narrative review on methods and new insights brought by artificial intelligence. J. Int. Med. Res. **52**(9) (2024). https://doi.org/10.1177/03000605241263170
6. Azimbagirad, M., et al.: Robust semi-automatic segmentation method: an expert assistant tool for muscles in CT and MR data. Comput. Methods Biomech. Biomed. Eng.: Imaging Vis. **11**(7) (2024). https://doi.org/10.1080/21681163.2023.2301403
7. Li, X., Wang, X., Dai, Y., Zhang, P.: Supervised recursive segmentation of volumetric CT images for 3D reconstruction of lung and vessel tree. Comput. Methods Programs Biomed. **122**(3), 316–329 (2024). https://doi.org/10.1016/j.cmpb.2015.08.014
8. Jimenez-Carretero D., et al.: A graph-cut approach for pulmonary artery-vein segmentation in noncontrast CT images. Med. Image Anal. **52**, 144–159 (2019). https://doi.org/10.1016/j.media.2018.11.011
9. Charbonnier, J.P., Brink, M., Ciompi, F., Scholten, E.T., Schaefer-Prokop, C.M., van Rikxoort, E.M.: Automatic pulmonary artery-vein separation and classification in computed tomography using tree partitioning and peripheral vessel matching. In IEEE Trans. Med. Imaging. **35**(3), 552–892 (2016). https://doi.org/10.1109/TMI.2015.2500279
10. Nam, J.G., Witanto, J.N., Park, S.J., et al.: Automatic pulmonary vessel segmentation on noncontrast chest CT: deep learning algorithm developed using spatiotemporally matched virtual noncontrast images and low-keV contrast-enhanced vessel maps. Eur. Radiol. **31**, 9012–9021 (2021). https://doi.org/10.1007/s00330-021-08036-z
11. Tan, W., Zhou, L., Li, X., Yang, X., Chen, Y., Yang, J.: Automated vessel segmentation in lung CT and CTA images via deep neural networks. J. Xray Sci. Technol. **29**(6), 1123–1137 (2021). https://doi.org/10.3233/XST-210955
12. Han, J., He, N., Zheng, Q., Li, L., Ma, C.H.:3D pulmonary vessel segmentation based on improved residual attention u-net. Med. Nov. Technol. Devices **20**, 100268 (2023). https://doi.org/10.1016/j.medntd.2023.100268

13. Cheung, W.K., Pakzad, A., Mogulkoc, N., et al.: Automated airway quantification associates with mortality in idiopathic pulmonary fibrosis. Eur. Radiol. **33**, 8228–8238 (2023). https://doi.org/10.1007/s00330-023-09914-4
14. van Iersel, C.A.: Risk-based selection from the generalpopulation in a screening trial: selection criteria recruitment andpower for the Dutch-Belgian randomised lung cancer multi-slice CTscreening trial (NELSON). Int. J. Cancer **120**, 868–874 (2006). https://doi.org/10.1002/ijc.22134
15. van Ginneken, B.: Comparing and combining algorithms for computer-aided detection of pulmonary nodules in computed tomographyscans: The ANODE09 study. Med. Image Anal. **14**, 707–722 (2010). https://doi.org/10.1016/j.media.2010.05.005
16. Harumi, I., Mizuki, N., Hitoto, H.: Architecture of the lung: morphology and function. J. Thorac. Imaging **19**(4), 221–227 (2004). https://doi.org/10.1097/01.rti.0000142835.06988.b0
17. Pieper, S., Lorensen, B., Schroeder, W., Kikinis, R.: The NA-MIC Kit: ITK, VTK, pipelines, grids and 3D slicer as an open platform for the medical image computing community, In 3rd IEEE International Symposium on Biomedical Imaging: Nano to Macro, pp. 698–701 (2006). https://doi.org/10.1109/ISBI.2006.1625012
18. Dijkstra, E.W.: A note on two problems in connexion with graphs. Numer. Math. **1**, 269–271 (1959). https://doi.org/10.1007/BF01386390
19. Poletti, J., et al.: Automated lung vessel segmentation reveals blood vessel volume redistribution in viral pneumonia. Eur. J. Radiol. **150** (2022). https://doi.org/10.1016/j.ejrad.2022.110259
20. Wasserthal, J., et al.: TotalSegmentator: robust segmentation of 104 anatomic structures in CT images. Radiol. Artif. Intell. **5**(5) (2023). https://doi.org/10.1148/ryai.230024
21. Taha, A.A., Hanbury, A.: Metrics for evaluating 3D medical image segmentation: analysis, selection, and tool. BMC Med. Imaging **15**(29) (2015). https://doi.org/10.1186/s12880-015-0068-x
22. Garg, S., Devi, B.: Shortest path finding using modified Dijkstra's algorithm with adaptive penalty function. In 14th International Conference on Computing Communication and Networking Technologies (ICCCNT), Delhi, India, pp. 1–9 (2023). https://doi.org/10.1109/ICCCNT56998.2023.10308130
23. Chung, M., Lee, J., Chung, J.W., Shin, Y.G.: Accurate liver vessel segmentation via active contour model with dense vessel candidates. Comput. Methods Programs Biomed. **166**, 61–75 (2018). https://doi.org/10.1016/j.cmpb.2018.10.010
24. Chen, W., Zhao, L., Bian, R., et al.: Compensation of small data with large filters for accurate liver vessel segmentation from contrast-enhanced CT images. BMC Med. Imaging **24**(129) (2024). https://doi.org/10.1186/s12880-024-01309-1

LightDenoise-HL: A Compact Attention-Based Denoising Framework with Dual-Frequency Filtering for Endoscopic Imaging

M. J. Aashik Rasool[1], Shabir Ahmad[2], M. J. Akeel Ahamed[1], and Taeg Keun Whangbo[1](✉)

[1] Department of IT Convergence Engineering, Gachon University, Sujeong-Gu, Seongnam-si, Gyeonggi-do 461–701, Republic of Korea
tkwhangbo@gachon.ac.kr

[2] Center of Artificial Intelligence for Medical Instruments, CAIMI Pvt., Ltd., 21982 Incheon Metropolitan City, Republic of Korea

Abstract. Endoscopic imaging is pivotal in the early detection and diagnosis of gastric cancer; however, noise introduced during image acquisition can significantly degrade fine structural details, limiting diagnostic reliability. Existing lightweight denoising methods, although suitable for deployment on resource-constrained endoscopic platforms, often compromise spatial fidelity or introduce perceptual artifacts. To address these challenges, we propose a novel lightweight denoising framework that incorporates a dual frequency-aware attention mechanism, enabling selective enhancement of high- and very high-frequency spatial components. Furthermore, we design compact residual blocks utilizing depthwise and pointwise convolutions to improve feature representation while maintaining low computational overhead. Extensive experiments on clinically relevant datasets demonstrate that our approach achieves superior performance in both quantitative metrics and perceptual quality, effectively preserving anatomical structures and enabling real-time applicability in medical environments.

Keywords: Image Denoising · Endoscopic Imaging · Dual-Frequency Filtering

1 Introduction

Endoscopic imaging is very important for finding and treating problems with the digestive system, such as gastric cancer, which is one of the main causes of cancer deaths around the world [8]. Doctors can see inside tissues in real time with endoscopy, which helps them find problems like mucosal patterns, lesion boundaries, and blood vessels. It also helps them plan procedures that aren't too invasive and keep an eye on how the condition is getting worse. Finding

U. Anazodo et al. (Eds.): MIRASOL 2025, LNCS 16398, pp. 124–132, 2026.
https://doi.org/10.1007/978-3-032-13654-1_13

gastric cancer early, especially early gastric cancer (EGC), is very important because it lets doctors treat it quickly and greatly improves the chances of a good outcome for the patient. Studies have shown that detection rates can be as low as 10% because of problems with picture quality. [11]. Adding to these problems with picture quality, noise in endoscopic imaging, especially for EGC diagnosis, sometimes obscures important structural elements like lesion boundaries and mucosal patterns. This makes it harder to find elements and makes clinical accuracy worse [5]. This problem is particularly noticeable in clinical settings with limited resources where endoscopic systems depend on embedded hardware with constrained computational power. There is a need for sophisticated solutions to improve diagnostic accuracy while maintaining computational efficiency because current lightweight denoising techniques, despite being made for such environments, frequently introduce visual artifacts or compromise spatial fidelity [11].

Several deep learning based denoising methods have been developed to address image noise [6,12], but their applicability to endoscopic imaging remains limited [2,9,10]. High-performance models like DnCNN and IRCNN use deep convolutional architectures and residual learning to remove noise from images at different levels of noise [13,14]. However, they are too complicated to use on endoscopic hardware with limited resources, where real-time processing is needed [1]. Lightweight methods like FFDNet and LPIENet, on the other hand, are made to be efficient. FFDNet uses downsampled sub-images and noise maps to lower the s reduce computational load, and LPIENet uses plug-and-play modules for low-power devices [3,15]. These methods are useful in situations where resources are limited, but they often lose spatial fidelity or add visual artifacts, especially when denoising complex endoscopic images with subtle lesions that are important for EGC diagnosis [3,15].

Existing denoising techniques' trade-offs between computational efficiency and detail preservation point to a serious weakness in endoscopic image analysis. While lightweight techniques like FFDNet and LPIENet frequently fail to preserve the fine structural details required for precise clinical diagnostics, high-performance models like DnCNN and IRCNN are not appropriate for embedded systems. This requires the development of an advanced denoising framework that combines the efficiency necessary for resource-constrained endoscopic systems with enhanced detail preservation.

To address this challenge, we introduce a novel lightweight denoising framework that incorporates a Dual-Frequency Filtering-aware Attention (DFFA) mechanism, specifically designed to enhance both high- and very high-frequency information in endoscopic images. The proposed model is further equipped with compact residual blocks constructed using efficient depth-wise and point-wise convolutions, enabling effective noise suppression while maintaining a low computational footprint. This design significantly improves the visual quality of EGC images, making it well-suited for real-time clinical deployment in resource-constrained environments. OurLightDenoise-HL offers three major contributions as follows:

- We propose a novel DFFA mechanism that effectively captures and enhances both high- and very high-frequency components, preserving crucial structural details in noisy endoscopic images.
- We design a compact residual architecture using efficient depth-wise and point-wise convolutions, enabling high-quality denoising while maintaining low computational complexity for real-time performance.
- We conduct extensive evaluations on synthetically deteriorated endoscopic images, demonstrating the framework's capabilities to improve image quality while preserving a compact architecture, indicating potential for future incorporation into real-time clinical operations.

2 Methods

In the EGC enhancement task, considering the set $\mathcal{I}$ denotes the collection of image samples, where each element $i \in \mathcal{I}$ comprises a noisy image x_i^{noisy} and its corresponding clean counterpart x_i^{clean}. In this formulation, a mapping function $x_i^{\text{clean}} = f_{\theta_i}(x_i^{\text{noisy}})$ is derived, where the parameters θ_i are optimized specifically for the noise characteristics and structural features of the task i. However, these approaches often result in substantial computational overhead, and their practicality in resource-constrained clinical settings.

2.1 Dual-Frequency Filtering Aware Attention Mechanism

In here, the DFFA block is used to enhance spatial features by sequentially collecting and presenting different frequency components of the given input feature map. The input feature map $\mathbf{X} \in \mathbb{R}^{C \times H \times W}$, the module utilizes a depth-wise Gaussian blur to extract a smoothed feature $\mathbf{LP}_1$. As a next step, subtracting this feature map from the original input, it derives the high-frequency (HF) feature residual $\mathbf{HP}_1 = \mathbf{X} - \mathbf{LP}_1$, which contains rich edge information. The second Gaussian blur is applied to $\mathbf{LP}_1$, resulting in $\mathbf{LP}_2$. The difference of $\mathbf{HP}_2 = \mathbf{LP}_1 - \mathbf{LP}_2$ captures very high-frequency (VHF) material, such as micro-spatial details. These residuals go through two distinct channels of attention. All branches individually consist of a global average pooling operation, two fully linked layers with ReLU activation, and a final sigmoid gate. Each branch includes a global average pooling operation, two fully connected layers with a ReLU activation, and a final sigmoid gate. This results in channel-wise attention maps $\mathbf{w}_1$ and $\mathbf{w}_2$ for HF and VHF components, respectively.

The final feature vector is created by tweaking the original feature map using residual modulation, guided by attention weights the model has learned. In simple terms, the process can be written as $\mathbf{Y} = \mathbf{X} \cdot (1 + \mathbf{w}_1) \cdot (1 + \mathbf{w}_2)$, where $\mathbf{X}$ is the input data, and $\mathbf{w}_1$ and $\mathbf{w}_2$ are the attention maps the model figures out. This approach lets the HFFA block focus on the most important parts of the image while keeping the computational load light, making it great for efficient EGC image restoration when resources are limited. The corresponding representation is outlined in Algorithm 1.

Algorithm 1 Dual-Frequency Filtering aware Attention mechanism

Require: Feature map $\mathcal{X} \in \mathbb{R}^{C\times H\times W}$, kernel size k, reduction ratio r
Ensure: Re-weighted output $\mathcal{Y} \in \mathbb{R}^{C\times H\times W}$
1: $p \leftarrow \lfloor k/2 \rfloor$ ▷ Padding size
2: $\mathcal{LP}_1 \leftarrow \text{DepthwiseGaussianBlur}(\mathcal{X}, k, p)$
3: $\mathcal{HP}_1 \leftarrow \mathcal{X} - \mathcal{LP}_1$ ▷ High-frequency residual
4: $\mathcal{LP}_2 \leftarrow \text{DepthwiseGaussianBlur}(\mathcal{LP}_1, k, p)$
5: $\mathcal{HP}_2 \leftarrow \mathcal{LP}_1 - \mathcal{LP}_2$ ▷ Very high-frequency residual
6: $\mathbf{w}_1 \leftarrow \sigma\left(\text{Conv}_2\left(\text{ReLU}\left(\text{Conv}_1\left(\text{AvgPool}(\mathcal{HP}_1)\right)\right)\right)\right)$
7: $\mathbf{w}_2 \leftarrow \sigma\left(\text{Conv}_2\left(\text{ReLU}\left(\text{Conv}_1\left(\text{AvgPool}(\mathcal{HP}_2)\right)\right)\right)\right)$
8: $\mathcal{Y} \leftarrow \mathcal{X} \cdot (1 + \mathbf{w}_1) \cdot (1 + \mathbf{w}_2)$
9: **return** $\mathcal{Y}$

2.2 LightDenoise-HL Framework

In order to perform lightweight EGC image enhancement, the proposed LightDenoise-HL uses an asymmetric encoder-decoder structure that has been enhanced with DFFA. The encoder has four stages, each with a compact residual block (CSB), DFFA, and a downsampling convolution to extract increasingly abstract features. The CSB consists of a depth-wise separable convolution followed by a point-wise convolution for reducing computation, as well as a compact squeeze-and-excitation (SE) [4] module for channel-wise attention. The SE block generates calibration weights, which control the resultant features, utilising two fully connected nonlinear layers. Finally, a scaled residual connection is established to the input, maintaining stability while facilitating efficient gradient flow. Afterward, the fifth residual block serves as the bottleneck, preserving architectural symmetry alongside the final DFFA.

The decoder uses transposed convolutions to gradually upsample information, with controlled skip connections from the encoder improving spatial detail at each step. The output is generated using residual learning by adding the final decoder output. The network can return both the recovered output and intermediate features when using the optional knowledge distillation mode. This architecture strikes a solid compromise between performance and efficiency, making it appropriate for real-time, resource-constrained EGC. The overview of our LightDenoise-HL framework is presented in Fig. 1.

3 Experimental Setup and Results Analysis

3.1 Dataset and Training Details

In this study, we employed the Curated Colon Dataset for Deep Learning [7], publicly available on Kaggle. This dataset contains high-resolution endoscopic images specifically curated for colonoscopy-based deep learning tasks. To imitate noisy imaging situations in clinical settings, we injected Gaussian noise with a standard deviation of 50 (σ) into the clean images. This technique allows for the evaluation of denoising performance in controlled yet demanding circumstances.

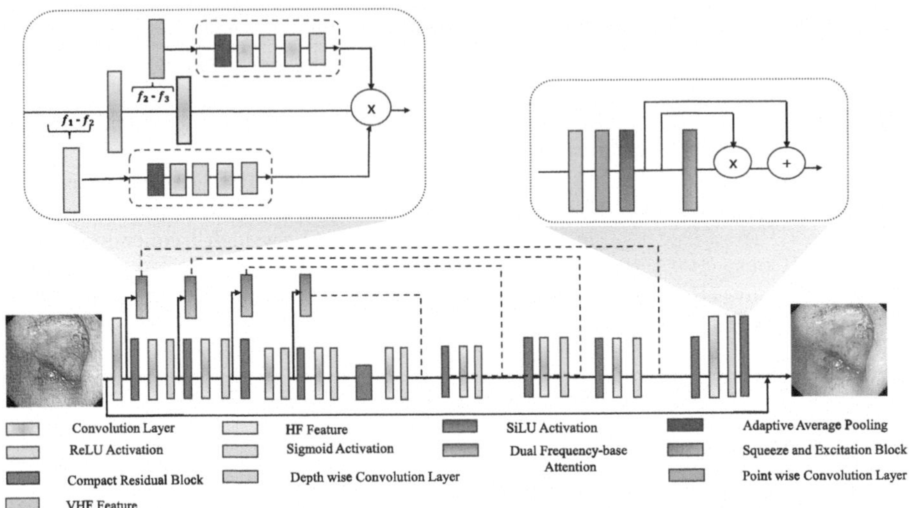

Fig. 1. The overview of our LightDenoise-HL framework.

The dataset was used for both training and testing, providing a thorough evaluation of the proposed framework's robustness and generalizability. During training, we utilized the L1 loss function with a learning rate of 0.001, a batch size of 128, and scaled all input images to 256*256 (Table 1).

Table 1. Quantitative comparison of different denoising methods

Method	PSNR	SSIM	LPIPS	FPS(CPU)
DnCNN [13]	23.45	0.8248	0.2116	16.00
IRCNN [14]	24.52	0.8362	0.1902	13.21
FFDNet [15]	24.28	0.8268	0.1994	10.45
LPEINet [3]	23.87	0.8276	0.2050	25.45
Our Approach	**25.14**	**0.8454**	**0.1858**	**23.21**

4 Results and Analysis

4.1 Quantitative Comparison

Figure 2 presents a scatter plot illustrating the relationship between peak signal-to-noise ratio (PSNR) and inference speed (measured in frames per second) for all evaluated denoising models. Among the compared methods, DnCNN and IRCNN achieve relatively high PSNR but suffer from limited inference speed, rendering them unsuitable for real-time deployment. On the other hand, LPEINet and FFDNet offer higher throughput but yield lower PSNR, compromising image quality.

The proposed LightDenoise-HL framework demonstrates the most favorable trade-off, achieving the highest PSNR of 25.14 dB while maintaining a competitive inference speed of 23.21 FPS on CPU. This balance confirms the model's practical applicability for real-time clinical environments without compromising denoising effectiveness.

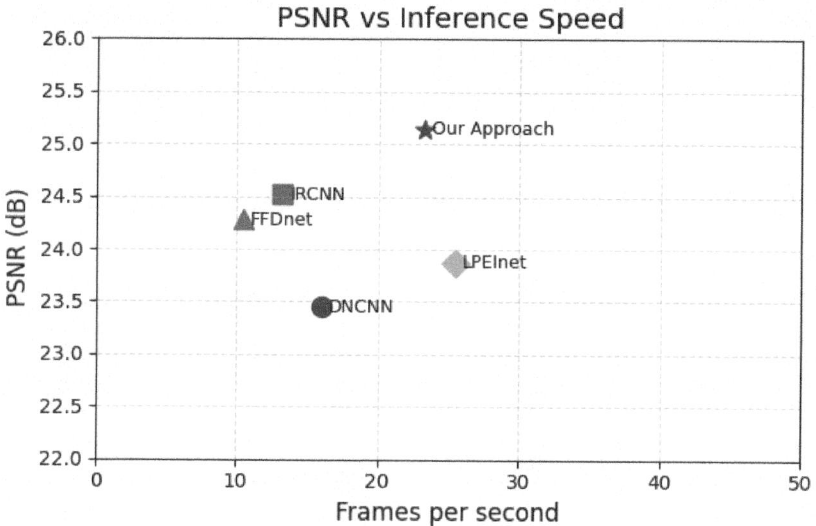

Fig. 2. PSNR vs. Inference Speed for various denoising methods.

4.2 Qualitative Evaluation

To assess visual quality and anatomical structure preservation, Fig. 3 and Fig. 4 present side-by-side comparisons of denoised outputs from all methods under evaluation. Figure 3 shows a colonoscopic image affected by Gaussian noise. Conventional methods such as DnCNN and IRCNN produce over-smoothed outputs, removing not only noise but also critical mucosal details. LPEINet and FFDNet, though computationally efficient, introduce visual artifacts and fail to restore textural fidelity. In contrast, the proposed method yields a sharper reconstruction with well-preserved tissue structures, closely resembling the clean ground truth.

Figure 4 displays results on a gastric image exhibiting fine vascular structures. While all methods reduce noise to some extent, only the proposed approach maintains vessel continuity and enhances local contrast without artificial

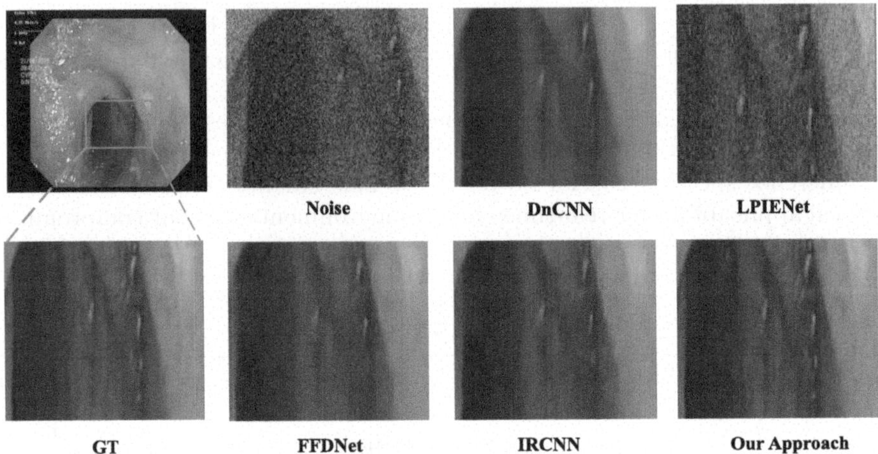

Fig. 3. Qualitative comparison on a colonoscopy image. The proposed method achieves superior visual restoration while preserving mucosal textures.

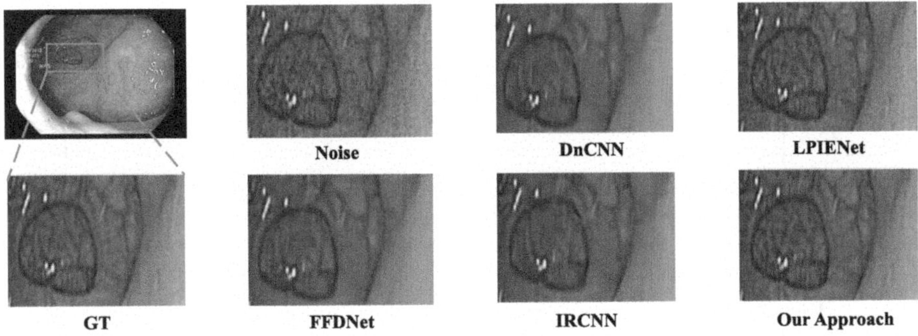

Fig. 4. Qualitative comparison on a gastric endoscopic image. The proposed method retains fine vascular structures and produces visually faithful reconstructions.

texture or over-smoothing. This demonstrates its ability to retain diagnostically important features even under challenging high-frequency noise conditions.

These results validate that LightDenoise-HL offers a compelling combination of high perceptual quality, structural integrity preservation, and computational efficiency suitable for real-time medical imaging scenarios.

5 Discussion

The results presented in Sect. 3 elucidate the effectiveness of the proposed LightDenoise-HL framework for real-time endoscopic image denoising under clinically realistic noise settings. In particular, the PSNR vs. inference speed analysis as portrayed in Fig. 2) reveals that the proposed model achieves a favorable balance between reconstruction quality and computational efficiency, surpassing both high-performance and lightweight baselines. This suggests that the integration of dual-frequency filtering with attention modulation and compact residual design enables effective noise suppression without compromising inference throughput. Qualitative comparisons further reinforce the quantitative findings. As shown in Fig. 3 and Fig. 4, existing methods such as DnCNN and IRCNN exhibit over-smoothing, leading to the loss of clinically relevant textural details. Lightweight alternatives, including LPEINet and FFDNet, tend to preserve more structure but often introduce residual noise and perceptual artifacts. In contrast, LightDenoise-HL consistently maintains anatomical fidelity across diverse anatomical regions, preserving mucosal textures and vascular boundaries that are critical for early-stage gastric cancer detection.

One of the key strengths of the proposed framework is the explicit modeling of high-frequency and very high-frequency components through the dual-attention mechanism. This design allows the network to selectively enhance fine-grained spatial patterns while suppressing irrelevant noise, aligning well with the diagnostic needs of clinical endoscopy. Furthermore, the compact residual blocks and pointwise-depthwise separable convolutions contribute to the overall efficiency of the model, making it suitable for deployment in embedded medical systems. Although the current evaluation demonstrates robust performance under synthetic noise conditions, future work should explore generalization to real-world intra-procedural noise, domain shifts between imaging systems, and the integration of domain adaptation techniques. Moreover, clinical validation with expert readers would be a critical step toward assessing diagnostic impact.

6 Conclusion

This paper introduces LightDenoise-HL, a lightweight denoising framework tailored for real-time endoscopic imaging under computational constraints. The model leverages a dual-frequency filtering-aware attention mechanism in conjunction with compact residual blocks to achieve high denoising quality and low latency. Extensive experiments demonstrate that the proposed method outperforms state-of-the-art baselines in both quantitative metrics and qualitative fidelity, while maintaining competitive runtime performance. The proposed approach addresses a pressing need in gastrointestinal diagnostics, where image quality is often compromised due to noise introduced during acquisition. By preserving fine structural details essential for clinical decision-making and ensuring compatibility with low-power hardware, LightDenoise-HL represents a promising step toward practical, deployable AI tools in endoscopic procedures. Future work will focus on expanding the dataset to include real-world clinical noise patterns, validating the model's performance on embedded hardware, and assessing its impact in prospective clinical workflows.

Acknowledgments. This work was supported by the GRRC program of Gyeonggi Province (GRRC- Gachon2023(B02), Development of AI-based medical service technology)

References

1. Ali, S.: Where do we stand in ai for endoscopic image analysis? deciphering gaps and future directions. npj Digit. Med. **5**(1), 184 (2022)
2. Arshaghi, A., Ashourian, M., Ghabeli, L.: Denoising medical images using machine learning, deep learning approaches: a survey. Current Med. Imaging Rev. **17**(5), 578–594 (2021)
3. Conde, M.V., Vasluianu, F., Vazquez-Corral, J., Timofte, R.: Perceptual image enhancement for smartphone real-time applications. In: Proceedings of the IEEE/CVF Winter Conference on Applications of Computer Vision, pp. 1848–1858 (2023)
4. Hu, J., Shen, L., Sun, G.: Squeeze-and-excitation networks. In: Proceedings of the IEEE conference on computer vision and pattern recognition, pp. 7132–7141 (2018)
5. Ko, W.J., An, P., Ko, K.H., Hahm, K.B., Hong, S.P., Cho, J.Y.: Image quality analysis of various gastrointestinal endoscopes: why image quality is a prerequisite for proper diagnostic and therapeutic endoscopy. Clin. Endosc. **48**(5), 374–379 (2015)
6. Kong, Z., Deng, F., Zhuang, H., Yu, J., He, L., Yang, X.: A comparison of image denoising methods. arXiv preprint arXiv:2304.08990 (2023)
7. Mon, F.: Curated colon dataset for deep learning (2022). https://www.kaggle.com/datasets/francismon/curated-colon-dataset-for-deep-learning, Accessed May 2025
8. Papp, A., Yang, H., Ottlakan, A., Vincze, Á.: Open access edited by. Surgical and Oncological Updates in the Management of Gastric Cancer: the Role of Neoadjuvant Therapy and Minimally Invasive Surgery, p. 69 (2023)
9. Su, X., Liu, Q., Gao, X., Ma, L.: Evaluation of deep learning methods for early gastric cancer detection using gastroscopic images. Technol. Health Care **31**(1_suppl), 313–322 (2023)
10. Tang, D., et al.: A deep learning-based model improves diagnosis of early gastric cancer under narrow band imaging endoscopy. Surg. Endosc. **36**(10), 7800–7810 (2022)
11. Yao, L., et al.: A gastrointestinal endoscopy quality control system incorporated with deep learning improved endoscopist performance in a pretest and post-test trial. Clin. Transl. Gastroenterol. **12**(6), e00366 (2021)
12. Yapici, A., Akcayol, M.A.: A review of image denoising with deep learning. In: 2021 2nd International Informatics and Software Engineering Conference (IISEC), pp. 1–6. IEEE (2021)
13. Zhang, K., Zuo, W., Chen, Y., Meng, D., Zhang, L.: Beyond a gaussian denoiser: Residual learning of deep CNN for image denoising. IEEE Trans. Image Process. **26**(7), 3142–3155 (2017)
14. Zhang, K., Zuo, W., Gu, S., Zhang, L.: Learning deep cnn denoiser prior for image restoration. In: Proceedings of the IEEE conference on computer vision and pattern recognition, pp. 3929–3938 (2017)
15. Zhang, K., Zuo, W., Zhang, L.: Ffdnet: Toward a fast and flexible solution for CNN-based image denoising. IEEE Trans. Image Process. **27**(9), 4608–4622 (2018)

Accessible Skin Analysis: A Low-Cost Multispectral Imaging System with Skin-Mimicking Phantoms

Nathan Shen Baldon[1], Ana Clara Caznok Silveira[1], Clarimar José Coelho[2], Bruna Alice Gomes de Melo[3], and Leticia Rittner[1](✉)

[1] School of Electrical and Computer Engineering, Universidade Estadual de Campinas (UNICAMP), Campinas, São Paulo, Brazil
lrittner@unicamp.br
[2] Polytechnic School, Pontifical Catholic University of Goiás, Goiânia, Goiás, Brazil
[3] Center for Biomedical Engineering, Universidade Estadual de Campinas (UNICAMP), Campinas, São Paulo, Brazil

Abstract. Skin abnormalities are a common factor among different pathological conditions, and there is always growing interest in the development of technologies that allow faster, more detailed, and less subjective skin evaluation. Among the technologies under investigation, multispectral imaging (MSI) has shown good potential, but has not yet been put to practical use, two of the reasons being the lack of clinical validation and high financial cost, which can be particularly limiting in low- to middle-income countries (LMICs). Aiming to develop a more accessible approach to MSI-based skin analysis, this paper presents a proof of concept for a low-cost solution. The contributions of this work are: (1) design and construction of a low-cost MSI acquisition system prototype, with a bill of materials of less than $200; (2) the development of a set of skin-mimicking phantoms with phototype diversity and emulated melanoma lesions; (3) a preliminary analysis of the designed prototype, using the skin phantoms to simulate possible clinical scenarios for the MSI system validation. Results demonstrate the potential of the low-cost MSI system prototype as a feasible solution for skin analysis in constrained resource settings, and how a more thorough MSI system validation can be performed using skin phantoms.

Keywords: skin analysis · multispectral imaging · skin phantoms · melanoma

1 Introduction

Skin abnormalities are a common factor among different pathological conditions (e.g., skin cancer, lupus). Preliminary skin lesion analysis is typically performed by visual inspection, which can be enhanced with a dermoscope; however, its accuracy is limited by doctor experience. A more detailed evaluation still requires

U. Anazodo et al. (Eds.): MIRASOL 2025, LNCS 16398, pp. 133–142, 2026.
https://doi.org/10.1007/978-3-032-13654-1_14

histological analysis, which can be costly and cause discomfort to the patient [2,7,19,22]. In response to these diagnostic challenges, there is growing interest in developing technologies that allow faster, more detailed, and less subjective skin evaluation. In this scenario, optical methods are preferred since they can be used noninvasively [7].

Among the technologies under investigation, multispectral imaging (MSI) has shown good potential. MSI can capture both spectral and spatial information from biological tissues, producing a unique spectral signature for each pixel, which portrays the light-tissue interaction at that point in the sample [7,11]. From the spectral information, the chromophore distribution and concentration can be estimated and correlated with different pathologies, being useful for diagnostics. In addition to that, these images can carry information at wavelengths in which light penetrates deeper in the skin, also allowing visualization of deeper structures [2,4,7,22]. Its potential has been attested for melanoma [8], dermatitis [6,9], burn wounds [1], skin ulcers, etc. [4,7,19]. However, despite its great potential in skin analysis and being described for medical use since the late 1980s [1], this technology has not yet been put to practical use, two of the reasons being the lack of clinical validation and high financial cost [4,7]. This latter disadvantage can be particularly limiting in low- to middle-income countries (LMICs) and the exploration of more affordable MSI hardware is desirable.

The most straightforward form of MSI acquisition is called area scanning, which is achieved by the acquisition of an image at each wavelength, stacked together to build the data cube [3]. Different low-cost systems have been reported using this method, differing in the way of selecting the desired spectral bands. One way is the selection of bands directly in the illumination using monochromatic light sources, such as light-emitting diodes (LEDs) [5,12,14,18].

When developing low-cost MSI hardware, one of the challenges is validating that relevant information—such as chromophore content—can be extracted, and identifying which spectral bands are optimal for doing so. Also, methods of comparing new systems with commercially available ones are desired [4]. All of the above demands can be achieved using phantoms, devices with well-known characteristics that, among their varied functions, can be used for calibration, validation, and comparison of different equipment or processes. In this way, a phantom that mimics the optical properties of the skin could be used.

With all this in mind, this paper presents a proof of concept for a low-cost solution to skin analysis using multispectral imaging (MSI). The contributions of this work are: (1) design and construction of a low-cost MSI acquisition system prototype, mainly composed of an RGB camera, a set of LEDs of six different wavelengths and microcontrollers; (2) the development of a set of skin-mimicking phantoms with phototype diversity and melanoma lesions; (3) a preliminary analysis of the designed system, using the skin phantoms to simulate possible clinical scenarios for the MSI system validation. Results demonstrate the potential of the low-cost MSI system prototype as a viable solution for skin analysis in constrained resource settings, and how a more thorough MSI system validation can be performed using skin phantoms. Also, the results make it clear why phototype diversity is important during the validation process, and why it should be a central concern.

2 Methods

The proposed proof of concept for a low-cost solution using MSI for skin analysis consisted of the development of two complementary tools: (a) a low-cost MSI acquisition system prototype; (b) and a set of skin phantoms. A preliminary validation of the MSI system prototype using the skin phantoms is also described.

2.1 Development of a Low-Cost MSI Acquisition System

Based on previous studies [5,12,14,18] and aiming for simple hardware, the method of area scanning was chosen for image acquisition. For band selection, LEDs of different wavelengths were employed, because of their ease of control by a microcontroller and inexpensive existing options. Six wavelengths in the visible spectrum were explored, ranging from 395 to 620 nm, referred to as: ultraviolet (UV, 395 nm); violet (V, 420 nm); cyan (C, 490 nm); green (G, 520 nm); yellow (Y, 590 nm); and red (R, 620 nm) (wavelengths informed by the manufacturer). Using a spectrometer, it was confirmed that the light emitted by each LED had narrow spectral band and peak in the expected wavelength.

A microcontroller-compatible RGB camera was chosen as the detector. To isolate the sample from external illumination, a box with black interior (referred to as 'dark chamber') was incorporated to the system. Microcontrollers were added for automatic sequential activation of the LEDs, camera triggering, and image storage on a memory card. A printed circuit board (PCB) was designed and manufactured to provide secure electronic connections. Finally, a mechanically robust 3D-printed enclosure was designed and fabricated to accommodate the PCB and the LEDs, which was attached to the dark chamber. In the end, the bill of materials was less than $200, a very affordable price compared to commercial spectral imaging systems (Fig. 1).

2.2 Development of Skin Phantoms

The second part of the proof-of-concept solution is the development of skin phantoms, seeking standardized comparison and validation of different spectral imaging systems and guarantee that it can serve any patient regardless of skin phototype. The developed phantoms were inspired by those designed by Yim et al. [21], with the aim of mimicking the optical properties of the skin in various wavelengths and its phototype diversity (different skin tones). Three different phototypes were developed (types 1, 3 and 5 in the Fitzpatrick scale - Fitz.1, Fitz.3 and Fitz.5 - according to Yim et al. [21]). In addition, aiming for the clinical use of the MSI system, a melanoma lesion was incorporated into the phantom, with depth variation (accomplished with a triangular cross-section) and higher absorbance relative to healthy skin, a melanoma feature [10]. The

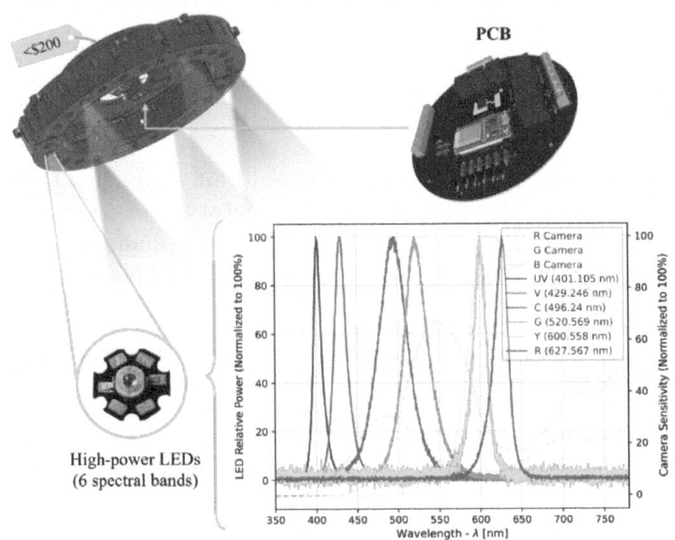

Fig. 1. Low-cost MSI system overview: 3D-printed ring-shaped enclosure (top-left); designed PCB, containing the microcontrollers and RGB camera (top-right); an LED used (bottom-left); LEDs' emission spectra and RGB camera sensitivity (bottom-right, adapted from [15]).

phantoms were composed of: epidermis (0.3 mm thick), dermis (2.5 mm thick), and the melanoma lesion, with total dimensions of 20 × 20 × 2.8 mm (Fig. 2, on the left).

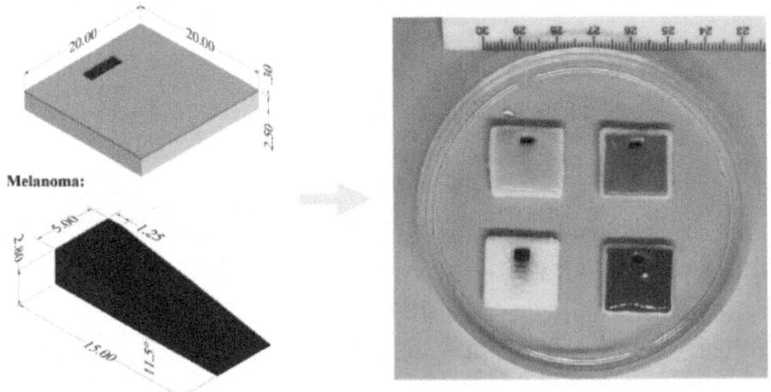

Fig. 2. The developed skin phantoms. On the left, the 3D-model and dimensions of the phantoms. On the right, the manufactured phantoms with three different skin phototypes: Fitz.1 (top-left), Fitz.3 (top-right) and Fitz.5 (bottom-right). An additional phantom ('control phantom') without epidermis was also manufactured (bottom-left).

Materials were chosen based on other studies [20,21,23]. More specifically, polidopamine (PDA, also known as synthetic melanin) was used to emulate the melanin absorption profile, and titanium dioxide powder (TiO_2), to emulate the scattering behavior of biological tissue. The dermis was fabricated using bioprinting, made of 10% (w/v) gelatin methacrylate (GelMA), 2% gelatin and 0.1131% TiO_2, being crosslinked in UV light (2 mWcm^{-2}) for 5 min in the presence of 1% Irgacure 2959. The epidermis was bioprinted using 5% sodium alginate, 2% gelatin, 0.1131% TiO_2 and PDA (3 mg/mL, 1,01 mg/mL and 0,338 mg/mL to achieve, respectively, phototypes Fitz.5, Fitz.3 and Fitz.1). Epidermis was crosslinked in a $1M$ solution of calcium chloride ($CaCl_2$) for 5 min. The emulated melanoma was fabricated in a mold, with 5% sodium alginate, 2% gelatin and PDA ($9mg/mL$ - higher concentration than used in the epidermises to achieve higher relative absorbance), being crosslinked in $CaCl_2$ solution for 10 min. The optical properties, absorption and reduced scattering coefficients (μ_a and μ_s), were estimated with an UV-Vis-NIR spectrophotometer and the Inverse Adding-Doubling algorithm [16] (Fig. 3). For each sample, total reflectance and total transmittance were measured twice, leading to four estimations of optical properties using the referred algorithm. The manufactured synthetic epidermis presented similar properties to those of real epidermis across all visible spectrum. Following Yim et al.'s storage method [21], the manufactured skin phantoms were stored at 4 °C in a Petri dish, sealed with Parafilm, then wrapped in plastic film and aluminum foil.

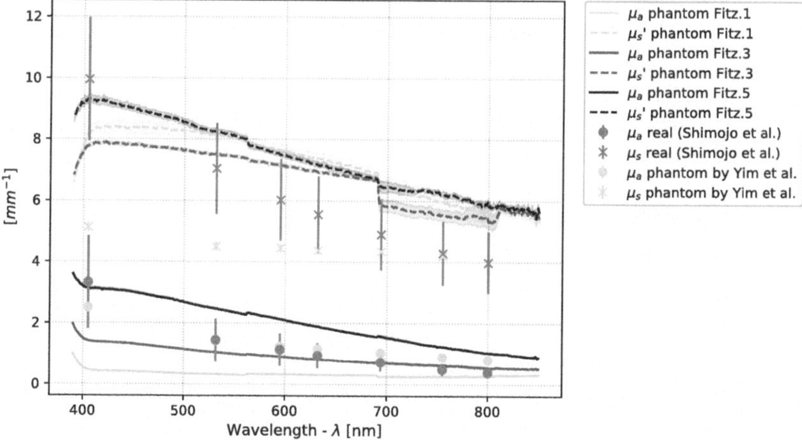

Fig. 3. Optical properties of the manufactured epidermises (mean, with the shaded area indicating ±1 standard deviation across four estimations), as well as of real epidermis (reported by Shimojo et al. [17]) and of mimicked epidermis by Yim et al. [21].

3 Experiments and Results

The phantoms were imaged using the low-cost MSI system prototype in three replicates. For each image, the acquisition protocol took approximately 20 s and consisted of the capture of eight pictures: (a) one picture with the light source deactivated (dark reference); (b) one picture in each different spectral band (six pictures); (c) and one picture with all LEDs on. Before the acquisitions, a color reference chart (ColorChecker Passport Photo) was imaged. In the color chart, patch number 19 ('white') was used as a white standard [13]. A common image processing pipeline was used [3,14]: (i) conversion from RGB to hue, saturation, and intensity (HSI color space); (ii) reflectance calibration for each pixel, using $I = \frac{I_0 - D}{W - D}$, where I_0 is the raw intensity value, W is the maximum intensity value of ColorChecker's patch 19 in the analyzed band and D is the intensity value measured in the same pixel without illumination. After processing, a calibrated image was obtained for each spectral band (Fig. 4, on the left). Three characteristics of the phantoms can be explored with the images: (a) phototype dependence in the analysis; (b) melanoma and epidermises differentiation potential, and determination of optimal bands for this task; (c) and capacity for evaluating structures beneath the skin tissue.

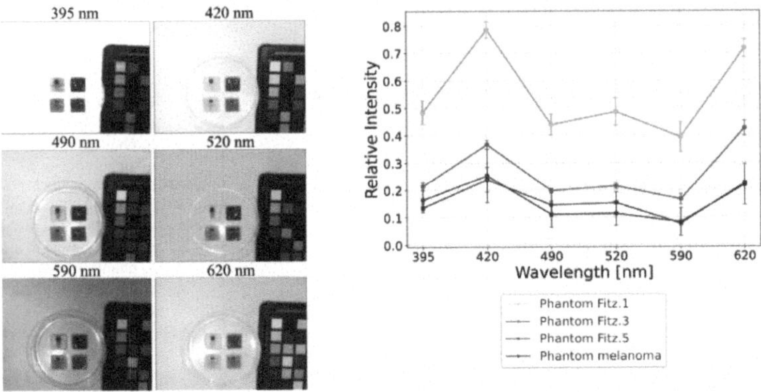

Fig. 4. On the left: multispectral image of the skin phantoms (six different wavelengths). On the right: spectral signatures of the mimetic epidermises and melanoma lesion (extracted from the control phantom). The error bars at each wavelength refer to the standard deviation calculated from values of 600 different pixels.

In the captured multispectral image, one can see how the different phantoms respond to radiation of different parts of the electromagnetic spectrum. At 395 nm, the three different epidermises appear very dark and it is difficult to distinguish the epidermis Fitz.5 from the melanoma. Also, only in the control phantom it is possible to see signs of melanoma underneath the mimetic tissue. This is due the high absorbance of UV light by PDA, as confirmed by the measured optical properties. A very different scenario is observed at 620 nm, where epidermises and melanoma seem lighter and more distinguishable from each other. The lesion beneath the tissue is also noticeable, which is expected, since light can penetrate deeper in this part of the spectrum (less absorption).

Further exploration was performed by plotting the spectral signatures of the three different epidermises and melanoma lesion. For each of these tissues, spectral signatures of 600 different pixels were extracted (200 pixels of each three replicate image), selecting pixels from each tissue using Multi-Otsu thresholding and distance transform (Fig. 4, on the right). The average spectral signatures of the exploited tissues are distinguishable from one another, and harder differentiation is noticeable between epidermis Fitz.5 and the melanoma lesion. For an analysis in which this differentiation is aimed, spectral signatures are further apart at 490 and 520 nm, and this bands could be more useful at this task. Further, as expected, at any wavelength, the higher the PDA concentration (Fitz.1 to Fitz.5), the smaller the measured relative intensity (less reflectance).

The deep tissue perception by the MSI acquisition system was explored by cropping each phantom image, isolating a strip containing the melanoma lesion (Fig. 5, on the top). The intensity values along the vertical axis (named 'vertical pixels') were plotted, by computing the weighted average of the three most frequent relative intensity values. Specular reflection was avoided by disregarding values 35% higher than the last measured average value.

For each phantom, three main levels of intensity are observable: at both ends of the graph, higher intensity values are found, due to the white background; lower intensity values are observed between pixels 50 and 75, where the fully exposed melanoma lesion lies; finally, the intermediate intensity level corresponds to epidermis regions (Fig. 5, on the bottom). The gradual change of intensity along the 'vertical pixels' axis is most noticeable in the control phantom images - since there is no coverage by any epidermis - and more evident at 395, 490, 520 and 590 nm. The same behavior is less prominent in phantoms with epidermis, because of higher light absorption. However, a slight slope can be seen between pixels 75 and 100, especially in phantom Fitz.1 at 490, 520, 590 and 620 nm.

4 Discussion

Experiments showed how MSI provides a more detailed understanding about a sample and how specific spectral bands can bring more valuable information for a specific task (e.g., distinction between melanoma lesion and healthy tissue). Also, it became clear that the analysis is dependent on skin phototype, depicting the importance of evaluating this aspect during the validation of MSI systems. This dependence was especially observed when evaluating structures beneath the mimetic tissue, in which the low-cost MSI system performance was only satisfactory in the analysis of the control phantom. One of the reasons was that only the visible spectrum was exploited. The use of light in the near-infrared (NIR) region would have improved the analysis in this scenario, since the absorber of

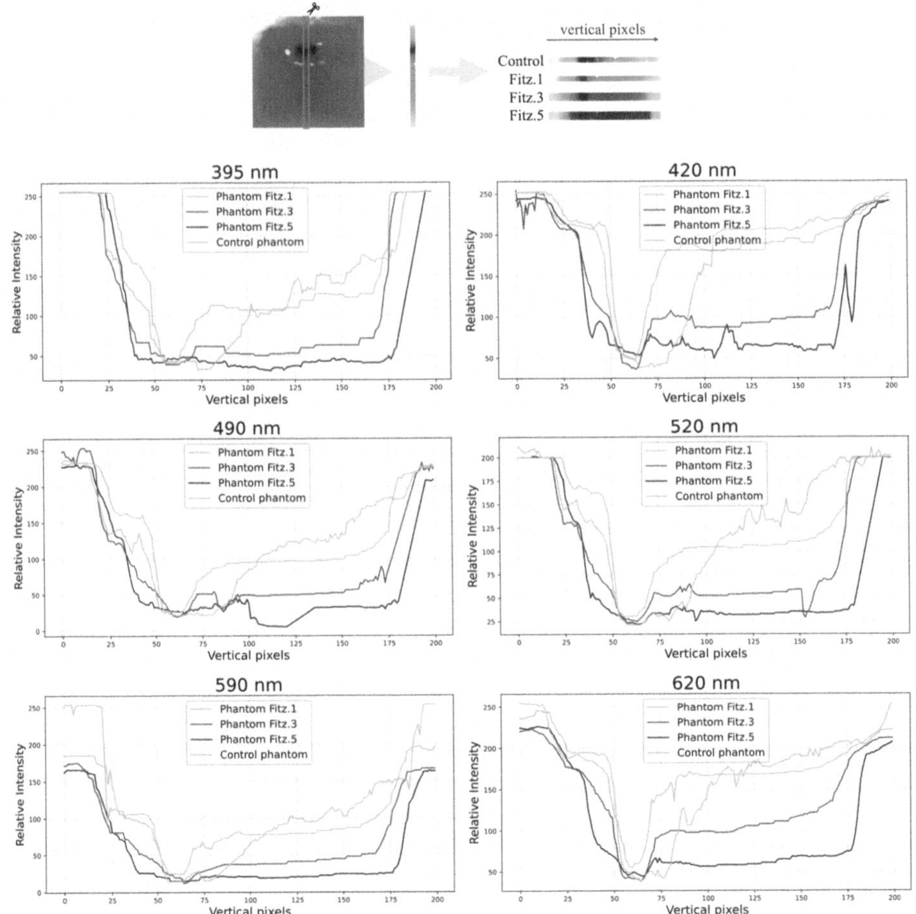

Fig. 5. On the top: the skin phantoms images were cropped to produce strips containing the melanoma lesions. On the bottom: relative intensity values measured along the 'vertical pixels' axis at each spectral band.

the mimetic tissue (PDA) interacts less with this type of radiation. In future versions of the low-cost MSI acquisition system, a detector sensitive to NIR could be used, as well as LEDs that emit light in that part of the spectrum.

These remarks about the MSI system were possible because of the prior knowledge about the samples imaged (i.e., skin phantoms) and the different possible scenarios created by them (e.g., phototype diversity, melanoma depth variation). In the future, other characteristics can be incorporated in the phantoms (e.g., uneven chromophore distribution in the melanoma lesion [10]) to enrich the analysis. Going further, the skin phantoms could also be imaged with other available spectral acquisition systems, for better comparison between them.

5 Conclusion

Aiming to develop a more accessible approach to MSI-based skin analysis, this study presents two complementary tools: a low-cost multispectral acquisition system prototype; and skin phantoms, for MSI systems' validation and comparison. The developed acquisition system was designed to capture images in six different wavelengths, distributed throughout the visible spectrum. Inexpensive and widely available components were used, with a bill of materials of less than $200, much more affordable than commercial options. Next, skin phantoms were designed and manufactured, with very similar optical properties to those of real skin, phototype diversity and a mimetic melanoma lesion, with depth variation and higher absorbance than that of the mimetic skin. The skin phantoms were imaged with the developed MSI system prototype and its potential for skin analysis was exemplified. Also, in the evaluation, the need to consider phototype diversity during validation became more evident. The prior knowledge of the skin phantoms, the variety of mimicked features, and their skin tone diversity were key in this investigation. These same characteristics would be valuable for future MSI equipment comparison, and validation of algorithms for estimation of chromophore content in the skin.

Acknowledgments. The authors would like to thank the São Paulo Research Foundation (FAPESP - grants #2022/11762-8 and #20/09838-0), the National Council for Scientific and Technological Development (CNPq - grant #317133/2023-3) and the Teaching, Research, and Extension Support Fund (FAEPEX - grants #3181/24 and #3434/24) for providing financial support.

Disclosure of Interests. The authors have no competing interests to declare that are relevant to the content of this article.

References

1. Afromowitz, M., Callis, J., Heimbach, D., DeSoto, L., Norton, M.: Multispectral imaging of burn wounds: a new clinical instrument for evaluating burn depth. IEEE Trans. Biomed. Eng. **35**(10), 842–850 (1988). https://doi.org/10.1109/10.7291
2. Aggarwal, L.P., Papay, F.A.: Applications of multispectral and hyperspectral imaging in dermatology. Exp. Dermatol. exd.14624 (2022). https://doi.org/10.1111/exd.14624
3. Amigo, J.M., Grassi, S.: Configuration of hyperspectral and multispectral imaging systems. In: Data Handling in Science and Technology, vol. 32, pp. 17–34. Elsevier (2019). https://doi.org/10.1016/B978-0-444-63977-6.00002-X
4. Bolton, F.J., Bernat, A.S., Bar-Am, K., Levitz, D., Jacques, S.: Portable, low-cost multispectral imaging system: design, development, validation, and utilization. J. Biomed. Opt. **23**(12), 1 (2018). https://doi.org/10.1117/1.JBO.23.12.121612
5. Bosse, J.L., Adhiwibawa, M.A.S., Brotosudarmo, T.H.: Multispectral imaging with raspberry pi for assessment of plant health status. Indon. J. Nat. Pigments **1**(2), 30 (2019). https://doi.org/10.33479/ijnp.2019.01.2.30
6. Du, T., et al.: Hyperspectral imaging and characterization of allergic contact dermatitis in the short-wave infrared. J. Biophotonics **13**(9), e202000040 (2020)

7. Ilisanu, M.A., Moldoveanu, F., Moldoveanu, A.: Multispectral imaging for skin diseases assessment-state of the art and perspectives. Sensors **23**(8), 3888 (2023). https://doi.org/10.3390/s23083888
8. Johansen, T.H., et al.: Recent advances in hyperspectral imaging for melanoma detection. Wiley Interdisc. Rev.: Comput. Stat. **12**(1), e1465 (2020)
9. Kim, E.B., Baek, Y.S., Lee, O.: A study on the classification of atopic dermatitis by spectral features of hyperspectral imaging. IEEE Access (2024)
10. Kuzmina, I., et al.: Towards noncontact skin melanoma selection by multispectral imaging analysis. J. Biomed. Opt. **16**(6), 060502 (2011). https://doi.org/10.1117/1.3584846
11. Li, Q., He, X., Wang, Y., Liu, H., Xu, D., Guo, F.: Review of spectral imaging technology in biomedical engineering: achievements and challenges. J. Biomed. Opt. **18**(10), 100901 (2013). https://doi.org/10.1117/1.JBO.18.10.100901
12. Lopez-Ruiz, N., Granados-Ortega, F., Carvajal, M.A., Martinez-Olmos, A.: Portable multispectral imaging system based on Raspberry Pi. Sens. Rev. **37**(3), 322–329 (2017). https://doi.org/10.1108/SR-12-2016-0276
13. McCamy, C.S., Marcus, H., Davidson, J.G., et al.: A color-rendition chart. J. App. Photog. Eng. **2**(3), 95–99 (1976)
14. McCarthy, A., Barton, K., Lewis, L.: Low-cost multispectral imager. J. Chem. Educ. **97**(10), 3892–3898 (2020). https://doi.org/10.1021/acs.jchemed.0c00407
15. OmniVision Technologies: OV2640 Color CMOS UXGA (2.0 MegaPixel) CameraChip^{TM}with OmniPixel2^{TM}Technology (2006). Version 1.6
16. Prahl, S.: Everything i think you should know about inverse adding-doubling. Oregon Med. Laser Cent. St. Vincent Hospit. **1344**, 1–74 (2011)
17. Shimojo, Y., Nishimura, T., Hazama, H., Ozawa, T., Awazu, K.: Measurement of absorption and reduced scattering coefficients in Asian human epidermis, dermis, and subcutaneous fat tissues in the 400- to 1100-nm wavelength range for optical penetration depth and energy deposition analysis. J. Biomed. Opt. **25**(04), 1 (2020). https://doi.org/10.1117/1.JBO.25.4.045002
18. Shrestha, R., Hardeberg, J.Y.: Multispectral imaging using LED illumination and an RGB camera. In: Color and Imaging Conference, vol. 21, no. 1, pp. 8–13 (2013). https://doi.org/10.2352/CIC.2013.21.1.art00003
19. Vasefi, F., MacKinnon, N., Farkas, D.: Hyperspectral and multispectral imaging in dermatology. In: Imaging in Dermatology, pp. 187–201. Elsevier (2016). https://doi.org/10.1016/B978-0-12-802838-4.00016-9
20. Verduzco-Grajeda, L.E., Solís-Delgadillo, N.V., Romo Castañeda, A.G., Ortíz-Martínez, M., Alfaro-Gómez, M.: Structural analysis and spectroscopic characterization of melanin-alginate films. Chem. Phys. Impact **9**, 100733 (2024). https://doi.org/10.1016/j.chphi.2024.100733
21. Yim, W., et al.: 3D-bioprinted phantom with human skin phototypes for biomedical optics. Adv. Mater. **35**(3), 2206385 (2023). https://doi.org/10.1002/adma.202206385
22. Zhang, Y., et al.: In vivo monitoring of rashes caused by systemic lupus erythematosus disease using snapshot spectral imaging. J. Biophotonics **13**(3), e201960067 (2020). https://doi.org/10.1002/jbio.201960067
23. Zhao, F., et al.: Reproducibility of identical solid phantoms. J. Biomed. Opt. **27**(07) (2022). https://doi.org/10.1117/1.JBO.27.7.074713

Development and Evaluation of an AI-Driven Telemedicine System for Prenatal Healthcare

Juan Barrientos[1](✉), Michaelle Pérez[1], Douglas González[1], Favio Reyna[2], Julio Fajardo[1], and Andrea Lara[1]

[1] BiomedLab, Biomedical Engineering Institute, Universidad Galileo, Guatemala City, Guatemala
{juan.barrientos,perezm7,duglasa,julio.fajardo,andrealh}@galileo.edu

[2] Faculty of Medicine, Universidad Francisco Marroquín, Guatemala City, Guatemala
freyna@ufm.edu

Abstract. Access to obstetric ultrasound is often limited in low-resource settings, particularly in rural areas of low- and middle-income countries. This work proposes a human-in-the-loop artificial intelligence (AI) system designed to assist midwives in acquiring diagnostically relevant fetal images using blind sweep protocols. The system incorporates a classification model along with a web-based platform for asynchronous specialist reviews. By identifying key frames in blind sweep studies, the AI system allows specialists to concentrate on interpretation rather than having to review entire videos. To evaluate its performance, blind sweep videos captured by a small group of soft-trained midwives using a low-cost Point-of-Care Ultrasound (POCUS) device were analyzed. The system demonstrated promising results in identifying standard fetal planes from sweeps made by non-experts. A field evaluation indicated good usability and a low cognitive workload, suggesting that it has the potential to expand access to prenatal imaging in underserved regions.

Keywords: Human-in-the-loop · Telemedicine · Obstetric Ultrasound · Blind sweeps

1 Introduction

Access to specialized healthcare services remains a major global challenge. In low- and middle-income countries (LMICs), limited resources and socioeconomic disparities, along with ongoing underinvestment in education and training, reduce the availability of medical care in rural and underserved areas [10]. While high-income countries average more than three physicians per 1,000 inhabitants, many LMICs remain at or below one physician per 1,000, highlighting a critical gap in healthcare capacity [22]. Telemedicine has emerged as a vital strategy for bridging healthcare capacity gaps by enabling remote access to specialized care. Since the COVID-19 pandemic, digital health has been increasingly emphasized

U. Anazodo et al. (Eds.): MIRASOL 2025, LNCS 16398, pp. 143–152, 2026.
https://doi.org/10.1007/978-3-032-13654-1_15

to expand medical expertise and relieve pressure on local health systems [17]. This approach is especially beneficial in regions with significant specialist shortages, where traditional healthcare models are inadequate. In LMICs, tailored telehealth systems have led to improvements in healthcare access [9,16]. Despite its potential, telemedicine implementation in LMICs remains limited, hindered by factors such as low digital literacy, insufficient training for providers, and sociocultural resistance [1,2].

Some telemedicine solutions now incorporate artificial intelligence (AI) tools to enhance diagnostic accuracy and scalability in resource-constrained settings. While these AI-based solutions show promising clinical results [11], their adoption in LMICs is still restricted due to a lack of user experience evaluation and cultural relevance. Additionally, most AI tools are developed within narrow clinical contexts, raising concerns about their generalizability across diverse health systems [3,5].

Obstetric ultrasound telemedicine tools are vital for prenatal care in rural and underserved areas; however, their use is limited by poor infrastructure, a lack of trained personnel, and insufficient technical support. Most ultrasound machines and AI solutions are designed for hospitals, making them unsuitable for community-based maternal care [6]. Many AI-driven tools provide diagnostic outputs without human oversight, complicating interpretation for healthcare providers, which is critical for effective patient communication [19]. Recent studies suggest human-in-the-loop frameworks as a solution, involving local providers in data collection and trained specialists for interpretation. This approach improves clarity, reduces the cognitive load on non-experts, and fosters trust, ultimately enhancing outcomes in maternal healthcare [21]. In Guatemala, healthcare is centralized in urban areas, restricting access to prenatal care, including obstetric ultrasounds, for rural and Indigenous populations [14,23]. Midwives and nurses typically provide primary care in these regions but often lack training and tools for obstetric imaging [12,15].

This work presents a pilot study on the NatalIA system conducted with midwives in rural Guatemala, as shown in Fig. 1. The system integrates portable

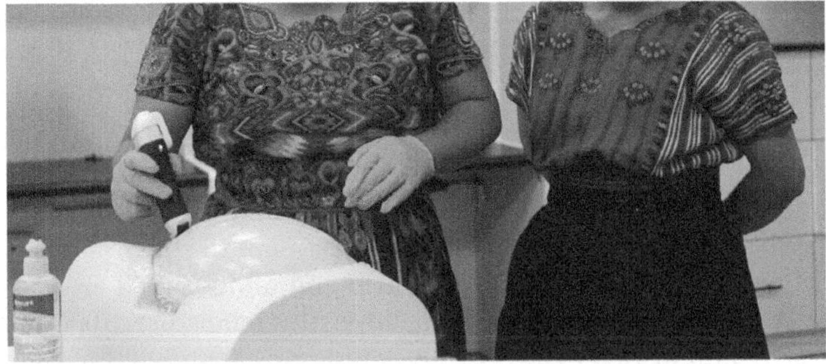

Fig. 1. Midwives from rural areas in Guatemala performing blind sweep ultrasound scans using a handheld POCUS device.

ultrasound technology, deep learning (DL), a subfield of AI, and community-based care, allowing midwives to perform image acquisition using blind sweep protocols without needing trained personnel [20]. Collected videos are analyzed by an DL model, which preselects relevant views, allowing remote specialists to focus on interpretation rather than data curation, thus supporting patient follow-up. The study evaluated the acceptance, usability and potential for adoption of the system in community-based maternal care settings. It tested the effectiveness of blind sweep protocols for capturing relevant fetal images and included clinical validation with specialists to assess the model accuracy in identifying standard fetal planes. User perception and cognitive load were analyzed from the midwife's perspective, covering tasks like performing the blind sweep, uploading studies, and interacting with NatalIA interface. Results showed good acceptance among midwives and specialists, indicating that the system is suitable for low-resource environments and has significant potential to connect specialists with rural populations.

2 Methods

2.1 System Architecture

The proposed system was designed following a client-server architecture with a clear separation between the frontend and backend components, as illustrated in Fig. 2. This web-based design minimizes computational demands on user devices and allows access from any browser without requiring a dedicated application. The frontend was implemented using the Angular framework, providing an interface for midwives to upload ultrasound studies and for specialists to review results. The backend was developed using Flask in Python to manage data processing and communication. Additionally, PyTorch was used to execute the DL model asynchronously, while MongoDB and a local image bucket were used for structured data and image storage, respectively.

Ultrasound videos acquired by midwives using a portable Point-of-Care Ultrasound (POCUS) device and a mobile interface are saved locally and uploaded to the system web platform directly from the same device. Once uploaded, the system processes the studies in the background and automatically identifies standard fetal planes. Ultrasound specialists review these predictions

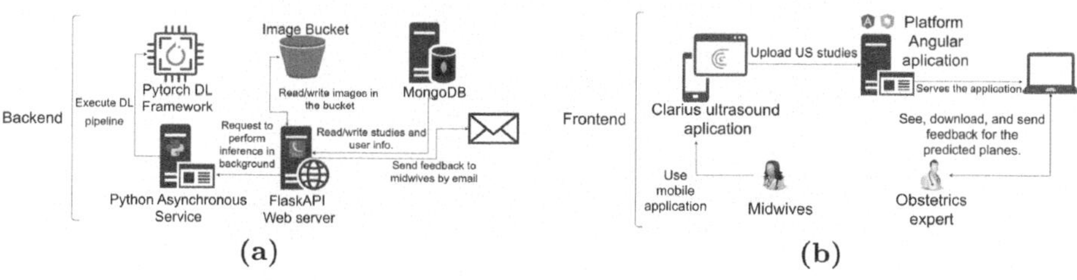

Fig. 2. System architecture composed of (a) backend and (b) frontend components.

asynchronously via the web interface and may also download the videos if necessary. Finally, clinical feedback is provided to the midwives through the email registered on the platform and also stored within the system for later access, ensuring a traceable and accessible diagnostic workflow.

2.2 Dataset

The dataset for this study, NatalIA: PBF-US1 was collected using a US-7a SPACE FAN obstetric phantom (Kyoto Kagaku, Japan) that simulates a 23-week pregnancy [7]. For its construction, blind sweep ultrasound videos were obtained using a Clarius C3 HD3 handheld POCUS device (Clarius, Canada) under standardized conditions. In addition, forty-five nursing students participated after receiving a one-hour training session on a blind sweep protocol adapted from Toscano et al. [20], which includes vertical and horizontal trajectories. We extended the protocol by adding two diagonal sweeps to capture greater anatomical variation. The four types of blind sweeps used in this study are illustrated in Fig. 3, leading to 90 ultrasound video recordings.

Annotations were performed based on the International Society of Ultrasound in Obstetrics and Gynecology (ISUOG) guidelines for fetal biometry [13,18]. The labeling scheme included five standard fetal planes: biparietal, abdominal, heart, spine, and femur, as well as a "no plane" label for frames that did not contain any of the relevant anatomical structures. The dataset comprises 42 biparietal, 63 abdominal, 61 heart, 134 spine, 46 femur plane frames, and 19,061 frames labeled as "no plane".

2.3 DL Models

Three different DL architectures were selected for this study in order to extract standard fetal planes from blind sweep ultrasound videos: SonoNet, ResNet-18, and ResNet-50 [4,8]. These architectures were chosen due to their relatively lightweight design, which makes them suitable for potential deployment

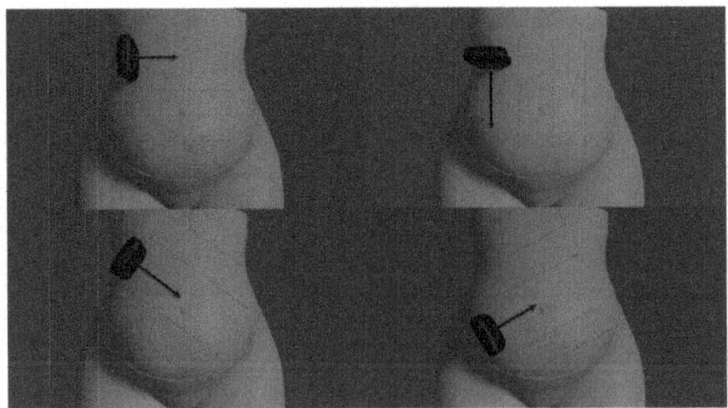

Fig. 3. Blind sweep acquisition protocol with vertical, horizontal, and two diagonal trajectories.

on portable devices commonly used in low-resource or remote environments. Additional frames were incorporated by adding the consecutive frames to the labeled ones. Frames showing high similarity, with both the Structural Similarity Index (SSIM) and normalized cross-correlation exceeding 0.90, were included and assigned the same label as the original frame. This approach expanded the dataset and allowed the model to learn from a wider set of clinically meaningful patterns. After this step, the dataset included 562 abdominal, 264 biparietal, 365 femur, 544 heart, and 1067 spine frames. For the "no plane" class, 30% of available frames were randomly selected, totaling 4,765 examples. The model was trained and validated using an 80/20 split of the dataset. Its performance was further assessed on blind sweep acquisitions performed by midwives, which specialists independently reviewed to confirm the presence of standard fetal planes.

3 Experiments and Results

To evaluate both the technical performance and field usability of the proposed system, a series of experiments were conducted comprising three main components: (1) the training and evaluation of DL models for fetal plane detection, (2) A clinical validation phase where specialists verified the accuracy of the fetal plane detections made by the NatalIA DL model based on blind sweep acquisitions performed by midwives, and (3) a usability study involving midwives and obstetric ultrasound specialists. The following section describes the study design and summarizes the key findings from all three components.

3.1 DL Model Evaluation

To evaluate the ability of architectures to classify standard fetal planes in blind sweep acquisitions, we first tested SonoNet, a widely used model for fetal plane detection. This initial experiment used the original pretrained weights without retraining to assess the model's out-of-domain performance. As expected, results were suboptimal due to domain shift, with an accuracy of 30% and an F1-score of 39%, confirming the need for adaptation to the characteristics of blind sweep acquisitions performed by non-experts.

Subsequently, SonoNet was fine-tuned on our dataset using the same training procedure as for other architectures. In parallel, ResNet-18 and ResNet-50 were tested using a transfer learning approach initialized with ImageNet weights. All models were trained with the AdamW optimizer, a learning rate of 0.000001, and a batch size of 64 for 300 epochs. The cross-entropy loss function was used, and data augmentation techniques included random horizontal and vertical flips and rotations up to 45°. As part of preprocessing, images were resized to 224×224 pixels and normalized according to the standard used by the original pretrained models.

The classification results are presented in the confusion matrices of Fig. 4. After fine-tuning, **SonoNet** achieved an accuracy of 75.5% (95% CI: 72.9–77.2), a weighted precision of 85.3% (95% CI: 83.9–86.7), a weighted recall of 75.1%

(95% CI: 72.7–77.2), and a weighted F1-score of 76.7% (95% CI: 74.7–78.6). **ResNet-18** reached an accuracy of 90.6% (95% CI: 88.9–91.9), a weighted precision of 91.6% (95% CI: 90.5–92.8), a weighted recall of 90.5% (95% CI: 88.9–91.8), and a weighted F1-score of 90.7% (95% CI: 89.3–92.0), while **ResNet-50** showed the best overall performance with 92.87%(95% CI: 91.5–94.1), a weighted precision of 93.5% (95% CI: 92.4–94.6), a weighted recall of 92.9% (95% CI: 91.6–94.1), and a weighted F1-score of 93.0% (95% CI: 91.8–94.2), demonstrating consistent improvements across fetal plane categories.

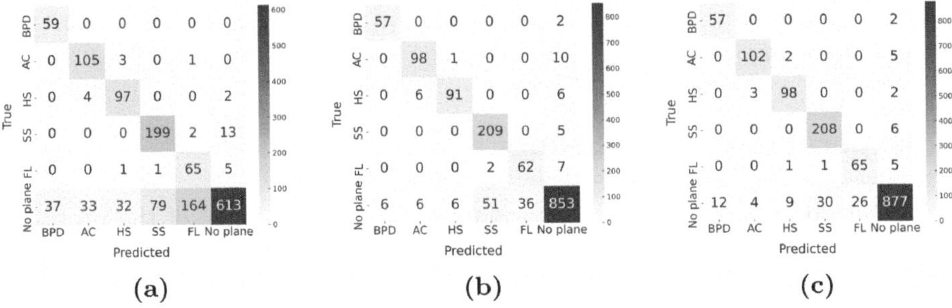

Fig. 4. Confusion matrices from the transfer learning experiment for the three architectures: (a) SonoNet, (b) ResNet-18, and (c) ResNet-50. Labels indicate fetal planes: BPD (biparietal diameter), AC (abdominal circumference), HS (heart standard view), SS (spine standard view), FL (femur length), and No plane (non-diagnostic frame).

3.2 Clinical Validation of System's Predictions by Specialists

A validation test was conducted using a fetal ultrasound phantom to assess the performance in real settings of the system. Eight midwives from rural regions of Guatemala each received a one-hour training session on performing blind sweep acquisitions. After training, each midwife independently performed two blind sweeps on the phantom. They then received a brief introduction to the digital platform and uploaded their recordings. The DL model processed these videos and returned the fetal planes it was able to detect. These outputs were reviewed by ultrasound specialists, who independently evaluated which fetal standard planes were present in the recordings. Table 1 shows a side-by-side comparison of the fetal standard planes identified by the DL model and those confirmed by the specialists for each midwife. In addition, false positives (FP) and false negatives (FN) were included to show the model performance.

3.3 Field Evaluation Study

To evaluate the usability and feasibility of the system in real settings, a field study was conducted involving eight midwives from rural regions of Guatemala

Table 1. Comparison of fetal planes identified by the system and confirmed by specialists for each midwife, including the number of false negatives (FN) and false positives (FP) per class.

Midwife	NatalIA System					Specialists				
	AC	BPD	HS	SS	FL	AC	BPD	HS	SS	FL
1	6	10	4	1	10	3	4	4	9	5
2	3	11	0	1	3	2	6	6	4	4
3	3	9	3	2	8	3	6	6	5	6
4	4	12	5	2	7	2	6	4	1	3
5	3	11	3	1	2	3	7	5	1	5
6	3	7	0	2	12	2	3	3	1	5
7	4	8	2	4	9	3	4	7	2	5
8	10	9	1	3	5	2	4	3	0	4
Metrics	FN (Underdetection)					FP(Overdetection)				
	AC	BPD	HS	SS	FL	AC	BPD	HS	SS	FL
	0	0	20	7	0	16	37	0	0	19

and three ultrasound specialists. Each participating midwife had between 5 and 15 years of experience in prenatal care, routinely performing procedures such as Leopold maneuvers and monitoring fetal heart rates with Doppler technology. As none of the midwives had previous training or experience with ultrasound techniques, they received prior training on the blind sweep protocol and the web-based platform. The evaluation focused on three main areas: blind sweep usability, system usability, and cognitive workload. Data for the first two areas were collected through structured questionnaires administered after completing both the training and the actual use of the system in the field. For cognitive workload assessment, the NASA Task Load Index (NASA-TLX) was used to quantify the mental, physical, and temporal demands perceived by the participants during use of the system.

Blind Sweep Protocol Usability. After the training session, all midwives were able to perform blind sweeps independently. Most participants reported that the protocol was easy to understand and required only one or two guided attempts to feel confident in its use. All completed two acquisition attempts using the ultrasound phantom. Several participants also mentioned limitations in their local infrastructure, such as unreliable access to electricity or the internet.

System Usability. Midwives found the web platform intuitive and easy to navigate. They successfully uploaded studies and accessed feedback using their mobile devices. Ultrasound specialists also reviewed the system interface and confirmed that it supported their clinical workflow. Suggestions for improvement included adding support for Mayan languages and incorporating a real-time help function.

Cognitive Workload Evaluation. NASA-TLX responses indicated that most participants experienced low to moderate levels of mental demand (M = 26.25, SD = 12.99) and very low physical demand (M = 12.5, SD = 4.14). Time pressure (M = 31.25, SD = 23.05), frustration (M = 22.5, SD = 20.62), and effort (M = 25.0, SD = 19.15) were generally rated as low, and the perceived performance score was consistently high (M = 80.0, SD = 21.60). Only a few participants reported isolated increases in workload dimensions. Figure 5 summarizes the average scores in the six dimensions of the NASA-TLX workload, including standard deviation bars to reflect the variability of the participants.

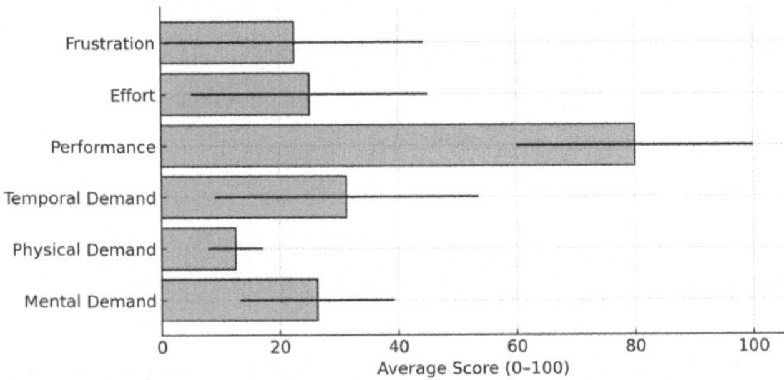

Fig. 5. Average scores reported by participants across the six NASA-TLX workload.

4 Discussion and Conclusion

This study suggest that a human-in-the-loop DL system can be effectively integrated into community-based maternal care workflows. The proposed system, demonstrated high usability and low cognitive demand when used by midwives with no prior experience in ultrasound. After brief instruction, participants successfully acquired ultrasound studies and interacted with the platform to upload and review findings. The overall process was positively received by both midwives and remote specialists, suggesting strong potential for adoption in low-resource settings, despite that the participants identified ultrasound as the least accessible diagnostic tool due to high equipment costs, limited availability in their communities, and its technical complexity.

The DL model evaluation showed that the original SonoNet architecture had limited performance under the new data conditions, while the adapted ResNet-based models, especially ResNet-50, achieved significantly better results. These findings highlight the importance of adapting models and using transfer learning when applying DL systems in clinical environments that differ from those used during initial training. The system was able to identify relevant fetal planes from blind sweep videos acquired by users without prior experience, and most of its predictions matched the evaluations made by independent specialists. However, performance still needs improvement for certain classes, particularly the HS and SS planes, which were often underdetected.

Despite the encouraging findings, this study presents several limitations. Data collection was conducted using a phantom model rather than real patients, and the usability assessment involved a small group of midwives. Moreover, while efforts were made to contextualize the usability findings, a review of the literature revealed a lack of published studies evaluating the usability of tele-ultrasound systems in comparable settings. This absence underscores the novelty and relevance of our work. Future research should include clinical validation with pregnant women, long-term follow-up, and participation from a broader range of primary healthcare providers. Beyond this study, the workflow combines blind sweep acquisition by non-experts, DL-based automatic analysis, and remote specialist review. This approach is valuable where training, time, or equipment are limited. Further evaluation in diverse clinical and cultural settings is needed.

Acknowledgments. This project was developed with the financial support of the Instituto de Efectividad Clínica y Sanitaria (IECS) trough a grant awarded by the International Development Research Centre (IDRC), Ottawa, Canada.

Disclosure of Interests. The views and opinions expressed in this document are those of the authors and do not necessarily represent those of Instituto de Efectividad Clínica y Sanitaria (IECS), International Development Research Centre (IDRC), or their governing bodies.

References

1. Al-Samarraie, H., Ghazal, S., Alzahrani, A.I., Moody, L.: Telemedicine in middle eastern countries: progress, barriers, and policy recommendations. **141**, 104232 (2020). https://doi.org/10.1016/j.ijmedinf.2020.104232
2. Alnasser, Y., Proaño, A., Loock, C., Chuo, J., Gilman, R.H.: Telemedicine and pediatric care in rural and remote areas of middle-and-low-income countries: narrative review. **14**(3), 779–78 (2024). https://doi.org/10.1007/s44197-024-00214-8
3. Assing Hvidt, E., et al.: Low adoption of video consultations in post–COVID-19 general practice in Northern Europe: barriers to use and potential action points. **25**, e4717 (2023). https://doi.org/10.2196/47173
4. Baumgartner, C.F., et al.: SonoNet: real-time detection and localisation of fetal standard scan planes in freehand ultrasound. IEEE Trans. Med. Imaging **36**(11), 2204–2215 (2017). https://doi.org/10.1109/TMI.2017.2712367
5. De La Torre, A., Diaz, P., Perdomo, R.: Analysis of the virtual healthcare model in Latin America: a systematic review of current challenges and barriers. **10**, 20 (2024). https://doi.org/10.21037/mhealth-23-47
6. Ginsburg, A.S., Liddy, Z., Khazaneh, P.T., May, S., Pervaiz, F.: A survey of barriers and facilitators to ultrasound use in low- and middle-income countries. **13**(1), 332 (2023). https://doi.org/10.1038/s41598-023-30454-w

7. González, D., Barrientos, J.P., Perez, M., Fajardo, J., Reyna, F., Lara, A.: NatalIA: PBF-US1 (phantom blind-sweeps for fetal ultrasound scanning) (2024). https://doi.org/10.5281/zenodo.14193949
8. He, K., Zhang, X., Ren, S., Sun, J.: Deep residual learning for image recognition (2015). https://arxiv.org/abs/1512.03385
9. Khurram, S.S., Aga, I.Z., Muzzamil, M., Karim, M., Hashmi, S.: Transitioning from crisis to continuity post-COVID-19 pandemic: adoption of telehealth by Sehat Kahani Healthcare Providers, Karachi, Pakistan. Telehealth Med. Today **9**(6) (2024). https://doi.org/10.30953/thmt.v9.541
10. Kruk, M.E., Gage, A.D., Joseph, N.T., Danaei, G., García-Saisó, S., Salomon, J.A.: Mortality due to low-quality health systems in the universal health coverage era: a systematic analysis of amenable deaths in 137 countries. Lancet **392**(10160), 2203–2212 (2018). https://doi.org/10.1016/S0140-6736(18)31668-4
11. Liu, X., et al.: Evaluation of an OCT-AI-based telemedicine platform for retinal disease screening and referral in a primary care setting. Transl. Vision Sci. Technol. **11**(3), 4 (2022). https://doi.org/10.1167/tvst.11.3.4
12. Martinez, B., et al.: mHealth intervention to improve the continuum of maternal and perinatal care in rural Guatemala: a pragmatic, randomized controlled feasibility trial. **15**(1), 12 (2018). https://doi.org/10.1186/s12978-018-0554-z
13. Carvalho, J.S., et al.: ISUOG Practice Guidelines (updated): sonographic screening examination of the fetal heart (2013)
14. Organization, P.A.H.: Guatemala - country profile | health in the Americas (2022). https://hia.paho.org/en/node/215, accessed: 2025-05-19
15. Ramos, E., et al.: Mobil Monitoring Doppler Ultrasound (MoMDUS) study: protocol for a prospective, observational study investigating the use of artificial intelligence and low-cost Doppler ultrasound for the automated quantification of hypertension, pre-eclampsia and fetal growth restriction in rural Guatemala. **14**(9), e09050 (2024). https://doi.org/10.1136/bmjopen-2024-090503
16. Rees, G.H., Peralta, F.: Telemedicine in Peru: origin, implementation, pandemic escalation, and prospects in the new normal. Oxford Open Digit. Health **2**, oqae002 (2024). https://doi.org/10.1093/oodh/oqae002
17. Saigí-Rubió, F.: Promoting telemedicine in Latin America in light of COVID-19. Rev. Panamericana Salud Pública **47**, e17 (2023). https://doi.org/10.26633/RPSP.2023.17
18. Salomon, L.J., et al.: Practice guidelines for performance of the routine mid-trimester fetal ultrasound scan. Ultras. Obstetr. Gynecol. **37**(1), 116–126 (2011). https://doi.org/10.1002/uog.8831
19. Stringer, J.S.A., et al.: Diagnostic accuracy of an integrated AI tool to estimate gestational age from blind ultrasound sweeps. JAMA **332**(8), 649–657 (2024). https://doi.org/10.1001/jama.2024.10770
20. Toscano, M., et al.: Testing telediagnostic obstetric ultrasound in Peru: a new horizon in expanding access to prenatal ultrasound. **21**(1), 328 (2021). https://doi.org/10.1186/s12884-021-03720-w
21. Vega, R., et al.: Overcoming barriers in the use of artificial intelligence in point of care ultrasound. **8**(1), 213 (2025). https://doi.org/10.1038/s41746-025-01633-y
22. World Bank: Physicians (per 1,000 people) - world bank data (2024). https://data.worldbank.org/indicator/SH.MED.PHYS.ZS. Accessed 19 May 2025
23. World Health Organization: Guatemala: Country health profile (2025). https://data.who.int/countries/320. Accessed 19 May 2025

CoMViT: An Efficient Vision Backbone for Supervised Classification in Medical Imaging

Aon Safdar(✉) and Mohamed Saadeldin

School of Computer Science, University College Dublin, Dublin, Republic of Ireland
aon.safdar@ucdconnect.ie, mohamed.saadeldin@ucd.ie

Abstract. Vision Transformers (ViTs) have demonstrated strong potential in medical imaging; however, their high computational demands and tendency to overfit on small datasets limit their applicability in real-world clinical scenarios. In this paper, we present **CoMViT** (Code available at: https://github.com/aonsafdar/CoMViT), a compact and generalizable Vision Transformer architecture specifically optimized for resource-constrained medical image analysis. CoMViT integrates a convolutional tokenizer, diagonal masking, dynamic temperature scaling, and pooling-based sequence aggregation to improve performance and generalization. Through systematic architectural optimization, CoMViT achieves robust performance across twelve MedMNIST datasets while maintaining a lightweight design with only ∼4.5M parameters. It matches or outperforms deeper CNN and ViT variants, offering up to 5–20× parameter reduction without sacrificing accuracy. Qualitative Grad-CAM analyses further reveal that CoMViT consistently attends to clinically relevant regions despite its compact size. Our findings highlight the potential of principled ViT re-design for developing efficient and interpretable models in low-resource medical imaging settings.

Keywords: Medical Imaging · Vision Transformers · Compact Models · Low-Resource · Tokenization

1 Introduction

Vision Transformers (ViTs) have emerged as powerful vision backbones due to their ability to model long-range dependencies via self-attention [3,17]. Their scalability has led to widespread adoption across classification, detection, and segmentation tasks [7]. However, their use in medical imaging remains limited, particularly in low-resource settings, where challenges such as small dataset sizes, scarce annotations, domain heterogeneity, and constrained compute persist [1,12].

Several strategies aim to mitigate ViTs' data inefficiency. Hybrid CNN-ViT models introduce inductive biases to preserve spatial locality [13,14], though at the cost of added architectural complexity. Others pursue transfer learning by distilling CNNs [17] or fine-tuning large ViTs pretrained on natural images [6].

U. Anazodo et al. (Eds.): MIRASOL 2025, LNCS 16398, pp. 153–163, 2026.
https://doi.org/10.1007/978-3-032-13654-1_16

These often struggle with domain shift, leading to sub-optimal transfer or negative knowledge transfer (NKT) [4].

Efforts to scale data via augmentation [18], unsupervised pretraining [2], or multi-domain training offer partial remedies but struggle with generalizability. Training across diverse domains may help one dataset while degrading performance on others. Domain-adaptive methods using CNN-based adapters [4] reduce this effect but add parameter overhead proportional to the number of domains.

We take a different approach. Rather than relying on scale, transfer, or domain-specific adaptation, we introduce **CoMViT**—a lean and universal ViT backbone for data- and compute-constrained medical imaging. CoMViT integrates several synergistic architectural innovations to boost efficiency and representational power. Specifically, (1) A lightweight transformer encoder with empirically chosen depth, embedding size, and MLP/head widths [8], balancing accuracy and efficiency. (2) A shallow convolutional tokenizer that encodes local context early, adding strong spatial priors. (3) Diagonal masking and learnable temperature scaling [11] to promote localized attention and gradient stability. (4) Sequence pooling in place of the classification token to improve aggregation and reduce redundancy.

We evaluate CoMViT across all twelve 2D MedMNIST datasets spanning multiple modalities (e.g., X-ray, OCT, microscopy) and diagnostic tasks. CoMViT consistently matches or exceeds the performance of significantly larger CNN and ViT baselines, using 5×–20× fewer parameters and FLOPs. Qualitative results confirm its focus on disease-relevant regions despite its compactness.

Our contribution are as under:

- We propose a compact ViT backbone optimized for data- and compute-limited medical imaging scenarios.
- We demonstrate that lightweight transformers can outperform deeper models across diverse medical modalities with superior accuracy-efficiency tradeoffs.
- We provide qualitative insights showing interpretable and diagnostically relevant attention maps.

Our results show that scale is not a prerequisite for success in medical imaging. Thoughtful architectural design can yield compact, generalizable ViTs suitable for real-world, resource-constrained deployment.

2 Related Work

The MedMNIST benchmark [21] introduced a suite of 2D biomedical datasets with low-resolution images across diverse modalities, enabling efficient model prototyping. Early baselines leveraged convolutional neural networks (CNNs) for their strong spatial priors, but CNNs require deep stacks to model long-range dependencies, increasing compute and memory cost [1].

Recent efforts have adapted ViTs for medical imaging. FPViT [13] integrates ResNet feature pyramids as token inputs to enhance patch encoding.

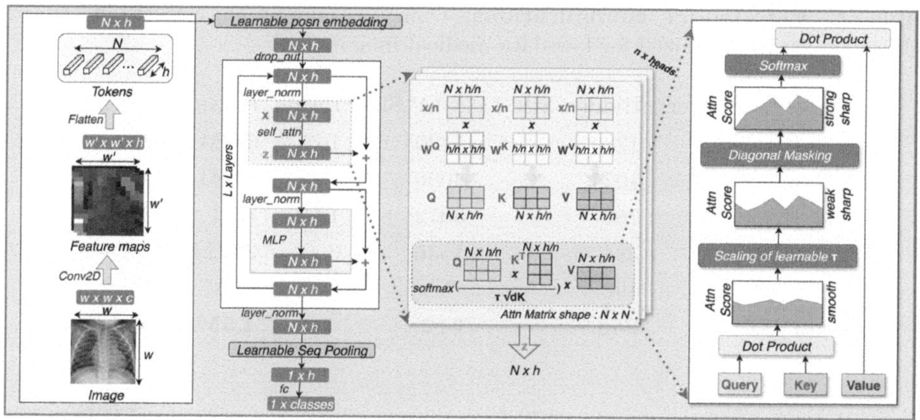

Fig. 1. CoMViT Architecture Overview. A light convolutional tokenizer produces tokens that are processed by a compact Transformer encoder with locality-promoting attention and learned sequence pooling.

MedViT [14] introduces local convolutional attention and depthwise MLPs for efficient context modeling. These methods improve locality but add architectural complexity. Fine-tuning large pretrained ViTs [6] yields competitive results but often suffers from domain mismatch and negative knowledge transfer [4]. Other works explore ViT robustness and domain alignment. PyramidAT [10] enhances resilience via multi-scale adversarial training; SelfCSL [15] leverages contrastive learning for semi-supervised domain adaptation; MedRDF [20] uses voting-based ensembling to defend against adversarial attacks. Despite these advances, lightweight ViT backbones tailored for low-resource medical imaging remain scarce. **CoMViT** addresses this gap by achieving competitive accuracy with ~4.5M parameters, surpassing deeper CNNs and larger ViTs—while maintaining practical efficiency for deployment in constrained clinical settings.

3 Method

We propose **CoMViT**, a compact Vision Transformer tailored to low-resource medical imaging. CoMViT incorporates five synergistic design choices and architectural modifications: (i) an *optimal compact ViT configuration* (depth/width/heads/MLP ratio); (ii) a *two-layer convolutional tokenizer* that injects spatial inductive bias; (iii) *diagonal masking* that removes self-loops in attention; (iv) *learnable per-head temperature* that sharpens attention distributions; and (v) *learned sequence pooling* that replaces the class token and learns to aggregate sementically-relevant tokens. Figure 1 summarizes the pipeline.

Model configuration (compact backbone). As shown in Table 1, CoMViT uses $L = 7$ layers, hidden size $h = 256$, $n_{\text{heads}} = 4$, and an MLP expansion of $2\times$ (about 4.5M parameters). This depth/width regime follows compact ViT design principles proposed by [9]. Shallow depth and moderate width reduce overfitting

Table 1. ViT model configurations. ComViT adopts an empirically tuned lightweight setup optimized for low-data medical imaging.

Model	Layers	Hidden Size D	MLP Size	Heads	Params	Source
ViT-H	32	1280	5120	16	632M	[3]
ViT-L	24	1024	4096	16	307M	[3,16]
ViT-B	12	768	3072	12	86M	[3,16]
ViT-S	12	384	1536	6	22.1M	[16]
ViT-Ti	12	192	768	3	5.8M	[16]
ComViT	**7**	**256**	**512**	**4**	**4.5M**	Ours

on small datasets; fewer heads avoid fragmenting capacity; and a 2× MLP limits redundancy while preserving nonlinearity.

Convolutional Tokenizer. Given an image $x \in \mathbb{R}^{w \times w \times c}$, we produce feature maps with two 7×7 convolutions followed by a 3×3 max-pool. Flattening yields tokens $T \in \mathbb{R}^{N \times h}$ and adding learnable positions P forms $Z_0 = T + P$. Learned tokenization injects spatial inductive bias and encodes boundaries and fine texture *before* attention. Empirically, such tokenizers improve data efficiency and reduce dependence on positional encoding, enabling smaller ViTs to train from scratch on limited data [9].

Locality-Promoting Attention. The sequence passes through L compact encoder blocks. Each of the L encoder blocks applies multi-head self-attention with two modifications that counteract the *over-smoothing* observed when many image tokens are present [11].

$$A = \frac{QK^{\top}}{\tau\sqrt{d_k}} + M, \quad \text{head}_k = \text{softmax}(A)\, V, \quad Z = \text{Concat}(\text{head}_1, \ldots, \text{head}_n)\, W^O \tag{1}$$

Here, τ is a learnable temperature and M is a diagonal mask with $Mii = -\infty$ (no self-attention). With many image tokens, vanilla attention tends to produce flat (over-smoothed) score distributions that dilute locality as attention is wasted on token self-correlation [5]. Firstly, *Diagonal masking* removes self-loops and suppresses self-attention, thus forcing tokens to attend to their neighbors and sharpening locality. Secondly, *Learnable temperature* lets each head adapt softmax sharpness to the layer and data; learned temperatures empirically converge below the conventional constant, and learning separate values per head outperforms sharing a single value [11]. Together, these changes increase the KL-divergence of the attention distribution relative to vanilla attention (less smoothing) and yield additive accuracy gains. Each attention sublayer is followed by a lightweight MLP (GELU) with residual connections and layer normalization.

Learned Sequence Pooling. Instead of a [CLS] token, we aggregate the final token sequence $O \in \mathbb{R}^{N \times h}$ with learned weights.

$$s = \sum_{i=1}^{N} w_i, O^i, \qquad w_i = \text{softmax}(W_p O^i), \tag{2}$$

and feed s to a linear classifier. Since tokens carry unequal diagnostic information, learning w_i lets the model emphasize informative regions and de-emphasize background or artifacts, improving stability on small datasets without adding parameters incurred by the class token [9].

With synergetic design choices and architectural modifications, **CoMViT** balances compactness with discriminative power to produce an efficient and interpretable ViT backbone that performs robustly in low-resource medical imaging settings.

4 Experiments and Results

4.1 Datasets

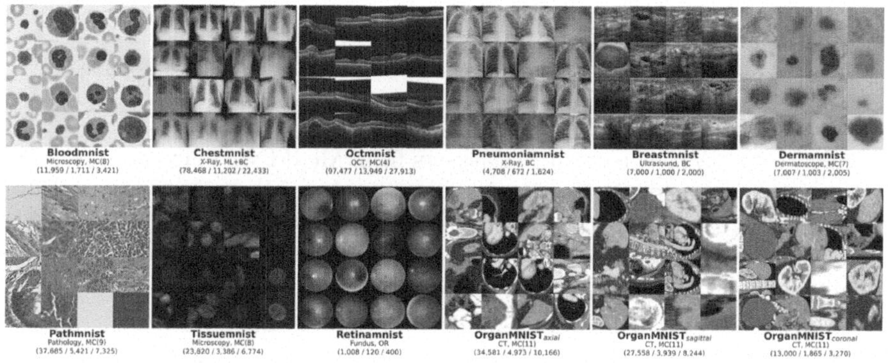

Fig. 2. Overview of the MedMNIST Dataset Family. Modality, Classification Task, and split (Train/Val/Test) are mentioned. Best viewed in color and zoomed in.

MedMNIST [21] used for evaluating CoMVit, is a curated suite of 12 biomedical imaging datasets covering diverse modalities such as X-ray, histopathology, dermatoscope, and microscopy (Fig. 2). All datasets are standardized to 224 × 224 resolution, enabling reproducibility. Unlike task-specific datasets, MedMNIST provides a modality-agnostic benchmark for evaluating generalizable tokenization and attention strategies. It supports multi-class, binary, and ordinal tasks, spanning texture- and structure-rich modalities under a unified evaluation protocol.

4.2 Experiment Protocol

Hyperparameters. ComViT is trained in PyTorch with `timm` [19], using AdamW for 300 epochs. The learning rate starts at 1.1×10^{-4} with cosine decay, warm-up (10 epochs), and cooldown (10 epochs). Regularization includes RandAugment, Mixup ($\alpha = 0.8$), CutMix ($\alpha = 1.0$), label smoothing, and drop-path (0.1). Mixup is probabilistically turned off after epoch 175. AMP and gradient clipping (max norm 1.0) are enabled. Batch size is 512 and input size is 224 × 224.

Table 2. Accuracy (%) comparison across MedMNIST2D datasets with 224 × 224 resolution. Red is best, blue is second-best.

Method	Path	Chest	Derma	OCT	Pneumonia	Retina	Breast	Blood	Tissue	OrganA	OrganC	OrganS
ResNet-18	90.9	94.7	75.4	76.3	86.4	49.3	83.3	96.3	68.1	93.6	92.0	78.5
ResNet-50	92.0	94.8	77.3	76.2	88.4	51.1	84.2	96.0	68.0	93.5	91.1	77.0
auto-sklearn	71.6	77.9	71.9	60.1	85.5	51.5	80.3	87.8	53.2	76.2	82.9	67.2
Google AutoML	72.8	77.8	77.8	76.8	91.6	53.1	86.1	90.8	67.3	88.6	87.7	74.9
MedViT-Tiny	95.6	95.6	76.8	76.7	94.9	53.4	89.6	95.0	70.3	93.1	90.1	78.9
CoMViT	91.08	95.12	77.0	84.6	92.14	53.9	83.97	98.04	69.79	95.15	92.84	80.44

Benchmarks and Metrics. We compare against strong ViT baselines (DeiT-Ti, PiT-Ti, PVT-Ti, RVT-Ti) and CNNs (ResNet-18, EfficientNet-B3) from [14]. Evaluation includes Top-1 test accuracy, parameter count, and GFLOPs per forward pass—capturing tradeoffs in accuracy, memory, and compute.

5 Results and Discussion

We evaluate CoMViT across all 12 datasets from MedMNIST2D and benchmark it against widely-used CNNs (ResNet-18/50), AutoML systems (AutoKeras, Google AutoML, auto-sklearn), and Transformer variants (MedViT-T). Table 2 shows that CoMViT achieves the best or second-best accuracy on 8 datasets, matching or outperforming much larger models. This highlights CoMViT's strong generalizability across diverse tasks and imaging modalities.

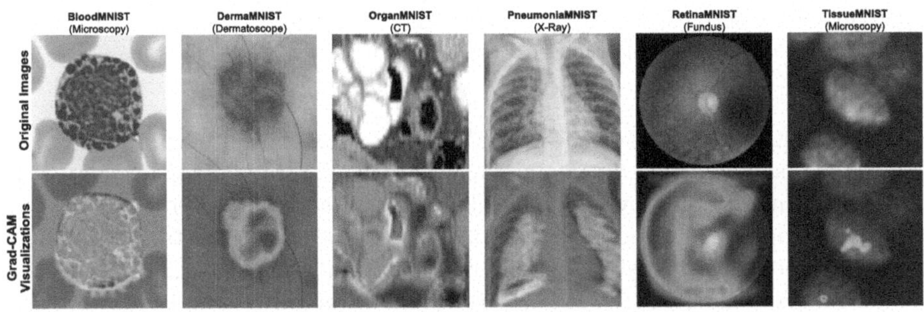

Fig. 3. Grad-CAM visualizations across six MedMNIST datasets highlighting the regions of interest that contribute most to the model's predictions. CoMViT accurately localizes relevant regions across diverse modalities.

Despite a compact footprint, CoMViT remains competitive with SOTA models. Table 3 shows average accuracy and parameter count. CoMViT achieves 84.5% accuracy using only 4.55M parameters, outperforming ResNet-50 (82.1%)

Table 3. Average accuracy comparison on MedMNIST-2D.

Methods	Params (M)	Avg. Top-1 Acc
ResNet-18 (224)	11.2	0.821
ResNet-50 (224)	23.5	0.821
auto-sklearn	–	0.722
AutoKeras	–	0.813
Google AutoML	–	0.809
MedViT-T (224)	10.2	0.840
MedViT-S (224)	23.0	0.851
MedViT-L (224)	45.0	0.842
CoMViT (Ours)	**4.55**	**0.845**

and matching MedViT-T (84.0%) which has over 2× parameters.Table 4 compares Tiny/Small/Large models on TissueMNIST. CoMViT leads among Tiny models (69.8%) while being the lightest. It rivals Small models like Swin-T and Twins-SVT-S with lower complexity, validating its architecture.

Table 4. Comparison with Tiny/Small/Large models on TissueMNIST.

Segment	Model	Img Size	Params (M)	FLOPs (G)	Top-1 (%)
Tiny	ResNet-18	224	11.7	1.8	68.1
	DeiT-Ti	224	5.7	1.3	59.5
	PiT-Ti	224	4.9	0.7	62.1
	PVT-T	224	13.2	1.9	63.4
	RVT-Ti	224	8.6	1.3	69.6
	CoMViT	224	**4.55**	**1.6**	**69.8**
Small	ResNet-50	224	25.6	4.1	68.0
	DeiT-S	224	22.0	4.6	67.0
	Swin-T	224	29.0	4.5	71.7
	Twins-SVT-S	224	24.0	2.9	72.1
	MedViT-S	224	23.6	4.9	73.1
Large	ResNet-152	224	60.2	11.3	67.5
	DeiT-B	224	87.0	17.5	66.9
	Swin-B	224	87.8	15.4	68.5
	MedViT-L	224	45.8	13.4	69.9

The parameter efficiency and accuracy trade-off are further visualized in Fig. 4. CoMViT appears near the Pareto frontier, striking a balance where further increases in model size yield only marginal accuracy gains. In contrast to

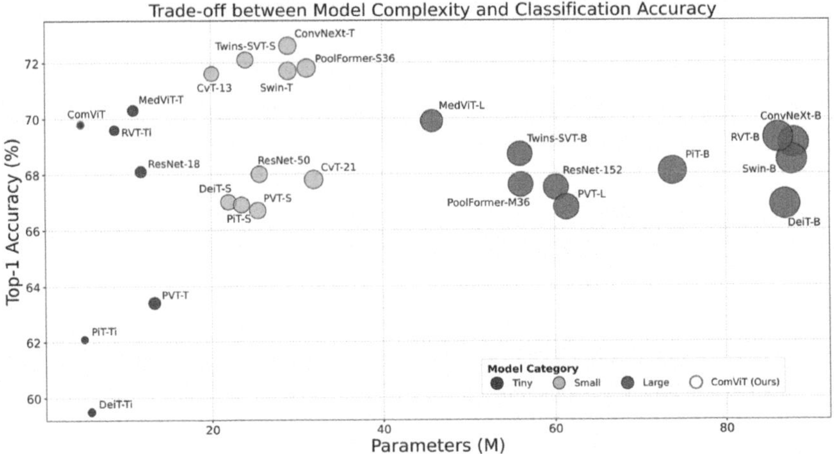

Fig. 4. Model Size vs. Accuracy on TissueMNIST. CoMViT achieves strong accuracy with minimal parameters, illustrating excellent efficiency.

Table 5. CoMViT ablation on *TissueMNIST* (Top-1%). Δ = difference with respect to CoMViT. ✓ = enabled.

	ConvTok	SeqPool	τ (learn.)	DMask	Top-1	Δ
ViT-Tiny (Patch 16, 12×192, 3h, MLP×4)	–	–	–	–	59.5	−10.3
	✓	✓	✓	✓	67.4	−2.4
CoMViT (full)	✓	✓	✓	✓	**69.8**	–
Patchify (no conv tokenizer)	–	✓	✓	✓	67.6	−2.2
CLS (no sequence pooling)	✓	–	✓	✓	68.3	−1.5
Fixed scale $\tau = \sqrt{d_k}$ (no learn.)	✓	✓	–	✓	68.9	−0.9
No DMask ($M = 0$)	✓	✓	✓	–	68.6	−1.2

other methods that require large parameter budgets to generalize well, CoMViT remains scalable and suitable for deployment across a variety of medical imaging tasks without retraining or architecture tuning. Figure 3 presents Grad-CAM visualizations to assess the interpretability of CoMViT. The model consistently highlights pathology-relevant regions across diverse modalities such as cell boundaries in BloodMNIST, lesion areas in DermaMNIST, and lung fields in PneumoniaMNIST, indicating that its predictions are grounded in clinically meaningful structures. Notably, in datasets with subtle or diffuse features (e.g., TissueMNIST), CoMViT still localizes relevant regions without relying on spurious artifacts. These results qualitatively support the model's capacity to extract semantically informative features while preserving spatial and structural priors (Table 5).

Ablations. We quantify the contribution of each component on TissueMNIST (Top-1% accuracy). A vanilla **ViT-Tiny** reaches **59.5%** Top-1. Keeping the

Tiny depth/width but swapping in CoMViT's modules (ConvTok, SeqPool, learnable τ, DiagMask) lifts accuracy to **67.4%** (+7.9), showing the modules drive most of the gain. Moving to our compact 7-layer/256-dim setup adds **+2.4** to **69.8%**. Within the CoMViT layout, ablating components hurts performance i.e. patchify (no ConvTok) −2.2, [CLS] (no SeqPool) −1.5, fixed $\tau = \sqrt{d_k}$ −0.9, no mask $M = 0$ −1.2. This supports that locality-aware tokenization and learned aggregation contribute most, while temperature and masking provide consistent, smaller gains.

In summary, CoMViT sets a strong baseline for low-resource medical image classification, offering competitive accuracy, lower inference cost, and improved scalability—making it a practical and robust choice for real-world medical AI systems. ComViT's improvements stem from a synergy of local bias, attention refinement, and compact architecture. The convolutional tokenizer extracts fine-grained local patterns absent in raw patch-based ViTs. Diagonal masking and dynamic temperature scaling complement this by enforcing spatial locality inside the attention heads, limiting unnecessary global mixing, and promoting robust local feature aggregation. Learnable pooling further allows flexible summarization of the token sequence without needing a rigid classification vector, improving adaptation to lesions of varying shapes and locations. Together, these balance capacity and generalization to yield high accuracy at minimal parameter cost, validating it as a robust strategy for low-resource medical imaging.

6 Conclusion

This work presented CoMViT, a compact Vision Transformer for low-resource medical imaging that combines a learned convolutional tokenizer, locality-promoting attention (diagonal masking with learnable temperature), and learnable sequence pooling within an empirically optimized lightweight configuration. This design injects strong spatial inductive bias while keeping the parameter count small. Across the 12 MedMNIST2D datasets, CoMViT matches or surpasses much larger CNN/ViT baselines with only 4.5M parameters, and Grad-CAMs confirm disease-relevant focus. CoMViT therefore offers a practical, deployable backbone for data and compute-constrained clinical settings.

Acknowledgments. This publication has emanated from research conducted with the financial support of Taighde Éireann – Research Ireland under Grant number 18/CRT/6183 (For the purpose of Open Access, the author has applied a CC BY public copyright licence to any Author Accepted Manuscript version arising from this submission).

Disclosure of Interests. The authors have no competing interests to declare that are relevant to the content of this article.

References

1. Aburass, S., Dorgham, O., Al Shaqsi, J., Abu Rumman, M., Al-Kadi, O.: Vision transformers in medical imaging: a comprehensive review of advancements and applications across multiple diseases. J. Imaging Inform. Med. 1–44 (2025). https://doi.org/10.1007/s10278-025-01481-y
2. Caron, M., et al.: Emerging properties in self-supervised vision transformers. In: 2021 IEEE/CVF International Conference on Computer Vision (ICCV), pp. 9630–9640. IEEE, Montreal (2021). https://doi.org/10.1109/ICCV48922.2021.00951, https://ieeexplore.ieee.org/document/9709990/
3. Dosovitskiy, A., et al.: An image is worth 16×16 words: transformers for image recognition at scale. arXiv:2010.11929 [cs] (2021)
4. Du, S., Bayasi, N., Hamarneh, G., Garbi, R.: MDViT: multi-domain vision transformer for small medical image segmentation datasets (2023). https://doi.org/10.1007/978-3-031-43901-8_43
5. Ferdous, G.J., Sathi, K.A., Hossain, M.A., Dewan, M.A.A.: SPT-swin: a shifted patch tokenization swin transformer for image classification. IEEE Access **12**, 117617–117626 (2024). https://doi.org/10.1109/ACCESS.2024.3448304, https://ieeexplore.ieee.org/document/10643534
6. Halder, A., Gharami, S., Sadhu, P., Singh, P.K., Woźniak, M., Ijaz, M.F.: Implementing vision transformer for classifying 2D biomedical images. Sci. Rep. **14**(1), 12567 (May2024). 10.1038/s41598-024-63094-9, https://www.nature.com/articles/s41598-024-63094-9
7. Han, K., et al.: A survey on vision transformer. IEEE Trans. Pattern Anal. Mach. Intell. **45**(1), 87–110 (2023). https://doi.org/10.1109/TPAMI.2022.3152247, https://ieeexplore.ieee.org/abstract/document/9716741
8. Hassani, A., Walton, S., Shah, N., Abuduweili, A., Li, J., Shi, H.: Escaping the big data paradigm with compact transformers. CoRR abs/2104.05704 (2021). https://arxiv.org/abs/2104.05704
9. Hassani, A., Walton, S., Shah, N., Abuduweili, A., Li, J., Shi, H.: Escaping the big data paradigm with compact transformers (2022).https://doi.org/10.48550/arXiv.2104.05704, arXiv:2104.05704 [cs]
10. Herrmann, C., et al.: CVPR 2022 open access repository. https://openaccess.thecvf.com/content/CVPR2022/html/Herrmann_Pyramid_Adversarial_Training_Improves_ViT_Performance_CVPR_2022_paper.html
11. Lee, S., Lee, S., Song, B.C.: Improving vision transformers to learn small-size dataset from scratch. IEEE Access **10**, 123212–123224 (2022). https://doi.org/10.1109/ACCESS.2022.3224044, https://ieeexplore.ieee.org/document/9957006/
12. Li, J., Chen, J., Tang, Y., Wang, C., Landman, B.A., Zhou, S.K.: Transforming medical imaging with Transformers? A comparative review of key properties, current progresses, and future perspectives. Med. Image Anal. **85**, 102762 (2023). https://doi.org/10.1016/j.media.2023.102762, https://www.sciencedirect.com/science/article/pii/S1361841523000233
13. Liu, J., Li, Y., Cao, G., Liu, Y., Cao, W.: Feature pyramid vision transformer for MedMNIST classification decathlon. In: 2022 International Joint Conference on Neural Networks (IJCNN), pp. 1–8 (2022). https://doi.org/10.1109/IJCNN55064.2022.9892282, https://ieeexplore.ieee.org/document/9892282. iSSN 2161-4407
14. Manzari, O.N., Ahmadabadi, H., Kashiani, H., Shokouhi, S.B., Ayatollahi, A.: MedViT: a robust vision transformer for generalized medical image classification. Comput. Biol. Med. **157**, 106791 (2023). https://doi.org/10.1016/j.compbiomed.2023.106791, arXiv:2302.09462 [cs]

15. Nguyen, N.Q., Le, T.S.: A semi-supervised learning method to remedy the lack of labeled data. In: 2021 15th International Conference on Advanced Computing and Applications (ACOMP), pp. 78–84 (2021). https://doi.org/10.1109/ACOMP53746.2021.00017, https://ieeexplore.ieee.org/document/9668240. iSSN 2688-0202
16. Steiner, A., Kolesnikov, A., Zhai, X., Wightman, R., Uszkoreit, J., Beyer, L.: How to train your ViT? Data, augmentation, and regularization in vision transformers (2022). https://doi.org/10.48550/arXiv.2106.10270, arXiv:2106.10270 [cs]
17. Touvron, H., Cord, M., Douze, M., Massa, F., Sablayrolles, A., Jegou, H.: Training data-efficient image transformers & distillation through attention. In: Proceedings of the 38th International Conference on Machine Learning, pp. 10347–10357. PMLR (2021). https://proceedings.mlr.press/v139/touvron21a.html. iSSN 2640-3498
18. Wang, W., Zhang, J., Cao, Y., Shen, Y., Tao, D.: Towards data-efficient detection transformers. In: Avidan, S., Brostow, G., Cissé, M., Farinella, G.M., Hassner, T. (eds.) ECCV 2022. LNCS, vol. 13669, pp. 88–105. Springer, Cham (2022). https://doi.org/10.1007/978-3-031-20077-9_6
19. Wightman, R.: PyTorch image model. https://doi.org/10.5281/zenodo.4414861, https://github.com/huggingface/pytorch-image-models
20. Xu, M., Zhang, T., Zhang, D.: MedRDF: a robust and retrain-less diagnostic framework for medical pretrained models against adversarial attack. IEEE Trans. Med. Imaging **41**(8), 2130–2143 (2022). https://doi.org/10.1109/TMI.2022.3156268, https://ieeexplore.ieee.org/document/9726228
21. Yang, J., et al.: MedMNIST v2 - a large-scale lightweight benchmark for 2D and 3D biomedical image classification. Sci. Data **10**(1), 41 (2023). https://doi.org/10.1038/s41597-022-01721-8, https://www.nature.com/articles/s41597-022-01721-8

Designing AI Algorithms to Suit Local Context

Yunusa G. Muhammed[1(✉)], Tarisiro Matiza[2], and Charles B. Delahunt[3]

[1] Gombe State University, Tudun Wada, Gombe, Nigeria
yunusa.mohammed@gsu.edu.ng
[2] Market Access Africa, Harare, Zimbabwe
tmatiza@marketaccess.africa
[3] Global Health Labs, Inc., Bellevue, WA, USA
delahunt@uw.edu

Abstract. Artificial Intelligence holds tremendous promise to transform healthcare, and if we can effectively convert technical advances in AI into robust deployments, then we can vastly improve quality of life for billions of currently underserved people. But a substantial barrier to this opportunity is a disconnect between AI developers and clinical realities: To deploy successfully, AI development must be shaped at every stage by the specifics of the local deployment context. This integration is complex, it's a necessary condition of success, and it falls to AI teams to carry out. However, because this task sits outside the traditional algorithm-centric focus of AI, it is poorly understood. Therefore this paper describes little-discussed but crucial aspects of an algorithm's path to deployment, each directly relevant to AI researchers and illustrated by concrete examples drawn from experiences in the African healthcare context.

We describe how AI is just one part of a much larger healthcare context, and how a core task of the AI team is to understand and translate this context into AI design choices from the very start of a project. We also include a set of actionable steps accessible to any researcher, which can markedly improve the fitness of an algorithm for future deployment.

Keywords: Healthcare · Domain expertise · Artificial Intelligence · Machine Learning

1 Introduction

AI has the potential to transform health care, especially for people who currently suffer large gaps in care delivery. However, a serious barrier to this opportunity is a disconnect between AI developers and clinical realities [13]. In this paper, we argue that to deploy successfully, AI development must be shaped, at every stage and in multiple ways, by an understanding of the specifics of the local deployment context. This integration of local context into AI work is complex and difficult,

Y. G. Muhammed, T. Matiza and C. B. Delahunt—Equal contributions.

U. Anazodo et al. (Eds.): MIRASOL 2025, LNCS 16398, pp. 164–174, 2026.
https://doi.org/10.1007/978-3-032-13654-1_17

and sits outside the traditional algorithm-centric focus of AI for health care. But it can also only be carried out by AI engineers - for example, while clinicians can specify real-world constraints, only AI engineers can translate these constraints into forces that shape an algorithm. Therefore, this integration of local context is an essential part of the technical workload of the AI researcher, just as important to success as architecture work. This expands the definition of technical AI work.

This topic has been addressed previously, for example by [15], which describes an excellent and comprehensive framework that is however necessarily general. In this paper, we extend the discussion by taking an "in the trenches" viewpoint. We describe some little-discussed but crucial aspects of the path to AI deployment, directly relevant to AI researchers, emphasizing how local context impacts even the earliest stages of algorithm development, with concrete examples drawn from experience in the African healthcare landscape. We hope that this approach highlights a crucial working rule of AI development, that it is just one part of a much larger puzzle, and that non-AI aspects of the local context must be integrated into AI work at every stage. Because not all AI researchers are positioned to fully address the factors we list, we also include a set of actionable steps, accessible to any AI researcher, which can markedly improve the fitness of an algorithm for future deployment.

2 Methods

The "local deployment context" is defined here as the clinical use case, location, and circumstances of the proposed deployment, including the needs of the patient population, infrastructure, medical personnel, and business models. In this section we describe some crucial factors of the local context, with examples to illustrate each. Since they all affect algorithm development, they need to be actively incorporated into AI work *from the start* to enable successful deployment:

1. Domain experts
2. Collaborative planning and goal alignment (includes defining AI goals)
3. Data collection
4. Algorithm design
5. Iterative co-design

2.1 Engage Local Domain Experts

This includes people who understand the specifics of the local context in which the AI algorithm will operate - clinicians, patients, policymakers, and researchers. This is a crucial first step because if the AI design does not incorporate the specific needs and constraints of each stakeholder, deployment will falter at some point due to an unaccounted-for factor. For example:

1. In India, revealing the sex of a fetus in ultrasound (US) is illegal [27] due to risk of female feticide. Thus an US application cannot provide sex-specific information about the fetus to patients, and the US devices themselves are strictly controlled. In Africa, by contrast, US images of the baby bring in the crowds, as parents want a glimpse of the new baby, and thus the images boost support of US in ante-natal care [17]. These contrasting local contexts will shape the basic design of AI applications for obstetric US.
2. Dr. Groesbeck Parham has led development of a successful cervical cancer screening program in Zambia [18]. Though an expert on cervical cancer in the USA, on arrival in Zambia he found a radically different clinical picture, and Zambian gynecologists educated him about the local context (e.g. its deadly combination with HIV infections). After developing a screening method on this basis, he found that conditions in outlying villages were radically different than in Lusaka. Village leaders and village clinic personnel educated him as to what could work. When the system deployed in these village clinics, he found that patient uptake depended on creating more trust, which required teams of local women recruited in each village. Thus, for this highly trained and experienced American gynecologist the journey to a successful deployment depended on aligning his skills with ever more local expertise.

2.2 Collaborative Planning and Goal Alignment

Because local experts have vital knowledge about their situation, without which a clinical problem cannot be even defined, much less solved, AI researchers must solicit collaborations with these experts to define problems and goals:

Jointly Define Use Cases with Impact. While we as AI researchers have ideas as to what AI can do, the local experts know what is needed locally. In the Zambian example above, a central factor was HIV as an accelerant of cervical cancer progression. Both parties have essential knowledge, but the local clinicians' knowledge has priority: applying AI to a real medical problem is difficult but can yield a useful tool, while an algorithm developed from internal AI perspectives is unlikely to have clinical utility.

Develop Shared Vocabulary. The AI field and the medical field each have a body of definitions and assumptions that can radically differ. Crucially, the fields disagree on what constitutes valid evidence. In medicine, the gold standard is a clinical trial, run by an independent party, which compares a therapeutic to the standard-of-care. In the field of AI, the standard is a comparison to other algorithms on benchmark datasets. By the medical field's standards, an AI-style comparison is inadequate to evaluate a diagnostic for safe deployment. So to deploy our algorithm in a clinic, the evidence must satisfy the medical field's rules, not those of AI. The WHO has a valuable document clarifying this distinction and describing the types of evidence required of AI for healthcare [25].

Understand the Local Context. The AI algorithm is only one piece of a complex puzzle, and the specific local context defines many of the other pieces, with which the AI piece must mesh for successful deployment. These pieces include constraints (and opportunities) of infrastructure, workforce, and logistics.

Price. Pricing is a major challenge. It must be sustainable for Ministries of Health (MoHs) or private markets to support an AI tool, and no AI algorithm will see deployment if the costs don't pencil out. African countries have limited budgets, especially with the recent withdrawal of U.S. and other funding. Products procured through non-local funding can be abandoned if the service and maintenance costs for equipment are unsustainable or too high for scale-up.

In Africa, most people seek care in public health facilities, a market with limited resources funded by governments, and the private sector market is very small. This strongly affects pricing and procurement mechanisms, which in turn affect basic design choices for algorithms and hardware. Pricing models vary:

(1) Upfront licensing involves one-time purchase of a perpetual license, often with accompanying hardware. However, the high initial investment often requires buy-in from international funding agencies.
(2) Subscriptions (monthly or annual) are used by many CAD vendors. The lower upfront cost for the AI portion is easier for governments to budget, but reducing high hardware cost impacts AI design choices (e.g. should the AI use dedicated scanning microscopes, or cellphone images from manual microscopes?).
(3) Pay per use (per scan analysed) has a similar low upfront cost to subscriptions. However, the long-term cost depends heavily on the volume of patients scanned. In Africa, the low cost per scan for CAD ($\approx$2 dollars/scan) requires high patient volumes and a pre-screening use case [10]. The choice of use case then drives how the algorithm is designed, trained, and evaluated.

Workforce Shortages. Africa has a shortage of specialized radiologists and radiographers (Table 1), most of whom work in the private sector and the rest in central hospitals, leaving remote areas with little coverage. In Zimbabwe, more than half of radiologists work in the capital city, the rest in two other cities, and none in rural districts. Thus, it is important to define the location of the use case and understand how the medical imaging product would integrate into the local workflow without dedicated and specialized healthcare professionals. Countries mitigate these shortages in various ways, affecting the potential users of AI tools: Lay cadres in Zimbabwe and Malawi have been trained to use digital X-rays and CAD, and radiographers in Kenya, review the CAD output and send reports, with complex cases sent to a radiologist in referral hospitals for review.

This labor scarcity means that AI can add tremendous value. For example, gestational age and number of fetuses can be accurately assessed with AI from untrained, blind ultrasound sweeps [20].

Supply chains can be hard to change. One malaria diagnostic system above found it useful, from an engineering/device viewpoint, to replace oil immersion

Table 1. Radiologist numbers by country. There is also great intra-country variation.

Country:	Population:	Radiologists:	Radiographers:	Approx population per radiographer:
UK	68,350,000	4,699	45,000	1,500
Zimbabwe	16,340,822	30	450	40,000
Nigeria	227,882,945	400	4,800	50,000
Cote d'Ivoire	31,165,654	13	200	150,000
Malawi	21,104,482	4	30	700,000

with coverslips for slide preparation. But due to the remote settings and difficult supply chains, requiring even a coverslip was a non-trivial barrier to acceptance.

Regulatory. AI researchers typically underestimate the importance and complexity of medical regulatory rules and thus omit the necessary groundwork. WHO has published a valuable framework [26], but countries are still assessing the regulation of AI tools. Kenya, South Africa and Uganda regulate these as software as a medical device with no clear framework and risk classification, but rather use ancillary laws like data protection. The EU's CE-mark and In Vitro Diagnostics Regulation (EU-IVDR), and the US FDA approval are currently the most common certifications for CAD software in the African market. But while these may fast-track in-country approval and introduction, local validation is still necessary for the specific location, use case, and imaging equipment. In addition to upfront regulation, it is important to proactively consider the local systems for post-marketing surveillance and quality assurance.

Co-create Success Metrics that Include Clinical Utility and Interpretability. This is a crucial (especially for AI researchers) subset of collaborative goal alignment since the AI project needs to track purely medical metrics of success.

1. In one project, hemozoin appeared to be a promising diagnostic biomarker for malaria. But it turned out to be absent in the blood samples of a crucial subset of cases, viz. synchronized *P. falciparum* infections most responsible for deaths [7]. A basic medical performance requirement was unattainable, so despite positive in-lab algorithm results the project was closed.
2. The intended public health impact may extend beyond standard AI notions, and must be clarified upfront. A study of CAD software for tuberculosis (TB) [5] defined these impacts as capacity creation, yield examination, efficiency gain and process optimisation. Identifying these non-AI goals early can inform algorithm and workflow design.

2.3 Data

AI researchers are familiar with how the quality and quantity of data can make or break an algorithm. Indeed, Andrew Ng has argued for a data-centric approach to AI optimization, because careful curation of training data can yield much greater improvements to performance than changes to architectures [19].

Suit the Data Collection and Annotations to the Use Case. The type of data and annotations required depend heavily on the particular use case and location. For example:

1. (Malaria) The diagnosis task requires species identification and thus data from multiple species, whereas drug resistance studies consider only *P. falciparum* [3], but post-drug-treatment malaria parasites are morphologically distinct due to damaged cytoplasm and thus require special post-treatment data. Also, the predominant (and thus clinically relevant) malaria species vary by region, so the required data for AI depends on locale.
2. In a dataset of African cervical images [9], the country-of-origin for many images could be visually identified by regional characteristics or comorbidities [4]. Data variability of this type can greatly impact an algorithm.
3. The expertise of field collaborators can greatly streamline data collection and annotation. For example, a *Loa loa* diagnostic required a dataset rich in very high parasitemia samples (due to the particular use case [12]). The field clinicians in Cameroon were able to efficiently and rapidly collect the required samples, because they knew exactly where the local hot spots were.

Local Data Effects. AI researchers must plan for the targeted region for deployment and account for disparities in data availability and representation, especially in under-resourced regions.

North-East Nigeria, home to ≈30 million people, has only eight functional CT scanners. The border state of Borno is a major medical hub, and patients from Chad and Cameroon often cross the border to access diagnostic services. These cross-border patients have distinct morphological characteristics due to genetic, nutritional, or environmental differences, skewing the dataset. Furthermore, diagnostic infrastructure in nearby states like Gombe and Bauchi is often unreliable, with machines frequently out of service. During such downtimes, patients are redirected to Maiduguri or vice versa, resulting in clusters of imaging data that reflect machine availability rather than true population distribution [1]. If AI models are trained on data collected during scanner downtime periods, the resulting models may reflect biased sampling and may fail to generalize.

Additionally, a supply of local data is required, at a minimum, to test and recalibrate models. For example, inter-clinic variability in slide preparation is perhaps the greatest AI challenge to automated malaria diagnosis [6,22]. As evident in many of these examples, local data may also be needed for training.

Trust. As the saying goes, "Deployment moves at the speed of Trust". A study of stakeholder perspectives [14] highlighted dataset representativeness and contextual equity as key determinants for trust and successful AI for healthcare. An approach to ethical AI development, emphasizing adaptation to local context, is described in [2]. Gaining trust requires a respect for patients' contribution, and often a way to share the benefit of the data. For more details see [11] and [23].

Data is central to AI success, and data collectors, who are usually the field clinicians, deserve more credit. The current model for AI conference papers gives precedence, even in dataset papers, to architectures, and thus AI researchers are typically the first authors. It requires something of a shift in thinking to highlight and reward the vital contributions of clinical partners who collect the data, e.g. by having them as first and anchor authors of dataset papers in AI venues.

2.4 Algorithm Design

This is the topic most familiar and dear to our AI hearts. It, of course, requires a great deal of straightforward AI-driven technical expertise. However, it also requires, as an equally necessary condition of success, an expanded set of uniquely healthcare-driven AI tools.

Suit the AI Architecture to the Local Needs. An inventory of the exact equipment available in healthcare facilities targeted by the AI project is important, as this can strongly affect architecture and model training decisions. In many African regions, analog X-ray machines have been retrofitted with computed radiography, while machines such as MRI, CT scans, and digital X-rays have limited availability. Most of the up-to-date medical imaging gear is found in the private sector; the public sector has limited functional equipment concentrated in district hospitals, and in rural areas (with over half the population), medical imaging equipment is scarce. Determining which type and brand of medical imaging equipment is available can help ensure integration of the software and minimize the need for countries to procure additional hardware.

Define Use Case-Relevant AI Metrics. AI is, by nature, quantitative, so the metrics chosen are central to algorithm development, whether for distance metrics, loss functions, or algorithm performance evaluation. We inherit many ready-made, easily applied metrics from the field of AI (e.g. loss functions in PyTorch). But in AI-for-healthcare, metrics must reflect the medical needs to guide the algorithm towards meeting the clinical performance requirements. For example, many AI papers evaluate algorithm performance at the object-level (e.g. parasite thumbnails), whereas clinicians evaluate performance at the patient level. This is a crucial distinction, and object-level metrics (whatever their intermediate value) are not adequate predictors of clinical performance [8,24]. For a comprehensive (and very readable) inventory of AI metrics for healthcare, see [21].

2.5 Iterative Co-design

A basic key to success is to interact with the local context early and often. Arriving at the clinic door at the project's end with a finished "solution" never works. Some useful techniques include:

1. Conduct feasibility studies with the clinical team for their input into the design and implementation of the product.
2. Conduct facility assessments to better understand the facility infrastructure, patient volumes, and available human resources.
3. Ensure that pilot deployments generate evidence acceptable to the medical world, not just to the AI world [25].
4. Work with MoHs or international programs for the specific disease areas, to get their buy-in and to understand the local context of deployment/use.
5. Keep clinicians looped in, for example, in full-team sync meetings, even when not in the field. A clinician may see elephants in pink tutus when the AI team sees nothing special and vice versa. This is a crucial way to head off pitfalls and, conversely, to leverage opportunities.

It is desirable to get into the field/clinic early and often to test prototypes, because important (even existential) reality checks always arise. For AI field tests, two common findings are (a) out-of-distribution data, and (b) unforeseen conditions in algorithm use. The earlier these situations are detected and incorporated into development, the better. Field tests always yield new, high-value information. For example:

1. In the *Loa loa* project described above, a field trial surfaced the issue that blood samples sometimes coagulated before being imaged, giving undercounts that would be disastrous in the targeted use case. This revealed the need for a built-in coagulation detector module.
2. A malaria project using Acridine Orange stain was highly effective, from a purely in-lab algorithm perspective. But during the first field test, the clinical teams made clear that AO stain was simply impractical for field use , an issue that led to shuttering the project.

3 OK, But Let's be Realistic

Addressing all the items listed above is ideal, but requires a team with substantial local connections, which most of us do not have. However, certain concrete actions, accessible to almost all researchers, can ensure that an AI-for-healthcare project is well-informed by local conditions and domain expertise:

1. At career transition points (e.g. choosing a PhD program), seek out labs and teams that have clinical connections: labs with adjoined research medical schools, with clinicians on the team, or with a track record of field work or advanced-stage projects.
2. Choose an AI problem based on your existing connections to local contexts: That is, let the AI project flow out of the needs of a local situation. This helps ensure that the problem being addressed is real, and that domain experts are in the loop from the start.
3. Subordinate algorithm design to the medical specifics. The best algorithmic solution might use off-the-shelf methods, in which case the novelty lies in its real-world utility.

4. Consult the *non-AI* medical literature - this is essential to understanding the AI task and is readily available. Sources include the medical literature on particular illnesses, and literature on health systems (e.g. local NGOs, PATH, CHAI, PLOS, ASTMH). Also, cite this literature in your AI paper to benefit other AI researchers.
5. Attend non-AI conferences, or those workshops at AI conferences with clinical presence, to meet collaborators, learn medical use cases, and the urgent problems AI can solve. E.g.: CLINICCAI at MICCAI; UnionConf (TB); ICASA (HIV); ASTMH (tropical medicine); IASLC Conference (lung cancer).
6. Define clinically relevant metrics for use in AI development for the targeted use case (an excellent resource for choosing appropriate metrics is [16]). This may require new construction or mathematical derivation, since these metrics often do not exist for a given medical use case in AI-usable form (for a concrete example from malaria, see [8]).

AI Lifecycle Mapping of Our Framework in problem definition, engage local experts to co-create clinically relevant goals (e.g., Zambia cervical cancer shaped by HIV); in data and model development, curate context-specific datasets, use patient-level metrics, and design for local hardware (e.g., CT downtime bias in North-East Nigeria); and in evaluation and deployment, validate with clinical standards, pilot through field testing, and adapt to pricing and regulatory systems (e.g., Loa loa trial revealing coagulation detector needs).

AI Expertise is a powerful and highly valued skill set. AI researchers can be welcome and high-value members of medical teams that accelerate progress in healthcare, if we carefully attend to the medical context.

Acknowledgments. Partially funded by Global Health Labs, Inc. (www.ghlabs.org).

Disclosure of Interests. The authors have no competing interests to declare that are relevant to the content of this article.

References

1. Adejoh, T., Onwujekwe, E., Chiegwu, H., et al.: Computed tomography scanner census and adult head dose in Nigeria. Egyptian J. Radiol. Nuclear Med. (2018)
2. Amugongo, L., Kriebitz, A., Boch, A., Lütge, C.: Operationalising AI ethics through the agile software development lifecycle: a case study of AI-enabled mobile health applications. AI Ethics (2025). https://doi.org/10.1007/s43681-023-00331-3
3. Ashley, E., Mehul Dhorda, M., White, N., et al.: Spread of artemisinin resistance in Plasmodium falciparum malaria. N. Engl. J. Med. (2014). https://doi.org/10.1056/NEJMoa1314981
4. Asiedu, D.M.: Personal communication (2024)

5. Creswell, J., Vo, L., Qin, Z., et al.: Early user perspectives on using computeraided detection software for interpreting chest x-ray images to enhance access and quality of care for persons with tuberculosis. BMC Glob. Public Health (2023)
6. Das, D., Vongpromed, R., Dhorda, M., et al.: Field evaluation of the diagnostic performance of EasyScan GO: a digital malaria microscopy device based on machine-learning. Malaria J. (2022)
7. Delahunt, C., Horning, M., Wilson, B., et al.: Limitations of haemozoin-based diagnosis of Plasmodium falciparum using dark-field microscopy. Malar J. (2014)
8. Delahunt, C., Gachuhi, N., Horning, M.: Metrics to guide development of machine learning algorithms for malaria diagnosis. Front. Malaria (2024)
9. Dotson, M.E., Asiedu, M., Ramanujam, N.: Speculum-free callascope for cervical self-visualization: acceptability, feasibility, and improved awareness of the reproductive system. JCO Glob. Oncol. (2020)
10. Du, Y., Greuter, M., Prokop, M., de Bock, G.: Pricing and cost-saving potential for deep-learning computer-aided lung nodule detection software in CT lung cancer screening. Insights Imaging (2023)
11. IDDO: https://www.iddo.org/governance/data-access-committee. Accessed 25 June 2025
12. Kamgno, J., Pion, S., Boussinesq, M., et al.: A test-and-not-treat strategy for onchocerciasis in Loa loa-endemic areas. New Engl. J. Med. (2017). https://doi.org/10.1056/NEJMoa1705026
13. Koller, D., Bengio, Y.: A fireside chat with Daphne Koller at ICLR (2018). https://www.youtube.com/watch?v=N4mdV1CIpvI
14. Kuo, R., Freethy, A., Furniss, D., et al.: Stakeholder perspectives towards diagnostic artificial intelligence: a co-produced qualitative evidence synthesis. eClinical Med. (2024)
15. Lekadir, K., Frangi, A., Porras, A., et al.: Future-AI: international consensus guideline for trustworthy and deployable artificial intelligence in healthcare. BMJ (2025). https://doi.org/10.1136/bmj-2024-081554
16. Maier-Hein, L., Reinke, A., et al.: Metrics reloaded: pitfalls and recommendations for image analysis validation. arXiv (2022). https://arxiv.org/abs/2206.01653
17. Mensah, Y., Nkyekyer, K., Mensah, K.: The Ghanaian woman's experience and perception of ultrasound use in antenatal care. Ghana Med. J. (2014)
18. Mwanahamuntu, M., Parham, G., et al.: Advancing cervical cancer prevention initiatives in resource-constrained settings: insights from the Cervical Cancer Prevention Program in Zambia. PLoS Med. (2011)
19. Ng, A.: Practical limitations of todays' deep learning in healthcare. ML4H conference (2020). https://nips.cc/virtual/2020/19127
20. Pokaprakarn, T., Stringer, J., et al.: AI estimation of gestational age from blind ultrasound sweeps in low-resource settings. NEJM Evidence (2022)
21. Reinke, A., Tizabi, M., et al.: Understanding metric-related pitfalls in image analysis validation. Nat. Methods (2024)
22. Torres, K., Bachman, C., Gamboa, V., et al.: Automated microscopy for routine malaria diagnosis: a field comparison on Giemsa-stained blood films in Peru. Malaria J. (2018)
23. Understanding Patient Data. https://understandingpatientdata.org.uk/. Accessed 25 June 2025
24. Varoquaux, G., Cheplygina, V.: Machine learning for medical imaging: methodological failures and recommendations for the future. NPJ Digit. Med. (2022). https://doi.org/10.1038/s41746-022-00592-y

25. WHO: Generating evidence for artificial intelligence-based medical devices: a framework for training, validation and evaluation: World Health Organization. Switzerland, Geneva (2021)
26. WHO: Regulatory considerations on artificial intelligence for health. World Health Organization (2023)
27. Wikipedia: Article on PCPND. https://en.wikipedia.org/wiki/PreConception_and_Pre-Natal_Diagnostic_Techniques_Act,_1994. Accessed 25 June 2025

Global South Health Practitioners' Awareness and Perceptions of Integrating Artificial Intelligence in Radiological Workflows: A Quantitative Nationwide Study in Health Facilities in Zambia

Peter Chibuta[1,3], Lighton Phiri[2,3](✉), Ernest Obbie Zulu[1,3], and Malaizyo Muzumala[2,3]

[1] Department of Radiology, University Teaching Hospitals, Lusaka, Zambia
[2] Department of Computing and Informatics, University of Zambia, Lusaka, Zambia
[3] DataLab Research Group, University of Zambia, Lusaka, Zambia
chibutapeter75@gmail.com, {lighton.phiri, malaizyo.muzumala}@cs.unza.zm, obbiernest@gmail.com

Abstract. Low- and Middle-Income Countries (LMICs) face a severe shortage of radiologists, which adversely affects medical imaging workflows. While Artificial Intelligence (AI) presents potential solutions, the awareness and perceptions of these technologies among healthcare practitioners in such settings are not well understood. This paper presents a nationwide study of medical practitioners in Zambia to investigate their awareness of and perceptions towards AI integration in radiology. The study utilised a quantitative, cross-sectional survey of 95 medical practitioners across Zambian health facilities. The findings reveal a significant gap in AI awareness, with practitioners being significantly more aware of AI for image interpretation than for radiological report writing ($p < .001$). Despite this, participants held positive perceptions of AI's potential to reduce workload and enhance efficiency. However, trust in AI was moderate and directly correlated with awareness levels (r_s = .21, p = .041), indicating that greater familiarity could potentially result in AI acceptance. While concerns about job displacement were present, these views did not significantly differ based on years of professional experience ($p = 0.18$). These results highlight a critical need for targeted education to bridge the awareness gap, build trust, and facilitate the effective integration of AI as a vital assistive tool in resource-constrained environments like Zambia.

Keywords: Artificial Intelligence · Global South · Generative AI

1 Introduction

Application of Artificial Intelligence (AI) in healthcare has rapidly increased in the recent decades globally and in all domains of healthcare [1]. In particular, more AI tools are being implemented in the realm of medical imaging, paralleling the rapid advancement in digital medical imaging technologies.

U. Anazodo et al. (Eds.): MIRASOL 2025, LNCS 16398, pp. 175–184, 2026.
https://doi.org/10.1007/978-3-032-13654-1_18

The integration of AI into medical imaging workflows has the potential to transform the practice of medical imaging, particularly in Low- and Middle-Income Countries (LMICs), which have been heavily impacted by the critical shortage of the radiology workforce and limited imaging infrastructure. AI potentially presents solutions to some of the challenges across the spectrum of the medical imaging workflow, from imaging examination requisition through to reporting, and by augmenting clinical decision-making and supporting non-specialist clinicians, thereby allowing the few radiologists more time to focus on more resource-intensive tasks [2]. Further, these tools offer the potential for improved diagnostic accuracy, efficiency, reduction in human errors and personalised radiological services [3]–[4].

However, the global south faces unique challenges to the adoption and integration of AI in medical imaging. RAD-AID has proposed a three-pronged approach to integration of AI in low-resource settings: simultaneous integration of clinical radiology education, infrastructure implementation, and phased AI introduction [5]. The successful implementation of AI-based technologies requires that radiology stakeholders are not only aware of these technologies but also perceive them as relevant, trustworthy, and beneficial to their clinical practice.

Multiple studies have examined stakeholder awareness, perception and attitudes towards adoption and integration of AI in medical imaging in the global south region, often revealing a mix of optimism and concerns [6, 7]. Like many global south countries, Zambia faces critical shortages of radiologists and other imaging professionals [8] and limited use of imaging technologies. Additionally, little is known about the knowledge, perceptions and attitudes towards the use of AI in medical imaging among healthcare professionals in Zambia. This study, therefore, investigates the challenges within current radiological workflows while assessing the awareness, perceptions, and attitudes of health practitioners in Zambia regarding the integration of AI.

The remainder of this paper is organised as follows: Sect. 2 describes existing studies related to this study; Sect. 3 details the methodological approach used to execute the study conducted; Sect. 4 is an outline of the findings and discussion of the findings and; finally, Sect. 5 presents concluding remarks.

2 Related Work

2.1 Challenges with Radiological Workflows in the Global South

There are a number of major challenges faced in radiological workflows in the Global South. A particular challenge is the severe shortage of radiologists. According to Tahir et al. [9] Sub Saharan Africa "… Averages fewer than 22 doctors per 100,000 people." They further mention that there's about 1 radiologist per 600,000 people in Nigeria for example. Because of this shortage experts are burdened with a high workload, especially as the number of images generated has increased.

Another challenge that affects the Global South particularly sharply is shortages of advanced imaging equipment with approximately 227,000 and 1,694,000 people served by 1 CT scanner in lower middle income and low income countries respectively [5] with other imaging equipment also being in short supply.

Furthering these challenges is the problem that much of the radiological equipment that is available has serious issues with maintenance causing a large amount of down time with 38.3% of donated equipment in developing countries being out of service for technical and non-technical reasons [10]. This regular downtime is very disruptive to workflows.

Poor integration and support systems also deeply affects the global south, with systems like Health Information Systems, Radiology Information Systems and Picture Archiving and Communication Systems often not being used at all or integrated directly with radiological workflows. Particularly in the public sector such as in Zambia [8].

These and other challenges deeply affect the radiological workflows in the Global South.

2.2 Awareness and Perceptions of Integrating Artificial Intelligence in Radiological Workflows

Rapid advancements in AI techniques such as generative AI, large language models and computer vision have made it so that the knowledge of AI's existence is generally mainstream.

Studies have shown that radiological experts generally hold positive attitudes towards AI [11, 12]. These studies also show that experts perceive AI will improve clinical quality. However, aspects of these studies had the radiologists showing uncertainty around the future, with studies having participants fearing AI may reduce demand for radiological experts.

Studies though have shown insistence on oversight in the use of these tools by human experts [13, 14].

3 Methodology

This study employed a quantitative approach, cross-sectional design conducted across selected health facilities in the Republic of Zambia. Ethical approval was obtained from The University of Zambia Biomedical Research Ethics Committee (Reference Number: 2731–2022) and The National Health Research Authority (Reference Number: NHRA000024/10/05/2022). In addition, formal permission to conduct the study was granted by the Republic of Zambia Ministry of Health.

3.1 Target Population and Sampling

The target population consisted of medical practitioners actively involved in radiological workflow activities within Zambian health facilities. Due to the geographical dispersion of this population and the absence of a centralized registry for recruitment, a non-probability sampling strategy was implemented.

Participants were recruited primarily through convenience sampling, leveraging professional WhatsApp[1] groups comprising medical practitioners. This sampling technique was supplemented by snowball sampling, where initial participants were encouraged to share the survey link with eligible colleagues.

[1] https://www.whatsapp.com

3.2 Measurement Instrument and Procedure

A self-administered Google Forms[2] online questionnaire was used to collect the responses from the participants. The questionnaire consisted of five sections, with questionnaire items for capturing participants' demographic factors; facility demographic factors; participants' experience interpreting medical images; participants' awareness of AI and; participants' perceptions of AI.

Prior to distribution, the questionnaire was pre-tested with a small group of medical professionals to ensure the clarity, relevance, and comprehensibility of the items.

3.3 Data Analysis

The data collected was pre-processed in Google Sheets and analysed in R[3] using descriptive statistics (frequencies, percentages, means) and non-parametric inferential tests. Specific analyses included a Wilcoxon Signed-Rank Test for paired awareness levels, a Spearman's rank-order correlation (r_s) to assess the relationship between awareness and trust, and a Kruskal-Wallis H test with Dunn's post-hoc comparisons to evaluate job loss perceptions across experience groups. A p-value of less than .05 was considered statistically significant.

4 Results and Discussion

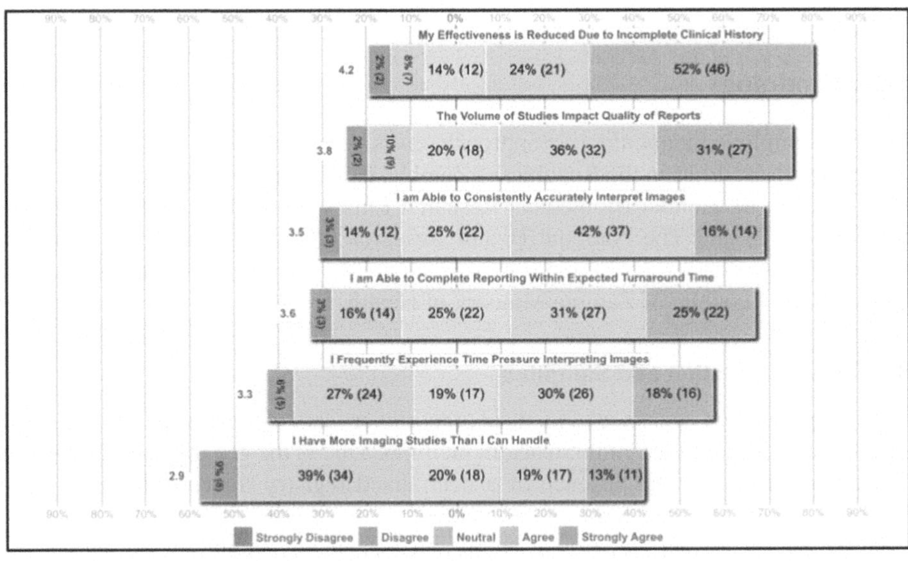

Fig. 1. Medical Imaging Interpretation Challenges.

[2] https://workspace.google.com/products/forms
[3] https://www.r-project.org

A total of 95 participants responded to the survey. The cohort was geographically diverse, with the highest representation from Lusaka Province (62.1%). The sample was balanced by gender (56.8% male), comprising various age groups (the largest being 31–40 years at 53.7%), and was well-educated, with a strong majority holding at least a Bachelor's Degree (68.4%). Experience levels varied, with the largest group (45.3%) having 2–5 years of practice.

4.1 Experience and Challenges Interpreting Medical Images

The vast majority of respondents (92.6%) interpret medical images—primarily X-rays (89.5%), ultrasound (54.7%), and CT scans (46.3%). This reliance on non-expert interpretation, confirming previous studies [15], is reflected in the finding that a majority (62.5%) reported only average confidence in their work. Practitioners also identified significant workflow challenges, as shown in Fig. 1. Notably, 73% confirmed their effectiveness is reduced by incomplete clinical history, a gap where AI-driven NLP techniques could assist [16]. While overall workload was perceived as manageable, significant time pressure during interpretation remained a key constraint, highlighting a primary bottleneck in the workflow.

4.2 Awareness of Integrating Artificial Intelligence Techniques in Radiological Workflows

The study participants' awareness levels regarding the use of AI in radiological workflows, by focusing on interpreting medical images and writing radiological reports. The participants' responses were categorised into five levels of awareness: "Not at all Aware", "Slightly Aware", "Somewhat Aware", "Moderately Aware" and "Extremely Aware". Figure 2 illustrates the varying degrees of AI awareness for the respondents.

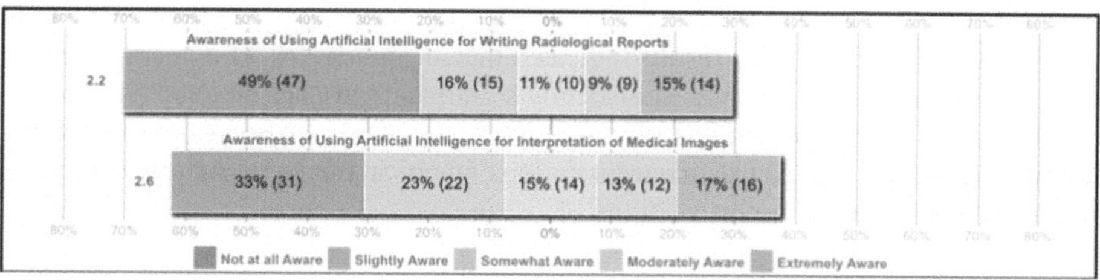

Fig. 2. Awareness Levels of Usage of Artificial Intelligence in Radiological Workflows.

Analysis 1. Awareness of Artificial Intelligence for Interpretation of Medical Images. The awareness levels for using AI in the interpretation of medical images suggest varying levels of awareness regarding AI's role in the interpretation of medical images.

While 33% (n = 31) of participants reported being "Not at all Aware"; this figure indicates a moderate baseline. In addition, 23% (n = 22) indicated being "Slightly Aware".

The proportions of higher awareness levels were also substantial, with 15% (n = 14) "Somewhat Aware"; 13% (n = 12) "Moderately Aware" and 17% (n = 16) "Extremely Aware". The mean awareness score for usage of AI in interpretation of medical images was 2.6, indicating a relatively low level of general awareness, positioned between "Slightly Aware" and "Somewhat Aware" within the surveyed population concerning AI in medical image interpretation.

The relatively low levels of awareness have significant implications for the adoption of AI in clinical practice. Despite the rapid advancements of AI in assisting with diagnostic tasks in healthcare, healthcare practitioners arguably do not fully grasp its potential and limitations. In order to address this deficiency, targeted skills building programmes focusing on AI applications in medical image interpretation could be introduced.

Analysis 2. Awareness of Artificial Intelligence for Writing Radiological Reports. The participants' levels of awareness for using AI to generate radiological reports were generally lower.

A significant portion of participants, 49% (n = 47), reported being "Not at all Aware" of using AI in this context, while 16% (n = 15) indicated being "Slightly Aware". Participants' levels of higher awareness were considerably lower: 11% (n = 10) were "Somewhat Aware", 9% (n = 9) were "Moderately Aware" and 15% (n = 14) indicated being "Extremely Aware". The mean awareness score was 2.2, suggesting an even more pronounced trend towards lower overall awareness. The particularly low awareness score highlights a missed opportunity to reduce the turnaround time associated with report writing through the use of AI techniques.

Analysis 3. Comparative Analysis of Awareness Levels. A comparative analysis of awareness levels of using AI for medical image interpretation and report writing reveals a consistent pattern of awareness, with a notable distinction in the "Not at all Aware" segment. The percentage of participants "Not at all Aware" was notably lower for AI in medical image interpretation (33%) than for AI in radiological report writing (49%). To determine if this observed difference was statistically significant, a Wilcoxon Signed-Rank Test was conducted. The results indicated that awareness for AI in medical image interpretation (Median = 2.0) was statistically significantly higher than for AI in radiological report writing (Median = 1.0), ($V = 568.5$, $p < .001$). This suggests that while general awareness is low, AI use in interpretation of medical images is significantly more recognised than its use in report writing.

4.3 Perceptions of Integrating Artificial Intelligence Techniques in Radiological Workflows

Participants' perceptions across several aspects of AI's potential impact were elicited using a five-point likert scale: "Strongly Disagree", "Disagree", "Neutral", "Agree" and "Strongly Agree". The results of the participants' ratings are shown in Fig. 3.

Analysis 1. Workload Reduction. The results suggest that the majority of participants perceive AI to have the potential to reduce the workload in radiological workflows, with a mean score of 4.0. Specifically, 37% (n = 35) "Agree" and 40% (n = 38) "Strongly

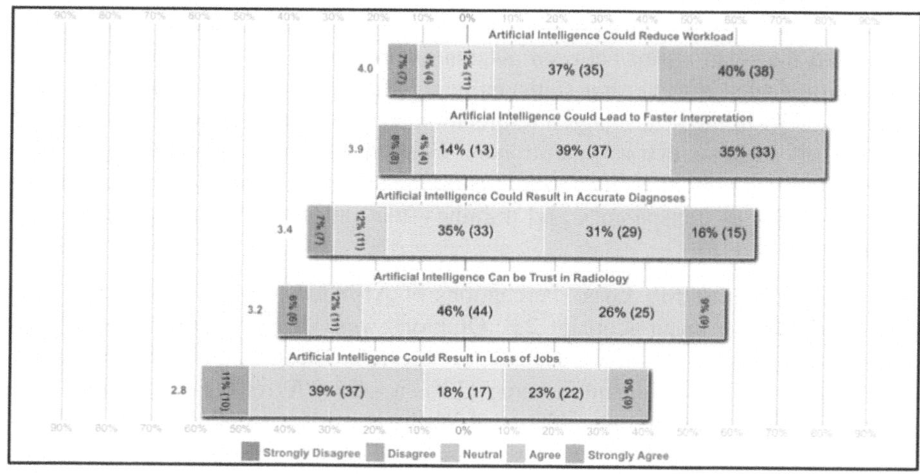

Fig. 3. Perceptions of Usage of Artificial Intelligence in Radiological Workflows.

Agree". Only a small minority—7% (n = 7) "Strongly Disagree" and 4% (n = 4) "Disagree"—while 12% (n = 11) remained "Neutral". This indicates a high level of optimism regarding AI's potential to reduce radiological workflow workloads. These results are aligned with our prior study [17] which demonstrated the reduction in the workload when using software to automatically interpret medical images.

Analysis 2. Efficient Medical Image Interpretation. There was also a clear positive perception that AI could lead to faster interpretation. This is indicated by the mean scores of 3.9. 39% (n = 37) and 35% (n = 33) for "Agree" and "Strongly Agree", respectively, that AI could accelerate the interpretation process. A small percentage "Strongly Disagree" (8%, n = 8) or "Disagree" (4%, n = 4), with 14% (n = 13) providing a "Neutral" rating. These findings with the perception of workload reduction, suggest a general expectation that AI will enhance the speed of diagnostic procedures, which could have significant implications for radiological workflow turnaround times.

Analysis 3. Accurate Diagnoses. The participants' perceptions regarding AI's ability to result in accurate diagnoses were moderately positive, with a mean score of 3.4. While 31% (n = 29) "Agree" and 16% (n = 15) "Strongly Agree", a notable 7% (n = 7) "Strongly Disagree" and 12% (n = 11) "Disagree"; interestingly enough, 35% (n = 33) providing a "Neutral" rating. These mixed responses suggest that while many acknowledge AI's potential, a significant segment are arguably uncertain about its direct contribution to diagnostic accuracy. This presents an opportunity to provide training sessions to demonstrate AI's role as an assistive tool rather than a replacement for human expertise.

Analysis 4. Trust in Radiological Workflows. The results suggest that Trust in AI is a controversial topic, with a mean score of 3.2. The majority of participants' provided a "Neutral" rating (46%, n = 44), indicating a large proportion of undecided participants. While 26% (n = 25) "Agree" and 9% (n = 9) "Strongly Agree", a combined 11% (n = 10) either "Strongly Disagree" or "Disagree". The high neutrality rate underscores that trust in AI is not yet fully established. To explore the factors influencing this, we investigated the relationship between awareness and trust. A Spearman's rank-order correlation revealed a weak, but statistically significant positive association between awareness of AI in medical image interpretation and trust in AI systems ($r_s = .21$, $p = .041$). This suggests that as practitioners' awareness of AI's capabilities increases, their trust in these tools also tends to rise. Building trust will arguably involve focusing on factors—such as transparency and training—that influence AI trust as reported in existing literature [18].

Analysis 5. Potential Job Loss. Perceptions of AI-induced job loss were the most negative, yielding a mean score of 2.8. Opinions were divided, with nearly half of the respondents (50%, $n = 47$) "Disagree"or "Strongly Disagree" with the idea of job displacement, while a significant minority (32%, $n = 31$) "Agree". To determine if this perception varied by experience, a Kruskal-Wallis H test was conducted but found no statistically significant difference across experience groups ($H(4) = 6.25$, $p = .181$), indicating that this concern is not unique to any particular seniority level. This aligns with existing studies where majorities of radiologists have also reported low concern about being replaced by AI [19, 20].

5 Conclusion

This study provides a quantitative snapshot of AI perceptions among healthcare practitioners in a resource-constrained setting. The study findings reveal a critical, statistically significant gap: while practitioners have some familiarity with AI for image interpretation, awareness of its use in reporting is nearly non-existent ($p < .001$). This is coupled with a key outcome: the high optimism for AI's potential to improve efficiency, but significant neutrality and skepticism regarding its trustworthiness and diagnostic accuracy. In addition, the results indicate a correlation between awareness and trust ($r_s = .21$), suggesting that familiarity is a key driver of acceptance. Finally, while job displacement fears exist, they were not tied to professional experience, indicating a general rather than a group-specific concern. These findings underscore the urgent need for targeted educational initiatives. To successfully integrate AI in the Global South, efforts must focus not only on demonstrating efficiency gains but also on building foundational knowledge to foster the trust necessary for clinical adoption.

Acknowledgement. This study was conducted as part of a much larger "Enterprise Medical Imaging in Zambia" project. The project is funded through generous grants from Google Research, Data Science Africa and The University of Zambia. In addition, this work was partly supported by the Italian Ministry of University and Research (MUR) under project PE0000013 – Future of Artificial Intelligence Research (FAIR). We are very appreciative of this funding.

References

1. Martinez-Millana, A., Saez-Saez, A., Tornero-Costa, R., Azzopardi-Muscat, N., Traver, V., Novillo-Ortiz, D.: Artificial intelligence and its impact on the domains of universal health

coverage, health emergencies and health promotion: an overview of systematic reviews. Int. J. Med. Inf. **166**, 104855 (2022). https://doi.org/10.1016/j.ijmedinf.2022.104855
2. Mekonen, K.A., Mohammed, S.H., Kebede, T., Bedane, A., Buser, A.A.: Artificial intelligence in radiology for ethiopia. Ethiop. J. Health Sci. **34**, 1–2 (2024). https://doi.org/10.4314/ejhs.v34i1.1S
3. Topol, E.J.: High-performance medicine: the convergence of human and artificial intelligence. Nat. Med. **25**, 44–56 (2019). https://doi.org/10.1038/s41591-018-0300-7
4. Pesapane, F., Codari, M., Sardanelli, F.: Artificial intelligence in medical imaging: threat or opportunity? Radiologists again at the forefront of innovation in medicine. Eur. Radiol. Exp. **2**, 35 (2018). https://doi.org/10.1186/s41747-018-0061-6
5. Mollura, D.J., et al.: Artificial intelligence in low- and middle-income countries: innovating global health radiology. Radiology **297**, 513–520 (2020). https://doi.org/10.1148/radiol.2020201434
6. Botwe, B.O., Antwi, W.K., Arkoh, S., Akudjedu, T.N.: Radiographers' perspectives on the emerging integration of artificial intelligence into diagnostic imaging: the Ghana study. J. Med. Radiat. Sci. **68**, 260–268 (2021). https://doi.org/10.1002/jmrs.460
7. Chinene, B., Mudadi, L.-S., Chitsva, W.: Patient perspectives on artificial intelligence in radiology: attitudes, knowledge, and concerns in public hospitals of Harare Metropolitan Province, Zimbabwe. MJZ **52**, 426–433 (2025). https://doi.org/10.55320/mjz.52.3.639
8. Zulu, E.O., Phiri, L.: Enterprise medical imaging in the global South: challenges and opportunities. In: 2022 IST-Africa Conference (IST-Africa), pp. 1–9. IEEE Explore (2022). https://doi.org/10.23919/ist-africa56635.2022.9845508
9. Tahir, M.Y., Mars, M., Scott, R.E.: A review of teleradiology in Africa – towards mobile teleradiology in Nigeria. S. Afr. J. Radiol. **26** (2022). https://doi.org/10.4102/sajr.v26i1.2257
10. Nigatu, A.M., et al.: Medical imaging consultation practices and challenges at public hospitals in the Amhara regional state, Northwest Ethiopia: a descriptive phenomenological study. BMC Health Serv. Res. **23**, 787 (2023). https://doi.org/10.1186/s12913-023-09652-9
11. Antwi, W.K., Akudjedu, T.N., Botwe, B.O.: Artificial intelligence in medical imaging practice in Africa: a qualitative content analysis study of radiographers' perspectives. Insights Imaging **12**, 80 (2021). https://doi.org/10.1186/s13244-021-01028-z
12. View of Radiographers' Perspectives on the Impact of Artificial Intelligence use on their future roles: A Qualitative Study. https://mjz.co.zm/index.php/mjz/article/view/363/425. Accessed 01 July 2025
13. Soleimantabar, H., Mahdavi, A., Qorbani, M., Madanipour, M., Tasharrofi, M.A., Okhovvat, B.: Artificial intelligence in radiology: Perceptions, adoption barriers, and trust among Iranian radiologists in a global context. InfoScience Trends **2**, 1–12 (2025). https://doi.org/10.61186/ist.202502.03.01
14. Cè, M., et al.: Radiologists' perceptions on AI integration: an in-depth survey study. Eur. J. Radiol. **177**, 111590 (2024). https://doi.org/10.1016/j.ejrad.2024.111590
15. Bwanga, O., Mulenga, J., Chanda, E.: Need for Image reporting by radiographers in Zambia. Med. J. Zambia **46**, 215–220 (2019). https://doi.org/10.55320/mjz.46.3.223
16. Chileshe, E., Phiri, L.: Large-scale analysis of medical image metadata. Zambia ICT J. **5**, 44–48 (2023)
17. Muzumala, M.G., Zulu, E.O., Chibuta, P., Nyirenda, M., Phiri, L.: Evaluating perceived workload, usability and usefulness of artificial intelligence systems in low-resource settings: semi-automated classification and detection of community acquired pneumonia. In: Wu, S., Shabestari, B., Xing, L. (eds.) Applications of Medical Artificial Intelligence, pp. 116–126. Springer, Cham (2025). https://doi.org/10.1007/978-3-031-82007-6_12
18. Henrique, B.M., Santos, E.: Trust in artificial intelligence: Literature review and main path analysis. Comput. Hum. Behav. Artif. Hum. **2**, 100043 (2024). https://doi.org/10.1016/j.chbah.2024.100043

19. Huisman, M., et al.: An international survey on AI in radiology in 1,041 radiologists and radiology residents part 1: fear of replacement, knowledge, and attitude. Eur Radiol. **31**, 7058–7066 (2021). https://doi.org/10.1007/s00330-021-07781-5
20. Chen, Y., et al.: Radiology residents' perceptions of artificial intelligence: nationwide cross-sectional survey study. J. Med. Internet Res. **25**, e48249 (2023). https://doi.org/10.2196/48249

Attention-Enhanced Deep Learning for Multi-class Alzheimer's Disease Classification Using Macular OCT Images in Low-Resource Settings

Edem Fiifi Dawson[1](✉), Louis Darko[1], Hephzi Tagoe[1], Edem Ahiabor[2], Kwame Oteng[2], and Samuel Gyamfi[3]

[1] Department of Biomedical Engineering, Academic City University, Haatso, Ghana
{edem.dawson,louis.darko,hephzi.tagoe}@acity.edu.gh
[2] Lions International Eye Centre, Korle-Bu Teaching Hospital, Accra, Ghana
[3] Emmanuel Eye Medical Centre, Accra, Ghana

Abstract. Alzheimer's Disease (AD) remains underdiagnosed in many low-resource settings, where access to neuroimaging tools is limited. Retinal imaging via Optical Coherence Tomography (OCT) offers a non-invasive window into neurodegenerative processes. This study investigates the use of deep learning models to detect and classify stages of AD using macular OCT images, with attention to explainability and resource accessibility. We evaluated two models: a baseline EfficientNet-B3 architecture and an attention-enhanced version incorporating Squeeze-and-Excitation (SE) blocks as an extra layer. Models were trained on 50% of a public dataset of over 80,000 OCT images and locally validated using anonymised scans from Ghanaian clinics. Grad-CAM was used to visualise model attention. The attention-enhanced model outperformed the baseline across all metrics, with improved classification of Mild Cognitive Impairment (MCI) and Alzheimer's Disease (AD) stages. Grad-CAM visualisations compared alongside ground truth, confirmed that attention mechanisms focused on clinically relevant macular regions better than the baseline model. This study demonstrates that attention-enhanced deep learning models, when validated with local clinical data, can reveal retinal structural and vascular features associated with Alzheimer's Disease (AD) and Mild Cognitive Impairment (MCI) using OCT. However, as OCT alone is not yet a validated standalone diagnostic test for Alzheimer's Disease (AD), this work highlights the potential of deep learning models in enhancing diagnostic capacity and accessibility in under-resourced health systems.

Keywords: Alzheimer Disease · Optical Coherence Tomography · Deep Learning · Low-Resource Settings · Medical Imaging

U. Anazodo et al. (Eds.): MIRASOL 2025, LNCS 16398, pp. 185–194, 2026.
https://doi.org/10.1007/978-3-032-13654-1_19

1 Introduction

Alzheimer's Disease (AD) is a progressive neurodegenerative disorder characterized by cognitive decline and memory loss [10]. Globally, it affects an estimated 50–55 million as of the early 2020s [12]. The burden of AD is increasing especially in low- and middle-income countries (LMICs), including regions like sub-Saharan Africa, which face constrained healthcare resources [19]. Traditional neuroimaging tools for diagnosis, such as MRI and PET scans, are often expensive, invasive, or inaccessible for widespread screening in these low-resource settings [2,11,13]. This creates an urgent need for affordable and non-invasive biomarkers to enable early detection and verification of AD in such settings.

Recent evidence suggests that the retina, being a developmental extension of the central nervous system can mirror neurodegenerative changes seen in the brain [14,20]. Optical Coherence Tomography (OCT) can detect structural alterations in the retinal nerve fiber layer and macular regions that correlate with AD progression [4,7]. The retina is unique as it is the only neural tissue that can be directly and non-invasively visualized, providing a window into central nervous system health. Both postmortem and clinical studies have shown that Alzheimer's Disease patients exhibit characteristic retinal abnormalities, including amyloid-β plaque accumulation and thinning of the retinal nerve fiber layer (RNFL) that parallel the expected pathological changes observed in the brain [18]. Prior studies have reported significant thinning of the RNFL and ganglion cell layer in AD, with retinal measures correlating to disease severity and brain changes like hippocampal atrophy [3]. This positions OCT as a promising, non-invasive biomarker tool for AD detection.

Deep learning approaches have recently achieved impressive success in medical image analysis, including analysis of retinal OCT images for various diseases. Convolutional neural networks (CNNs) have demonstrated expert-level accuracy in detecting retinal pathologies such as diabetic macular edema (DME) and age-related macular degeneration (AMD) from OCT scans [17]. For instance, Kermany et al. [9] showed that a CNN could distinguish Normal, drusen (dry AMD), choroidal neovascularization (CNV, wet AMD), and DME OCT images with over 95% accuracy. Subsequent studies have refined OCT classification with advanced architectures and larger datasets, even approaching near-perfect performance on public benchmarks [5]. Attention mechanisms incorporated into deep models have further enhanced performance by focusing on clinically relevant image regions. In particular, the Squeeze-and-Excitation (SE) module introduced by Hu et al. [6] adaptively recalibrates CNN feature maps, boosting important feature channels and suppressing less useful ones. Such attention modules can improve feature selection and have been applied in retinal OCT models to emphasize subtle disease patterns, leading to more robust feature extraction.

Despite these advances, most prior work has been developed and validated on datasets from high-resource settings. This study addresses these gaps by developing and evaluating an attention-enhanced deep learning model for multi-class classification of Alzheimer's stages using macular OCT images. Unlike previous works, we validate our models on both a large global OCT dataset and a new Ghanaian clinical OCT dataset, to ensure generalizability across patient populations. We integrate Squeeze-and-Excitation (SE) attention blocks into a state-of-the-art CNN architecture to enhance the model's ability to focus on retinal features relevant to Alzheimer's Disease. In addition, we employ explainability techniques to identify and visualize the image regions most influential in the model's predictions. By bridging advanced deep learning with local clinical data and interpretability, this work aims to advance accessible early diagnosis and human validation of AD through retinal imaging in under-served regions.

2 Materials and Methods

2.1 Dataset and Ethical Approval

This study used a combination of public and clinical datasets for training. The primary dataset was a publicly available collection of 83,482 retinal Optical Coherence Tomography (OCT) images from Kermany et al. [9]. The images were classified into four pathological categories—Normal, Drusen, Choroidal Neovascularisation (CNV), and Diabetic Macular Edema (DME)—following the original schema. These images were further divided into training, validation, and test sets, ensuring no patient overlap across subsets.

For external validation and clinical relevance, additional anonymized OCT images were obtained from two Ghanaian medical institutions: Korle-Bu Teaching Hospital and Emmanuel Eye Medical Centre. Ethical approvals were secured from both institutions before data collection. All data were de-identified in compliance with local research-ethics guidelines and international standards for medical data handling.

The dataset was reduced to serve two purposes: to meet computational constraints for training a deep learning neural network, and to test the reliability of the models under data-scarce conditions, often reflecting the reality in low-resource clinical settings. This reduction may introduce bias if the excluded images contain systematic differences (Table 1).

Table 1. Summary of Dataset Size Reduction for OCT Categories

Data Size	CNV	Drusen	DME	Normal	Total
100%	37,204	8,616	11,348	26,314	83,482
50%	18,602	4,308	5,674	13,157	41,741

2.2 Preprocessing

To ensure compatibility with EfficientNet-B3 and enable batch processing, OCT images were resized to 300×300 and pixel intensities normalised to (0,1) using a 1./255 factor. Data augmentation with Keras' ImageDataGenerator included

±20° rotation, ±10% width/height shifts, ±20% zoom, horizontal flips, and nearest fill mode. We address the class imbalance of fewer DME and Drusen samples by applying inverse frequency class weighting.

2.3 Label Mapping for Alzheimer's Stage Classification

To adapt the OCT pathology labels to investigate cognitive status, we remapped the original OCT categories into three cognitive group as follows:

Normal class matched to Cognitively Normal (CN), Drusen to Mild Cognitive Impairment (MCI), and CNV and DME to Alzheimer's Disease (AD). This relabelling is a heuristic adopted to reuse the annotated public OCT dataset. It is important to emphasize that the strongest, reproducible retinal biomarkers reported in the literature for AD and MCI are quantitative OCT/OCT-A measures such as retinal nerve fibre layer (RNFL) and ganglion-cell/inner plexiform layer (GC-IPL) thinning and changes in retinal microvascular density. Systematic reviews and meta-analyses support using layer thickness and microvascular metrics as candidate biomarkers for cognitive impairment [1,8,15].

2.4 Model Architecture

Baseline Model (EfficientNet-B3). The EfficientNet-B3 model was selected for its balance of accuracy and computational efficiency [16]. Pre-trained on the ImageNet dataset and implemented using TensorFlow and Keras, the model leveraged transfer learning by freezing the early convolutional blocks and fine-tuning blocks 5 through 7 to adapt to domain-specific features. The custom classification head consisted of a Global Average Pooling (GAP) layer, followed by a dense layer with 512 units, batch normalization, and ReLU activation. This was followed by a dropout layer with a rate of 0.4. A subsequent dense layer with 128 units and ReLU activation was included, followed by an additional dropout layer with a rate of 0.3. The final layer was a dense layer with 3 units and softmax activation to enable multi-class output.

Attention-Enhanced Model EfficientNet-B3 + SE Block. To improve feature selection, a Squeeze-and-Excitation (SE) block was inserted immediately after the convolutional backbone. The SE block recalibrates channel-wise feature responses by first applying a squeeze operation, which uses global average pooling across spatial dimensions. This is followed by the excitation step, involving bottleneck fully connected layers with a reduction ratio of 16, utilising ReLU and Sigmoid activations. Finally, the block performs scaling by reweighting input channels based on learned importance. This attention mechanism was lightweight and introduced minimal computational overhead while improving discriminative focus.

2.5 Training Protocol

Both models were compiled with the Sparse Categorical Cross-Entropy loss function and optimised using Adam with a learning rate of 1×10^{-5}. Training was performed with a batch size of 32 for 20 epochs. The training process incorporated several callbacks to ensure optimal performance and prevent overfitting. Early stopping was implemented with a patience of 12 epochs, monitoring validation loss. The learning rate was reduced by a factor of 0.5 if validation loss plateaued for 4 epochs. Model checkpointing was employed to retain the best model based on the highest validation AUC.

3 Evaluation Metrics

Model performance was evaluated using accuracy, precision, recall (sensitivity), F1-score, area under the curve (AUC), and top-3 accuracy. Additionally, confusion matrices and precision-recall curves were generated for each class. This allowed for a detailed assessment of misclassification trends and provided insight into the model's ability to distinguish between adjacent disease stages, such as mild cognitive impairment (MCI) versus Alzheimer's disease (AD).

3.1 Explainability via Grad-CAM

To enhance clinical trust in predictions, Gradient-weighted Class Activation Mapping (Grad-CAM) was used to generate heat maps indicating which regions of the image contributed most to the model's classification decision. These were implemented by back propagating gradients to the final convolutional layer and applying a weighted combination of feature maps. Visual overlays of Grad-CAM on input images were used to assess alignment with known pathological regions.

4 Results

4.1 Baseline Performance

The baseline EfficientNet-B3, was trained for 20 epochs and achieved 75.2% training accuracy, 90.3% validation accuracy, and an AUC of 0.978, indicating strong generalisation but with signs of overfitting due to the large train–validation gap. Precision-recall analysis showed a bias towards the AD class, with weaker recall for MCI. The model correctly classified 242/249 CN, 220/242 MCI, and 446/484 AD cases. Misclassifications were mostly between MCI and AD, highlighting the clinical difficulty of early-stage differentiation. F1-scores were 0.93 (CN), 0.87 (MCI), and 0.91 (AD). Calibration analysis gave a Brier score of 0.0330, with predicted probabilities generally well aligned but showing slight underestimation (0.1–0.3) and mild overestimation (0.7–0.8).

4.2 SE Attention-Enhanced Performance

The attention-enhanced model with Squeeze-and-Excitation (SE) blocks improved stability, interpretability, and calibration. It achieved 80.1% training accuracy, 91.4% validation accuracy, and an AUC of 0.982, with smoother convergence, especially in the first 10 epochs. Compared to the baseline, it showed stronger generalisation and reduced train–validation divergence.

For classification, the SE model correctly identified 246/249 CN, 232/242 MCI, and 460/484 AD cases—a net gain of 12 correct predictions, most notably in MCI, critical for early Alzheimer's detection. F1-scores were 0.95 (CN), 0.91 (MCI), and 0.93 (AD). The top-3 accuracy reached 99.2%, supporting clinical use where differential diagnosis is key.

Precision-recall analysis showed consistently high performance, with MCI recall improving from 0.86 to 0.91, reducing false negatives in early disease detection. The model also achieved a lower Brier score (0.0208) and a well-calibrated probability curve, with only slight deviations with underestimation below 0.2, and overestimation in the 0.7–0.8 range (Table 2).

Table 2. Comparative performance of baseline and SE-enhanced deep learning models across key evaluation metrics.

Model	Validation Acc.	AUC	Top-3 Acc.	Brier
Baseline	90.3%	0.978	97.6%	0.0330
SE-Enhanced	91.4%	0.982	99.2%	0.0208
Gain	+1.1%	+0.004	+1.6%	-0.0122

4.3 Grad-CAM Visualization and Explainability

To ensure that our method uses clinically relevant features in the OCT for predictions, Grad-CAM heatmaps were generated for sample predictions across all classes. Visual comparison between the baseline and SE-enhanced models revealed the following:

Normal (CN): Both models focused correctly on the central foveal depression, but the SE model exhibited more confined activation, reducing noise in peripheral layers.

Drusen (MCI): The SE model produced sharper activations in regions corresponding to retinal pigment epithelium irregularities, aligning with known pathophysiology.

DME/CNV (AD): The attention-enhanced model highlighted vascular distortions and subretinal fluid regions more distinctly than the baseline.

4.4 Specialist Description

Specialist ophthalmic interpretation of representative OCT scans is provided below to contextualize the model outputs.

Normal (CN): SD-OCT B-scan of the macula in the left eye showing a normal foveal depression (yellow arrow) with preserved retinal and choroidal structures.

Drusen (MCI Proxy): SD-OCT B-scan of dry age-related macular degeneration in the left eye showing small, discrete elevations of the hyper-reflective retinal pigment epithelium (blue arrow) from Bruch's membrane, associated with moderately reflective sub-RPE material (red arrow). The overlying retinal layers remain preserved.

DME (AD Proxy): SD-OCT B-scan of diabetic macular edema in the right eye demonstrating retinal thickening with multiple intraretinal cystoid spaces (green arrow).

CNV (AD Proxy): SD-OCT B-scan of neovascular AMD in the right eye showing variable intraretinal cystoid spaces (red star), retinal pigment epithelial detachment (yellow arrow), undulating RPE elevations, and hypertransmission into the choroid (green arrow), reflecting outer retinal and RPE atrophy (Fig. 1).

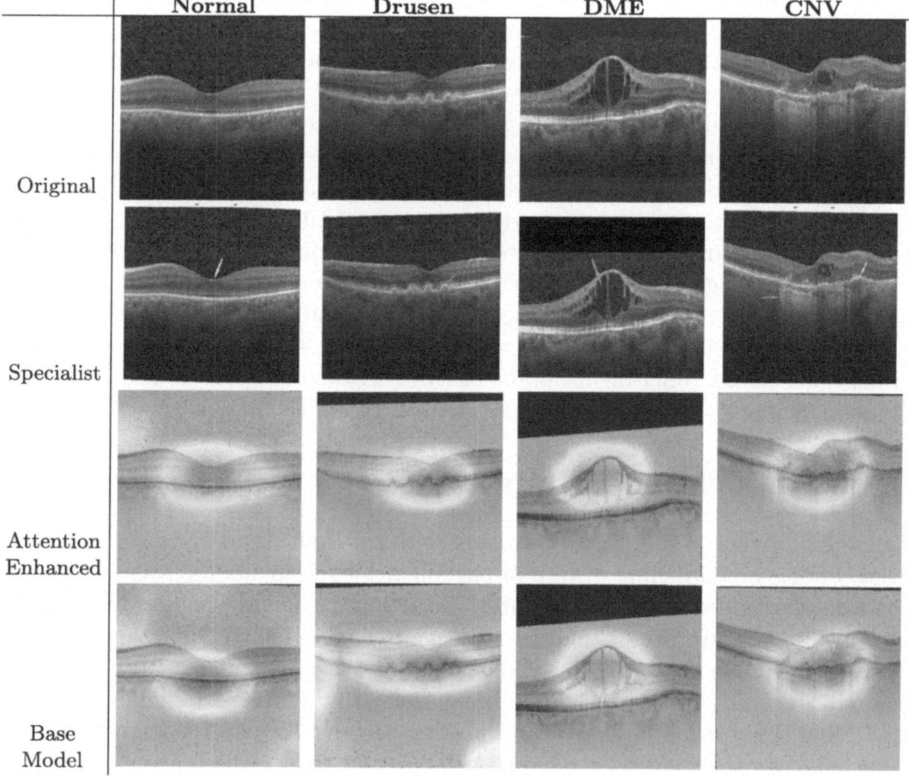

Fig. 1. Visual comparison across Normal, Drusen, DME, and CNV for Original input, specialist description, attention-enhanced model, and base model, output. (Color figure online)

5 Discussion

This study demonstrates the potential of attention-enhanced deep learning models, specifically EfficientNet-B3 with a Squeeze-and-Excitation (SE) block, for early-stage Alzheimer's disease classification using retinal OCT images. The SE-enhanced model outperformed the baseline in accuracy, AUC, and F1-score, while also showing greater stability and improved explainability through Grad-CAM visualisations that aligned with clinical ground-truth. These results are especially relevant for deployment in low-resource settings such as sub-Saharan Africa.

The integration of SE blocks allowed dynamic recalibration of feature responses, improving differentiation between Mild Cognitive Impairment (MCI) and Alzheimer's Disease (AD), a clinically challenging boundary. Enhanced classification of MCI is particularly important, as it represents a window for early intervention before irreversible neurodegeneration. The model's gains in F1-score and reduced misclassification support the hypothesis that attention mechanisms strengthen both performance and interpretability, consistent with prior work in retinal imaging and neurodegenerative disease prediction.

A key contribution of this study is the inclusion of anonymised OCT scans from Ghanaian patients, moving toward validation across more diverse populations. While most deep learning research in ophthalmology relies on high-income country datasets, our preliminary results suggest that local data improve robustness. Broader validation, however, is still required to confirm cross-population generalisability. Grad-CAM visualisations also helped bridge the "black box" problem by highlighting clinically relevant regions, reinforcing their role in building clinician trust and aiding diagnostic reasoning. Still, interpretability methods require further standardisation for reliable integration into clinical practice.

Finally, the lightweight nature of our model enables real-time inference on modest infrastructure, which is vital for low- and middle-income countries (LMICs) where access to advanced neuroimaging or specialists is limited. Such applications could function as point-of-care decision-support tools, extending diagnostic capacity in ophthalmic and neurology clinics.

6 Limitations and Future Work

While the current results are promising, several limitations must be acknowledged. First, the relabeling of retinal pathologies to Alzheimer's disease stages is based on associative evidence and may oversimplify complex neuro-ophthalmic relationships. Future work should integrate longitudinal clinical data or biomarkers (CSF tau, Aβ-42 levels) to validate the predicted cognitive stages. Second, although the study incorporates Ghanaian data, the sample size remains limited. Expanding the clinical dataset and validating across multiple African institutions will strengthen the model's ecological validity and reduce the bias in identifying unique cases. In addition, the study used a 3-class classification schema. More granular (early vs. late MCI) and multi-modal data integration (retinal thickness, patient history) could refine predictions. Future research may also explore federated learning approaches to train models across institutional boundaries without compromising patient privacy.

7 Conclusion

This study highlights the feasibility and clinical relevance of applying attention-enhanced deep learning to retinal OCT images for the classification of Alzheimer's disease stages. By incorporating Squeeze-and-Excitation mechanisms into an EfficientNet-B3 backbone, the model achieved superior diagnostic accuracy, particularly in distinguishing Mild Cognitive Impairment which is a critical target for early intervention. The use of Grad-CAM-based visual explanations further enhances the model's interpretability, offering transparency essential for clinical trust. Importantly, the integration of anonymised Ghanaian clinical data contributes to addressing the equity gap in AI healthcare tools, demonstrating potential for localisation and deployment in under-resourced healthcare environments. Taken together, these findings highlight the potential of OCT-based AI models as scalable, non-invasive, and interpretable screening tools for neurodegenerative diseases in settings with limited access to advanced neuroimaging. Future work should focus on expanding clinical validation, integrating multimodal data, and collaborating across institutions to ensure ethical, inclusive, and contextually relevant AI deployment.

References

1. Ashraf, G., McGuinness, M., Khan, M.A., Obtinalla, C., Hadoux, X., van Wijngaarden, P.: Retinal imaging biomarkers of Alzheimer's disease: a systematic review and meta-analysis of studies using brain amyloid beta status for case definition. Alzheimer's Dementia: Diagnosis Assess. Dis. Monit. **15**(2) (2023). https://doi.org/10.1002/dad2.12421
2. Bour, B.K., et al.: National inventory of authorized diagnostic imaging equipment in Ghana: data as of September 2020. Pan Afr. Med. J. **41**(1) (2022)
3. Chen, S., et al.: The association between retina thinning and hippocampal atrophy in Alzheimer's disease and mild cognitive impairment: a meta-analysis and systematic review. Front. Aging Neurosci. **15**, 1232941 (2023a)
4. Chen, S., et al.: Ultrahigh resolution oct markers of normal aging and early age-related macular degeneration. Ophthalmol. Sci. **3**(3), 100277 (2023b). https://doi.org/10.1016/j.xops.2023.100277
5. Gencer, G., Gencer, K.: Advanced retinal disease detection from OCT images using a hybrid squeeze and excitation enhanced model. PLoS ONE **20**(2), e0318657 (2025)
6. Hu, J., Shen, L., Sun, G.: Squeeze-and-excitation networks. In: Proceedings of the IEEE Conference on Computer Vision and Pattern Recognition, pp. 7132–7141 (2018)
7. Huang, D., et al.: Optical coherence tomography. Science **254**(5035), 1178–1181 (1991)

8. Jin, Q., Lei, Y., Wang, R., Wu, H., Ji, K., Ling, L.: A systematic review and meta-analysis of retinal microvascular features in Alzheimer's disease. Front. Aging Neurosci. **13** (2021). https://doi.org/10.3389/fnagi.2021.683824
9. Kermany, D.S., et al.: Identifying medical diagnoses and treatable diseases by image-based deep learning. Cell **172**(5), 1122–1131.e9 (2018). https://doi.org/10.1016/j.cell.2018.02.010
10. Kumar, A., Sidhu, J., Lui, F., Tsao, J.W.: Alzheimer disease. In: StatPearls [internet]. StatPearls Publishing (2024)
11. Musah, T., Adjei, P.E., Otoo, K.O.: Automated segmentation of ischemic stroke lesions in non-contrast computed tomography images for enhanced treatment and prognosis. In: Meets Africa Workshop, pp. 73–80. Springer, Cham (2024)
12. Nichols, E., et al.: Estimation of the global prevalence of dementia in 2019 and forecasted prevalence in 2050: an analysis for the global burden of disease study 2019. Lancet Public Health **7**(2), e105–e125 (2022)
13. Piersson, A., Gorleku, P.: Assessment of availability, accessibility, and affordability of magnetic resonance imaging services in Ghana. Radiography **23**(4), e75–e79 (2017)
14. Sadeghi, E., et al.: Choroidal biomarkers in age-related macular degeneration. Surv. Ophthalmol. **70**(2), 167–183 (2025). https://doi.org/10.1016/j.survophthal.2024.10.004
15. Song, A., Johnson, N., Ayala, A., Thompson, A.C.: Optical coherence tomography in patients with Alzheimer's disease: what can it tell us? Eye and Brain **13**, 1–20 (2021). https://doi.org/10.2147/eb.s235238
16. Tan, M., Le, Q.V.: Efficientnet: rethinking model scaling for convolutional neural networks (2019). https://doi.org/10.48550/ARXIV.1905.11946
17. Viedma, I.A., Alonso-Caneiro, D., Read, S.A., Collins, M.J.: Deep learning in retinal optical coherence tomography (OCT): a comprehensive survey. Neurocomputing **507**, 247–264 (2022)
18. Wang, L., Mao, X.: Role of retinal amyloid-β in neurodegenerative diseases: overlapping mechanisms and emerging clinical applications. Int. J. Mol. Sci. **22**(5), 2360 (2021)
19. Zencir, M., Kuzu, N., Beşer, N.G., Ergin, A., Çatak, B., Şahiner, T.: Cost of Alzheimer's disease in a developing country setting. Int. J. Geriatric Psychiatry: J. Psychiatry Late Life Allied Sci. **20**(7), 616–622 (2005)
20. Zhang, Y., Wang, Y., Shi, C., Shen, M., Lu, F.: Advances in retina imaging as potential biomarkers for early diagnosis of Alzheimer's disease. Transl. Neurodegeneration **10**(1) (2021). https://doi.org/10.1186/s40035-021-00230-9

Resource-Efficient Glioma Segmentation on Sub-Saharan MRI

Freedmore Sidume[1(✉)], Oumayma Soula[2,3], Joseph Muthui Wacira[4], YunFei Zhu[5], Abbas Rabiu Muhammad[6,7], Abderrazek Zeraii[2], Oluwaseun Kalejaye[8], Hajer Ibrahim[2,9], Olfa Gaddour[10], Brain Halubanza[11], Dong Zhang[12,13], Udunna C. Anazodo[5,12,13,14,15,16,17], and Confidence Raymond[5,12,14]

[1] BAC School of Computing and Information Systems, Gaborone, Botswana
freedmores@gmail.com
[2] Faculty of Medicine of Sfax, University of Sfax, Sfax, Tunisia
[3] Laboratory of Biophysics and Medical Technologies, Higher Institute of Medical Technologies of Tunis, University ElManar, Tunis, Tunisia
[4] London School of Hygiene and Tropical Medicine, London, UK
[5] Lawson Health Research Institute, London, ON, Canada
[6] Department of Radiology, Bayero University, Kano, Nigeria
[7] Department of Radiology, Aminu Kano Teaching Hospital, Kano, Nigeria
[8] Department of Anatomy,Federal University Lokoja College of Health Sciences, Lokoja, Kogi, Nigeria
[9] Department of Computer Science and Information, College of Science, Majmaah University, AL-Majmaah 11952, Saudi Arabia
[10] CES Laboratory, National School of Engineers of Sfax, Sfax, Tunisia
[11] Department of Computer Science and IT, Mulungushi University, Kabwe, Zambia
[12] Multimodal Imaging of Neurodegenerative Diseases (MiND) Lab, Department of Neurology and Neurosurgery, McGill University, Montreal, QC, Canada
[13] Department of Electrical and Computer Engineering, University of British Columbia, Vancouver, Canada
[14] Medical Artificial Intelligence Laboratory (MAI Lab), Lagos, Nigeria
[15] Montreal Neurological Institute, McGill University, Montréal, Canada
[16] Department of Medicine, University of Cape Town, Cape Town, South Africa
[17] Department of Clinical and Radiation Oncology, University of Cape Town, Cape Town, South Africa

Abstract. Gliomas are the most prevalent type of primary brain tumors, and their accurate segmentation from MRI is critical for diagnosis, treatment planning, and longitudinal monitoring. However, the scarcity of high-quality annotated imaging data in Sub-Saharan Africa (SSA) poses a significant challenge for deploying advanced segmentation models in clinical workflows. This study introduces a robust and computationally efficient deep learning framework tailored for resource-constrained settings. We leveraged a 3D Attention U-Net architecture augmented with residual blocks and enhanced through transfer learning from pre-trained weights on the BraTS 2021 dataset. Our model was evaluated on 95 MRI cases from the BraTS-Africa dataset, a benchmark for

F. sidume, O. Soula and J. M. Wacira—these author are equally contributed.

U. Anazodo et al. (Eds.): MIRASOL 2025, LNCS 16398, pp. 195–205, 2026.
https://doi.org/10.1007/978-3-032-13654-1_20

glioma segmentation in SSA MRI data. Despite the limited data quality and quantity, our approach achieved Dice scores of 0.76 (Enhancing Tumor - ET), 0.80 (Necrotic and Non-Enhancing Tumor Core - NETC), and 0.85 (Surrounding Non-Functional Hemisphere - SNFH). These results demonstrate the generalizability of the proposed model and its potential to support clinical decision making in low-resource settings. The compact architecture ($\sim$ 90 MB) and sub-minute per-volume inference time on consumer- grade hardware, further underscores its practicality for deployment in SSA health systems. This work contributes toward closing the gap in equitable AI for global health by empowering underserved regions with high performing and accessible medical imaging solutions.

Keyword: Glioma, BraTS-Africa, Low Resource setting, MRI, 3D attention, Residual blocks, Transfer learning.

1 Introduction

Brain tumors are a heterogeneous group of neoplasms that include gliomas, meningiomas, and pituitary adenomas, and can severely affect cognition, motor function, and quality of life. Gliomas are the most aggressive and account for the majority of malignant brain tumors in the central nervous system, with nearly 80% of patients dying within two years of diagnosis [3]. Magnetic resonance imaging (MRI) is central to glioma diagnosis and treatment planning, providing high-resolution, multi-parametric images across modalities such as T1-weighted (T1w), contrast-enhanced T1w (T1CE), T2-weighted (T2w), and FLAIR, which together capture tumor location, infiltration, and morphology [11]. However, accurate delineation still relies on expert radiologists, who are scarce in resource-limited settings such as sub-Saharan Africa (SSA) [9], [?]. Manual interpretation is further complicated by the large data volume, multiple modalities, and frequent presence of low-quality images.

Deep learning (DL) methods, particularly convolutional neural networks (CNNs), have advanced automated brain tumor segmentation by capturing complex spatial patterns and providing accurate delineation of tumor subregions. Architectures such as DeepMedic [6] and 3D U-Net [16], [?] have achieved state-of-the-art performance, while transfer learning has improved generalization on smaller or domain-specific datasets [14]. Yet, most approaches require substantial computational resources, limiting deployment in low- and middle-income countries (LMICs).

The Brain Tumor Segmentation (BraTS) Challenge, established in 2012, has driven progress by providing annotated datasets and standardized benchmarks. However, models trained on BraTS data often underperform in SSA due to differences in imaging protocols, resolution, and annotation scarcity [15]. Recent initiatives, such as BraTS-Africa [2], address this gap by curating SSA-specific datasets. Studies have shown that fine-tuning pre-trained models on local data improves segmentation, underscoring the need for domain adaptation [4], [?].

Building on these efforts, we propose **BrainUNet**, a lightweight 3D U-Net model with residual connections and attention mechanisms. Pre-trained on the BraTS 2021 dataset [3] and fine-tuned on BraTS-Africa [2], BrainUNet is optimized for the challenges of SSA imaging, including low resolution and modality variation. Its efficient design balances accuracy and computational cost, making it suitable for deployment in resource-constrained clinical environments.

2 Methods

2.1 The Dataset

The BraTS-Africa dataset comprises multi-parametric MRI (mpMRI) scans from 95 glioma patients (60 training, 35 validation) [2]. Each case includes four structural modalities: native T1-weighted (T1w), post-contrast T1-weighted (T1CE), T2-weighted (T2w), and T2 Fluid Attenuated Inversion Recovery (FLAIR). In this study, the native T1w modality was excluded. All scans were reviewed by board-certified neuroradiologists, pre-processed, and manually annotated using standardized BraTS protocols. Ground-truth tumor subregions—Enhancing Tumor (ET), Non-enhancing Tumor Core (NETC), and Surrounding Non-enhancing FLAIR Hyperintensity (SNFH)—were delineated and verified by expert neuroradiologists in accordance with BraTS 2023 challenge criteria [3], [?].

2.2 Proposed Approach

2.3 Proposed BrainUNet Framework

We propose **BrainUNet**, a 3D Attention U-Net with residual blocks as its backbone (Figs. 1, 2). The framework integrates transfer learning with tailored preprocessing to enhance MRI quality and minimize information loss. The U-shaped encoder-decoder structure extracts multi-scale features while skip connections preserve spatial detail. Attention gates at skip connections focus the network on clinically relevant regions, improving tissue discrimination critical for diagnosis [5]. Residual connections in both encoder and decoder blocks stabilize training, reduce computational complexity, and enhance feature representation.

Data Preprocessing. MRI scans were cropped to $128 \times 128 \times 128$ to reduce computational demand. Three modalities (FLAIR, T1CE, T2W) were stacked, providing optimal tumor contrast [1]. Percentile clipping mitigated outliers, while intensity normalization harmonized modalities and masks [10]. These steps improved efficiency and preserved critical information for robust segmentation.

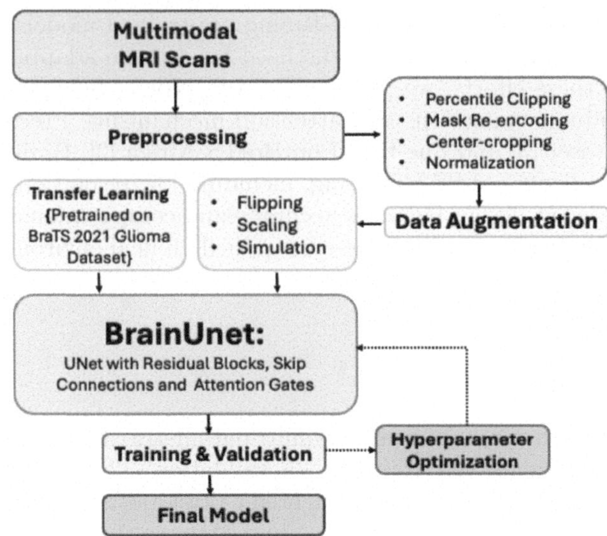

Fig. 1. Overview of the proposed BrainUNet framework.

Data Augmentation. To enhance generalization on the BraTS-Africa dataset, we applied augmentations including flipping, scaling, and gamma adjustment to account for variability in orientation and brightness. Motion and ghosting artifacts were introduced to simulate common MRI imperfections, improving robustness to real-world clinical imaging noise [12].

The Model Design:

2.4 Model Architecture

The proposed BrainUNet architecture (Fig. 2) builds on the U-Net framework to improve 3D medical image segmentation. It integrates two key components designed to enhance learning and focus within volumetric data:

1. **Residual Blocks:** Instead of standard convolutional blocks, the model employs residual blocks (Fig. 3a), each consisting of two 3D convolutional layers, ReLU activation, batch normalization, and a skip connection. Residual connections mitigate the vanishing gradient problem, preserve identity information, and facilitate deeper network training. This improves convergence speed, feature learning, and the ability to capture complex patterns.
2. **Attention Mechanisms:** Attention blocks (Fig. 3b) enable the network to focus on salient regions by adaptively re-weighting feature maps. Each block receives x (encoder features via skip connections) and g (decoder features from deeper layers), which are processed through 3D convolutions, batch

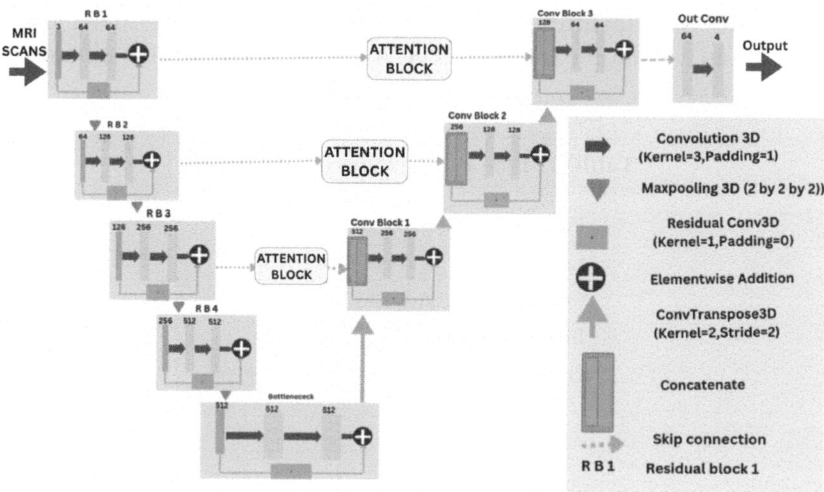

Fig. 2. 3D Attention U-Net with residual connections.

normalization, and activations. The gating mechanism emphasizes informative regions, improving discrimination of subtle tissue differences critical for clinical decision-making.

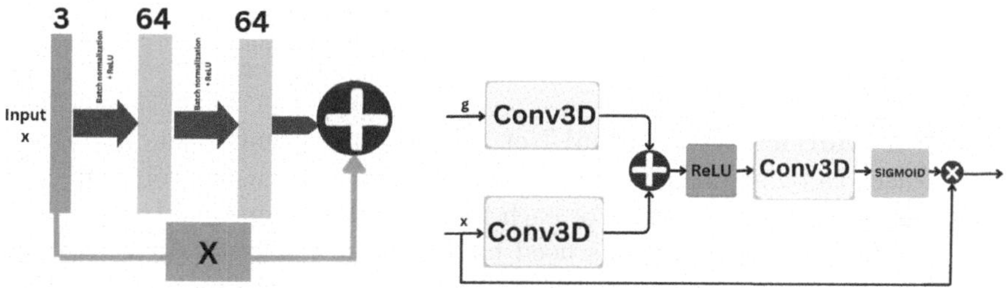

Fig. 3. Core components of BrainUNet: (a) residual block with skip connections, and (b) attention block with gating for region-specific focus.

Transfer Learning: Before fine-tuning on BraTS-Africa, the model was pre-trained on the larger BraTS-GLI 2021 dataset ($n = 1251$ 3D MRI scans). Training used the parameters in Table 1, with epochs and learning rate set to 30 and 1×10^{-3}, respectively. This pre-training phase exposed the model to diverse tumor types, shapes, and stages, enabling it to learn robust, generalized features while mitigating overfitting. Such a foundation improved transferability and provided a strong basis for fine-tuning on BraTS-Africa, enhancing clinical applicability.

Training and Validation: The pre-trained model was fine-tuned using 50 training and 10 validation samples. The hyperparameters are summarized in Table 1.

Table 1. Fine-tuning hyperparameters.

Parameter	Choice
Loss	Tversky loss [13]
Optimizer	Adam [7]
Learning rate	1×10^{-4}
Epochs	50

Rationale. Adam optimizer combines the advantages of AdaGrad and RMSProp, enabling efficient training with adaptive learning rates [7,8]. Tversky loss addresses class imbalance by weighting false positives and false negatives through α and β. It is defined as:

$$\text{Tversky Loss} = 1 - \text{TI}, \quad \text{TI} = \frac{\sum_i y_i \hat{y}_i}{\sum_i y_i \hat{y}_i + \alpha \sum_i \hat{y}_i (1 - y_i) + \beta \sum_i y_i (1 - \hat{y}_i)} \tag{1}$$

where y_i and $\hat{y}_i$ denote ground-truth and predicted labels. This formulation improves robustness by emphasizing minority tumor classes.

Evaluation Metrics: The BrainUNet model was evaluated on the BraTS-Africa dataset ($n = 95$) using 5-fold cross-validation to ensure robustness and mitigate bias from limited data. For each fold, one partition served as validation ($n = 35$) while the remainder was used for training, and results were averaged across folds. Performance was assessed using Dice score, Intersection over Union (IoU), and Hausdorff95 distance, tracked over 100 epochs. Predicted tumor sub-regions from the validation set were further benchmarked on the BraTS Challenge Synapse platform www.synapse.org/brats, with and without fine-tuning, to obtain standardized metrics [3]. Finally, runtime efficiency was measured by model size (91.4 MB) and per-volume inference time on CPU and GPU (Tesla P100, 16 GB VRAM, Kaggle environment).

3 Results

3.1 Cross-Validation Results for BrainUNet with SSA Dataset

The performance of the BrainUNet model before fine-tuning is summarized in Figs. 4, and shows the average training and validation Dice Score, IoU, and Loss across the folds. The results indicated a consistent improvement in training metrics over epochs, where the average training Dice score progressively improved, reaching 0.7, and the IoU increased to an average of 0.62. The validation metrics (Dice score stabilized at 0.55 and IoU averaged 0.47) showed a steady convergence, reflecting the model's ability to generalize across different data splits. Similarly, the loss curves exhibited steady convergence, with training loss decreasing to 0.39 and validation loss plateauing at 0.62.

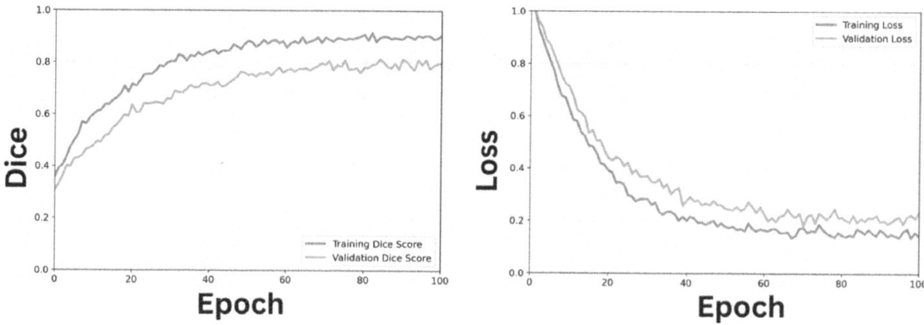

Fig. 4. Average training and validation metrics across folds during cross-validation. (Left) Average Dice Score, and (Right) Average Loss. These figures reflect the performance of the BrainUNet model before fine-tuning.

3.2 Synapse Evaluation: BrainUNet Before and After Fine-Tuning

We evaluated BrainUNet on the BraTS-Africa dataset before and after fine-tuning with BraTS-GLI 2021. As summarized in Table 2, fine-tuning markedly improved Dice scores across all tumor subregions (ET, NETC, SNFH) and reduced Hausdorff distances, indicating more precise boundary delineation. Figure 5 highlights a representative case, showing improved segmentation after fine-tuning.

Table 2. Performance of BrainUNet before and after fine-tuning, reported as Dice Score and Hausdorff Distance (95%) for Enhancing Tumor (ET), Non-Enhancing Tumor Core (NETC), and Surrounding Non-Enhancing FLAIR Hyperintensity (SNFH).

Model	Dice Score			Hausdorff Distance (95%)		
	ET	NETC	SNFH	ET	NETC	SNFH
BrainUNet$_{SSA}$	0.52	0.47	0.62	74.8	77.48	23.78
BrainUNet$_{SSA}$ fine-tuned	0.76	0.80	0.85	30.00	30.55	18.72

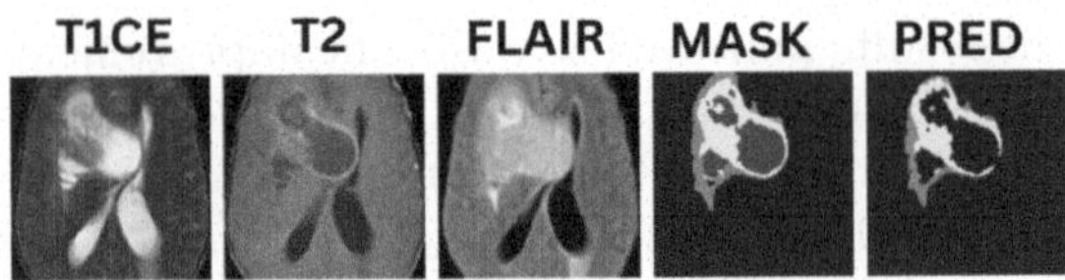

Fig. 5. Segmentation results with BrainUNet before fine-tuning. T1CE, T2W, and FLAIR modalities are shown with ground-truth (Mask) and predicted segmentation (Pred).

To benchmark BrainUNet, we compared it against state-of-the-art baselines (nnUNet and MedNeXt). Results in Table 3 show that although BrainUNet reports slightly lower Dice scores for ET and TC, it achieves competitive WT performance. Importantly, its compact size (~91 MB) and lower parameter count make it well suited for low-resource settings (Fig. 6).

Table 3. Lesion-wise and legacy Dice scores for BraTS-Africa segmentation. BrainUNet demonstrates competitive WT Dice while being significantly more lightweight (22M parameters, ~91 MB), underscoring its suitability for deployment in resource-limited clinical environments.

Model	Lesion-wise Dice			Avg	Legacy Dice			Avg	Params
	ET	TC	WT		ET	TC	WT		
nnUNet	0.797	0.786	0.846	**0.810**	0.850	0.853	0.913	**0.872**	30M+
MedNeXt Ens.	0.793	0.786	0.893	**0.824**	0.878	0.876	0.924	**0.893**	50M+
BrainUNet	0.684	0.714	0.831	**0.743**	0.759	0.791	0.869	**0.806**	22M+

3.3 Inference Analysis

To assess the runtime performance of our model (91.4 MB), we measured per-volume inference time on both GPU and CPU. On a Tesla P100 (16 GB VRAM) in the Kaggle environment, the model processed each $3 \times 240 \times 240 \times 155$ voxel case in an average of 30.7 s. When run on CPU only (Intel® Xeon® @ 2.20 GHz), average time rose to 56.3 s per case. Despite the roughly 1.8× slowdown, the sub-minute CPU runtime and compact weight file underline the model's portability for low-resource settings. Table 4 summarizes the detailed timings for the five validation cases.

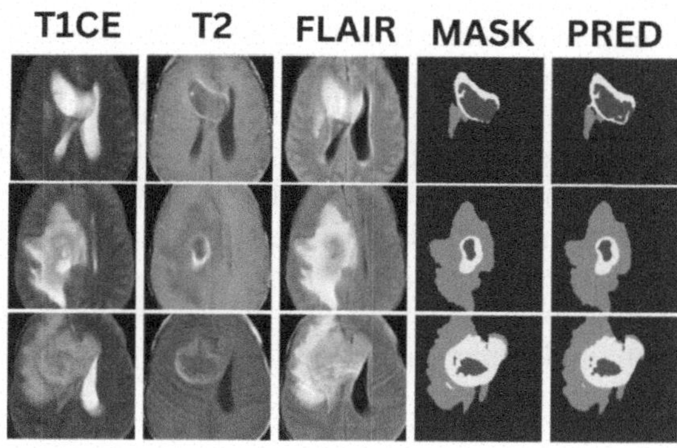

Fig. 6. Final segmentation results from BrainUNet fine-tuned on BraTS-Africa validation data.

Table 4. Inference time per case on GPU (Tesla P10) and CPU (Intel Xeon @ 2.20GHz)

Case ID	GPU (P100) Time (s)	CPU Time (s)
BraTS-SSA-00126-000	30.69	55.82
BraTS-SSA-00129-000	30.19	54.63
BraTS-SSA-00132-000	30.28	55.43
BraTS-SSA-00139-000	30.73	60.66
BraTS-SSA-00143-000	31.41	54.80
Average	**30.66**	**56.27**

4 Discussion and Conclusion

We introduced **BrainUNet**, a 3D U-Net variant enhanced with residual connections and attention mechanisms, tailored for glioma segmentation in sub-Saharan Africa (SSA). Using 5-fold cross-validation on the BraTS-Africa dataset, the model showed consistent improvements in Dice and IoU scores, though generalization to unseen samples remained challenging due to low resolution and imaging noise typical of SSA scans.

To address these limitations, we adopted a two-stage training strategy: pretraining on the large, high-quality BraTS 2021 dataset followed by fine-tuning on BraTS-Africa. This transfer learning approach improved segmentation performance, as validated on the Synapse platform using Dice scores and Hausdorff distances. While BraTS-Africa alone yielded reasonable results, fine-tuning substantially enhanced robustness against scanner variability and data artifacts.

In terms of deployment, BrainUNet demonstrated sub-minute inference times and compact memory size (91.4 MB), supporting feasibility in low-resource settings where GPU access is limited. These characteristics, combined with improved accuracy, highlight the model's potential for clinical integration in LMICs.

Nonetheless, several limitations remain. The small size of the BraTS-Africa dataset may restrict generalizability across diverse SSA populations and scanner types. Persistent MRI quality issues, including motion artifacts and resolution variability, continue to affect prediction consistency. Runtime was tested only on moderate hardware, and broader evaluation on low-spec devices is needed. Finally, external validation was limited to a single dataset; multi-center studies would strengthen robustness claims.

Future work should focus on enhancing interpretability, improving cross-dataset generalization, and exploring integration into real-time diagnostic workflows to maximize clinical impact.

Acknowledgments. The authors gratefully acknowledge the faculty and instructors of the SPARK Academy 2024 Summer School for their invaluable insights on deep learning in medical imaging, particularly brain tumors. We also thank the Digital Research Alliance of Canada for computational infrastructure, the McGill University Doctoral Internship Program for student exchange support, and the BrainUNet team for their guidance during model development. This work was supported by the Lacuna Fund for Health and Equity (PI: Udunna Anazodo, grant #0508-S-001), the Natural Sciences and Engineering Research Council of Canada (NSERC) Discovery Launch Supplement (PI: Udunna Anazodo, grant #DGECR-2022-00136), and partly by the RSNA Research & Education Foundation Derek Harwood-Nash International Education Scholar Grant (Investigators: Farouk Dako and Udunna Anazodo). This work was partly supported by the Italian Ministry of University and Research (MUR) under project PE0000013 - Future of Artificial Intelligence Research (FAIR).

Disclosure of Interests. The authors have no competing interests to declare that are relevant to the content of this article.

References

1. Abidin, Z.U., Naqvi, R.A., et al.: cc. Front. Bioeng. Biotechnol. **12** (Jul 2024). https://doi.org/10.3389/fbioe.2024.1392807, https://www.frontiersin.org/journals/bioengineering-and-biotechnology/articles/10.3389/fbioe.2024.1392807/full
2. Adewole, M., Rudie, J.D., et al.: The brain tumor segmentation (BraTS) challenge 2023: glioma segmentation in sub-saharan Africa patient population (BraTS-Africa) (2023). https://doi.org/10.48550/arXiv.2305.19369, arXiv:2305.19369 [physics]
3. Baid, U., Ghodasara, S., et al.: The RSNA-ASNR-MICCAI BraTS 2021 benchmark on brain tumor segmentation and radiogenomic classification (2021). https://doi.org/10.48550/arXiv.2107.02314, arXiv:2107.02314 [cs]
4. Barakat, M., Magdy, N., et al.: Towards SAMBA: segment anything model for brain tumor segmentation in sub-sharan african populations (2023). https://doi.org/10.48550/arXiv.2312.11775, arXiv:2312.11775 [cs, eess]
5. Hassanin, M., Anwar, S., Radwan, I., Khan, F.S., Mian, A.: Visual attention methods in deep learning: an in-depth survey (2024). https://doi.org/10.48550/arXiv.2204.07756, arXiv:2204.07756 [cs, eess]

6. Kamnitsas, K., Ledig, C., et al.: Efficient multi-scale 3D CNN with fully connected CRF for accurate brain lesion segmentation. Med. Image Anal. **36**, 61–78 (2017). https://doi.org/10.1016/j.media.2016.10.004, https://www.sciencedirect.com/science/article/pii/S1361841516301839
7. Kurbiel, T., Khaleghian, S.: Training of deep neural networks based on distance measures using RMSProp (2017). https://doi.org/10.48550/arXiv.1708.01911, arXiv:1708.01911 [cs, stat]
8. Mzoughi, H., Njeh, I., et al.: Deep multi-scale 3D convolutional neural network (CNN) for MRI gliomas brain tumor classification. J. Digit Imaging **33**(4), 903–915 (2020). https://doi.org/10.1007/s10278-020-00347-9, https://www.ncbi.nlm.nih.gov/pmc/articles/PMC7522155/
9. Nigatu, A.M., et al.: Medical imaging consultation practices and challenges at public hospitals in the Amhara regional state, Northwest Ethiopia: a descriptive phenomenological study. BMC Health Serv. Res. **23**, 787 (2023)
10. Ranjbarzadeh, R., Bagherian Kasgari, A., et al.: Brain tumor segmentation based on deep learning and an attention mechanism using MRI multi-modalities brain images. Sci. Rep. **11**, 10930 (2021). https://doi.org/10.1038/s41598-021-90428-8, https://www.ncbi.nlm.nih.gov/pmc/articles/PMC8149837/
11. Sahayam, S., Jayaraman, U.: Integrating edges into U-net models with explainable activation maps for brain tumor segmentation using MR images (2024). https://doi.org/10.48550/arXiv.2401.01303, arXiv:2401.01303 [cs, eess]
12. Sajjad, M., Khan, S., et al.: Multi-grade brain tumor classification using deep CNN with extensive data augmentation. J. Comput. Sci. **30**, 174–182 (2019). https://doi.org/10.1016/j.jocs.2018.12.003, https://www.sciencedirect.com/science/article/pii/S1877750318307385
13. Salehi, S.S.M., Erdogmus, D., Gholipour, A.: Tversky loss function for image segmentation using 3D fully convolutional deep networks (2017). https://doi.org/10.48550/arXiv.1706.05721, arXiv:1706.05721 [cs]
14. Tajbakhsh, N., Shin, J.Y., et al.: Convolutional neural networks for medical image analysis: full training or fine tuning? IEEE Trans. Med. Imaging **35**(5), 1299–1312 (2016). https://doi.org/10.1109/TMI.2016.2535302, arXiv:1706.00712 [cs]
15. Zhang, D., Confidence, R., Anazodo, U.: Stroke lesion segmentation from low-quality and few-shot MRIs via similarity-weighted self-ensembling framework. In: International Conference on Medical Image Computing and Computer-Assisted Intervention, pp. 87–96. Springer (2022)
16. Çiçek, î, Abdulkadir, A., Lienkamp, S.S., Brox, T., Ronneberger, O.: 3D U-net: learning dense volumetric segmentation from sparse annotation (2016). https://doi.org/10.48550/arXiv.1606.06650, arXiv:1606.06650 [cs]

TB Screening App: Smartphone Imaging of Tuberculin Skin Test Indurations for Latent Tuberculosis Screening in Low-Resource Settings

Ajibola S. Oladokun[1,2](✉), Bessie Malila[1,3], Yohhei Hamada[4], Watjana Lilaonitkul[5], Molebogeng X. Rangaka[4], and Tinashe E. M. Mutsvangwa[1,2]

[1] Division of Biomedical Engineering, University of Cape Town, Cape Town, South Africa
oldaji001@myuct.ac.za

[2] Data Science Department, IMT Atlantique, Brest, France

[3] Department of Electrical and Electronic Engineering Science, University of Johannesburg, Johannesburg, South Africa

[4] Institute for Global Health, University College London, London, UK

[5] Global Business School for Health, University College London, London, UK

Abstract. Latent tuberculosis infection (LTBI) is the precursor to active TB. Africa, South-east Asia, and Western-Pacific account for about 80% of global TB infections. The tuberculin skin test (TST) is the most common test for LTBI in these regions due to its low cost and simplicity. It involves the subdermal injection of tuberculin which may trigger the formation of an induration 48 to 72 h later. Patients often don't return to the clinic to read the induration diameter to determine LTBI diagnosis. Researchers in Cape Town, South Africa, developed the TB Screening App to enable automated LTBI diagnosis from home to prevent the need for secondary clinical visits. The app utilized a photogrammetry-based backend which did not yield reliable results. In this paper, we propose a vision transformer-based pipeline to replace the backend of the app and enable accurate LTBI diagnosis. We developed and tested the pipeline to classify 1,026 images collected in previous studies using the app. The pipeline enabled a mean balanced accuracy of 82.3% $\pm$ 2.9 for LTBI classification, and produced attention maps that visually correlate with a clinician's readings. This shows the potential of the pipeline to facilitate automated LTBI screening in low resource settings and eliminate the need for secondary clinical visits by patients.

Keywords: Tuberculin skin test · TST · Latent tuberculosis infection · LTBI · TB Screening App · Induration · Vision transformer · Mobile health · AI

1 Introduction

About two billion people globally are estimated to carry latent tuberculosis infection (LTBI) [1]. Caused by the pathogenic bacterium Mycobacterium tuberculosis (Mtb), LTBI is an asymptomatic and non-transmissible state where the host's immune system

U. Anazodo et al. (Eds.): MIRASOL 2025, LNCS 16398, pp. 206–216, 2026.
https://doi.org/10.1007/978-3-032-13654-1_21

contains Mtb replication below a symptomatic threshold [2]. It serves as a precursor and a major reservoir for the onset of active tuberculosis (TB) which is the biggest comorbidity for people living with HIV/AIDs [3]. With over 10 million new cases and 1.4 million TB-related deaths reported in 2020, Mtb remains the leading global cause of death from a single infectious agent [4]. Regions like Sub-Saharan Africa, Southeast Asia, and Western-Pacific bear the highest TB burden, accounting for approximately 80% of global TB infections [4, 5]. Notably, over 95% of TB cases and 99% of deaths occur in resource-limited settings [6]. The prevalence of LTBI in low-resource regions poses a significant challenge to the WHO's End TB strategy goals of reducing global TB incidence by 90% and deaths by 95% by 2035 [1].

Given that LTBI acts as a reservoir for future active TB, early detection and management are crucial components of TB control [7]. Diagnosing LTBI is becoming increasingly important as the World Health Organization (WHO) expands its target groups beyond those at highest risk—such as people living with HIV and children under five—to include others such as adult household contacts. LTBI testing enables targeted preventive treatment for individuals more likely to progress to active TB and provides an opportunity for education on maintaining health to prevent disease progression. Despite the availability of preventive TB treatment, low uptake of LTBI testing in Africa and other low-resource regions remain a significant challenge [8, 9].

Currently, LTBI is detected using a skin test, such as the Tuberculin Skin Test (TST), or the Interform Gamma Release Assay (IGRA) [10, 11]. The TST is the most commonly used test for LTBI in low- and middle-income countries (LMICs) due to its low cost and ability to be administered without the need for laboratory infrastructure [10]. On the other hand, the IGRA is a blood test which requires extensive laboratory infrastructure and skilled personnel, which makes it more expensive to administer [7]. While IGRA is more specific than TST—particularly because TST can yield false positives in individuals vaccinated with Bacillus Calmette–Guérin (BCG)—newer skin tests recently endorsed by the WHO offer similar accuracy to IGRA while retaining the operational advantages of the TST. The TST is administered by injecting tuberculin into the dermis of a patient's forearm. An induration (a subdermal wheal or lump) may form at the injection site 48 to 72 h after the test. Using the Mantoux reading method, a clinician measures the diameter of the induration and classifies patients with diameters above a consensus threshold as LTBI positive [7]. The requirement of patients to make secondary visits to the clinic for induration reading 48 to 72 h after initial injection leads to prevalence of invalid TST results LMICs as indurations are considered "expired" beyond 72 h. Patients cite time, transportation cost, and distance to the clinic as reasons for missing their secondary clinical visits [12]. Despite the low cost, and ease of administration of the TST and the newer skin tests, the requirement for a second visit to read the TST and newer skin tests limits their scale-up in LMICs.

Recent rise in the accessibility of smartphones and internet in LMICs inspired the development of the TB Screening App [13–15] by researchers in South Africa. The aim of the mobile app was to facilitate LTBI screening using a smartphone's camera. The goal of the app was to enable patients to capture their TST injection site and obtain LTBI diagnosis from home, 48 to 72 h after the initial test. This was a viable solution to increase the uptake of LTBI testing in LMICs. However, the backend diagnostic model of the app was plagued by some challenges [16]. In [13, 14, 16], the backend diagnostic model was implemented using structure-from-motion. This process involved capturing multiple images of the test site from multiple smartphone orientations to reconstruct a three-dimensional (3D) representation of the site, as shown in Fig. 1. The front end of the app was designed in [14, 15] to guide a patient on the phone orientations from which to capture the images before transmitting them to the diagnostic model, as shown in Fig. 1.

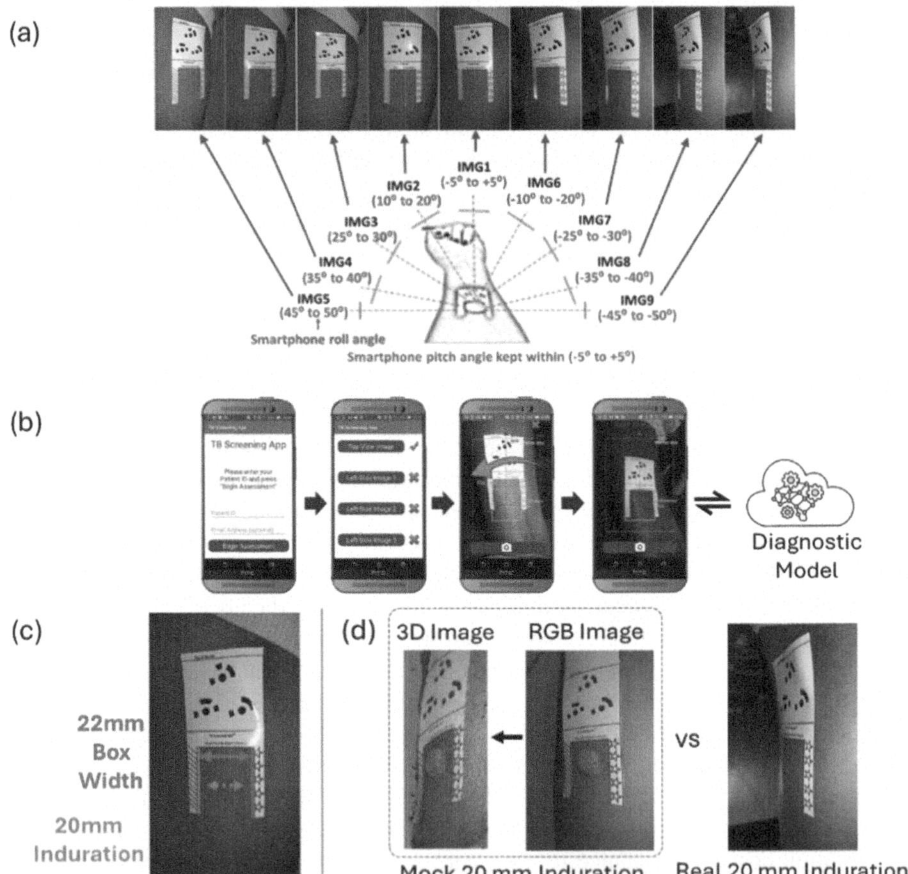

Fig. 1. (A) Imaging orientations of the TB Screening App, (b) the workflow of the app, (c) illustration of the sticker bounding box width, and (d) mock versus real induration comparison

Coded stickers were applied prior to imaging to provide fiducial markers across the images to facilitate the 3D reconstruction, as well as to provide a 22 mm bounding box around the site. Cosmetically generated mock indurations were utilized in [13, 14] to test the model and generate reliable 3D reconstructions from which digital diameter readings were obtained. However, the app was unable to generate good 3D reconstructions when tested on real TST indurations, as reported in [16]. This was due to the subtility of the geometric features of real indurations compared to the mock indurations.

Indurations form sub-dermally and their boundaries are difficult to visualize in dark-skinned participants [17, 18], which contributed to the failure of the 3D reconstruction approach. Due to the limited visual features of indurations, clinicians rely on tactile cues and palpations to identify induration boundaries during diameter readings [17]. Hyperspectral imaging (HSI) was proposed by [19] to capture subdermal features of indurations that are not accessible to smartphone cameras or the human eye. However, the limited availability of HSI precludes it from directly addressing the secondary visit problem. Thus, smartphone imaging of indurations remains a practical option.

About 1,026 images (nine images each from 114 participants in Cape Town, South Africa) of real TST sites were captured in [15, 16] using the app but the 3D reconstruction model failed to obtain consistent LTBI diagnosis [16]. The availability of such a unique dataset presents an opportunity to apply recent models, such as vision transformer models, to obtain LTBI diagnosis from the limited geometric features in the images – thereby facilitating the initial goal of the App to enable LTBI screening. The ability of these models to capture global relationships and long-range dependencies in an image suggests that they may capture subtle features in the TST images of dark-skinned participants. The attention maps of vision transformers offer an opportunity to visualize these subtle features and compare them to clinician diameter readings.

Thus, the contributions of this paper are: (1) the development of a vision transformer-based pipeline to classify the TST images of dark-skinned participants, (2) the identification of subtle image features that correlate with clinician readings using transformer attention maps, and (3) the provision of preliminary evidence of the validity of the TB Screening App to facilitate LTBI screening in a low-resource setting and mitigate the secondary clinical visit requirement for patients.

2 Methodology

The proposed pipeline for generating the LTBI classification and attention maps of the TST images, as shown in Fig. 2, is informed by the peculiarities of the images. The images were collected in [16] under the ethics approval number HREC 319/2018 granted by the University of Cape Town human research ethics committee. All the images had coded stickers around the skin test site. There is a possibility that a classification model may attempt to learn features from the sticker rather than only the reaction site. Thus, it is desired to segment the skin reaction site, which is enclosed within the sticker, and feed only the reaction site to the classification model. The uniqueness of the sticker makes it relatively easy to detect its position in an image. YOLO-v11 [20], one of the most recent models for real-time object detection, is utilized in the first stage of the pipeline to predict the coordinates for the bounding box and bounding polygon that encapsulates the sticker in each image. The bounding box enables cropping a rectangular region around the sticker while the polygon enables a close-fitting segmentation of the sticker and its embedded TST reaction site.

We annotated the sticker position in 207 (nine images from 23 subjects) out of the 1,026 images using polygon vertex points. We finetuned YOLO-V11 using these 207

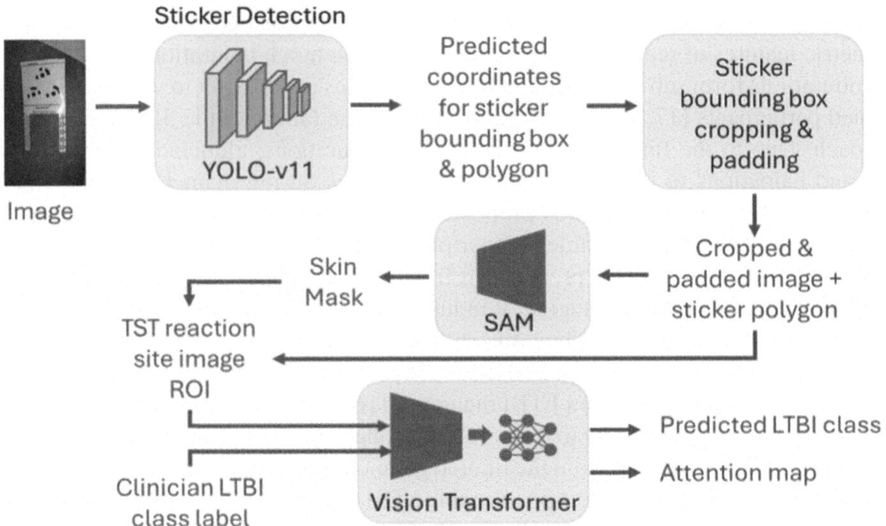

Fig. 2. Schematic of the vision transformer-based proposed pipeline

annotated images (153 for training and 54 for validation) to learn how to recognize the sticker in the nine images per subject. An input TST image is then cropped to the coordinates of the bounding box and padded, with black pixels, on the shorter box side to make the cropped image shape a square. At the next stage of the pipeline, the cropped image and the predicted polygon coordinates are fed to an instance of the segment anything model (SAM) [21] for zero-shot segmentation of the embedded skin site from the enclosing sticker. The SAM, pretrained on 11 million natural images and 1.1 billion masks by its authors, can segment objects in an image, without finetuning or training, by only using a few points that belong to the object of interest. The robustness of SAM suggests it can separate the TST reaction site, of varying skin tones, from the sticker. The predicted polygons from YOLO were such that their center always falls within the reaction site. Three points are randomly sampled near the center of the polygon and fed as point prompts to SAM. This prompts SAM to segment the object (TST reaction site) that produced the points. The output of the SAM stage is an image comprising only the skin region of interest (ROI) of the TST reaction site. This is such that the width of the region corresponds to the 22 mm width of the inner bounding box of the sticker. This enables the correlation of the attention maps to real-world measurement in mm. At the final stage of the pipeline, the TST reaction site image is fed to a classification model. The model generates a class prediction which is compared to a clinician's LTBI class label (LTBI positive (induration $\geq$ 10 mm) or negative ($<$ 10 mm)) during training.

Three vision transformer models: ViT [22], PVTv2 [23], and SwiftFormer [24], were pretrained on ImageNet-21k [25] and finetuned at the final stage of the pipeline. The 1,026 TST images were split into the ratio 60:20:20 for training:validation:test subsets. The split was done by subject, such that all nine images belonging to a subject are in the same split, to prevent data leakage between splits. The validation subset was utilized to implement early stopping [26] during training. The percentage of all images in the

LTBI positive class was approximately 33%. To minimize the effect of class imbalance, data augmentation was performed on the positive class using random color jitter within a narrow range of perturbations (± 0.2). Geometric transformations such as flipping or rotation were avoided as they might alter the implicit orientations captured by the smartphone camera. All 1,026 images were captured using the HTC One 8 smartphone [27]. A small in-house dataset of 38 overhead images of TST reaction sites of 38 dark skinned subjects, captured using the RGB viewfinder camera of Specim IQ [28], were included only during training to introduce some camera variability.

3 Results

The proposed pipeline was applied to all 1,026 images captured by the TB Screening App and Fig. 3 shows the outputs of the first three stages of the pipeline. Figure 3(a) shows the nine orientation images of a subject, which are inputs into the pipeline, as captured by the app. Figure 3(b) shows the output of the sticker detection stage with the predicted bounding boxes and polygons overlaid on the input images. The mean average precision of YOLO-v11 in predicting the sticker polygons of the validation subset at an IoU threshold of 0.5 (mAP@0.5) [29] was 0.995, and the mAP@0.5–0.95 [29] was 0.98351 for the IoU range of 0.5 to 0.95. This shows accurate detection of the sticker. The output of the sticker bounding box cropping and padding are as shown in Fig. 3(c). Guided by the polygons from YOLO-v11, the zero-shot segmentation masks predicted by SAM for the TST reaction site are as shown in Fig. 3(d). Comparing the SAM masks with expert-annotated masks for the reaction site for 54 images reveals a mean IoU of 0.98. This highlights the ability of SAM to perform segmentation of the test site without training or finetuning on TST images. Figure 3(e) shows the segmented TST reaction site for each image orientation after the application of the SAM masks. These TST reaction site images were fed to the classification models to produce the results in Table 1. After training the models, bootstrap resampling with replacement [30] was performed on the test dataset for 1,000 iterations. In each iteration, the balanced accuracy [31] was computed on the resampled test dataset to account for class imbalance. The mean and the standard deviation of the balanced accuracies for each model is as reported in Table 1. The ViT model achieved the highest accuracy of 82.3% ± 2.9. The attention maps of the trained ViT model for selected TST images are as shown inFigs. 4, 5 and 6.

Fig. 3. The nine orientation images of a subject showing (a) original images, (b) outputs of the sticker detection stage, (c) sticker region cropping and padding, (d) SAM zero-shot mask prediction for reaction sites, and (e) reaction site segmentation

Table 1. Bootstrap Performance of the Vision Transformer Models on Test Subset

	ViT	PVTv2	SwiftFormer
Balanced Accuracy	**82.3% ± 2.9**	73.7% ± 3.2	79.2% ± 2.9
Sensitivity	73.5% ± 5.3	65.3% ± 5.5	**76.4% ± 4.9**
Specificity	**91.1% ± 2.3**	82.1% ± 3.3	82.1% ± 3.2
ROC-AUC	**82.3% ± 2.9**	73.7% ± 3.2	79.2% ± 2.9

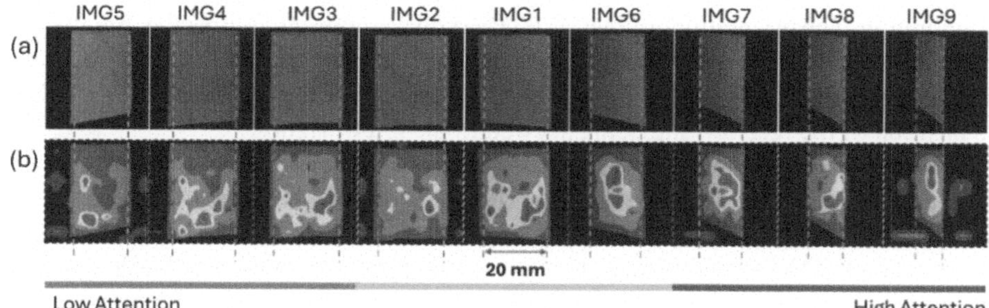

Fig. 4. Visualization of the attention map of the fourth attention head of the last attention layer of ViT for the nine TST image orientations

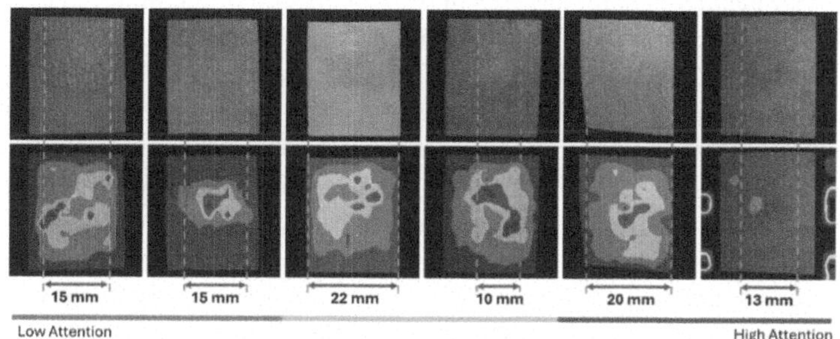

Fig. 5. Comparing the attention map of ViT's fourth attention head with the clinician diameter for some LTBI positive overhead (IMG1) orientation images in the test set

Fig. 6. Visualizing the attention map of ViT's 12th attention head for some LTBI negative overhead (IMG1) orientation images in the test set

4 Discussion

The fourth attention head of the final attention layer of ViT generated maps that appear to visually correlate to the induration diameter reading of a clinician. This suggests that the model was able to learn subtle features in the images that define the presence and extent of an induration. Visualizing the attention maps for the nine orientation images of a subject, as shown in Fig. 5, depicts that the model pays attention to some common regions across orientations (particularly IMG1 to IMG5 and IMG6 to IMG9). The regions of significant attention in the overhead images (IMG1) appear to correlate with the extent of an induration. The orientations of IMG2 to IMG5 do not appear to mirror the orientations of IMG6 to IMG9, despite the smartphone orientations being symmetric during imaging. This is likely due to the curvature of the ventral side of the forearm which may result in the position of the reaction site being skewed to the left or right.

The 13 mm induration image in Fig. 6 is an example of a case where the model wrongly classifies a positive LTBI image. The attention map here shows that the model was unable to recognize some induration features. For LTBI negative images, the 12th attention head of ViT generate maps that appear to suggest that this head has learned the characteristics of images with no significant induration. The image and corresponding attention in the last column of Fig. 7 is an example of a case where the model wrongly classified the image as LTBI positive when it should be negative. The attention map showing high focus on the background for this image highlights the failure of the model to identify LTBI negative characteristics in the foreground of the image. The generation of attention maps for TST indurations to offer explainability of model decision, sets this paper apart from studies such as [32–34] which attempt to directly segment the boundaries of indurations from RGB images. All three studies generated ground truths by annotating regions of an induration that is visually observable in an RGB image and training a model to generate a comparable segmentation mask. With this approach a model may only learn induration features that are easily visible in the image and ignore the subtle ones. Conversely, we focus on utilizing the more clinically valid Mantoux reading labels to classify TST images and utilize attention maps to show qualitative evidence of trustworthiness of the classification. Furthermore, we utilized the largest dataset, to date, of real RGB images of TST reaction sites. Conversely, the studies [32–34] utilized 66, 26, and 56 TST RGB images respectively, which suggests that our findings may be more reliable.

5 Conclusion

This paper shows that the proposed vision transformer-based pipeline for the backend of the TB Screening App is capable of consistently detecting the coordinates of the App's sticker in an image and segment the TST reaction site embedded within the sticker. The classification accuracy achieved suggests that implementing the pipeline as the backend of the app could enable patients to obtain around 82.3% accuracy of diagnostic prediction when they use the app. The provisioning of the attention map overlay by the pipeline could give more confidence to a patient that the model is focusing on the right areas of the reaction site and that the width of the attention cluster correlates with the predicted induration classification. A limitation of this paper is the limited number of unique subjects (114) in the dataset. Future work will include the acquisition of more TST images from more diverse populations and using other smartphones to improve the robustness of the classification model. The results of the application of this pipeline on the TST induration images of African subjects serve as preliminary evidence of the validity of the TB Screening App to facilitate automated LTBI screening in a low-resource setting and to potentially reduce follow-up visits.

Acknowledgments. This document has been produced with the financial assistance of the European Union (EU) and co-funding from the Carnegie Foundation of New York (Grant no.

DCI-PANAF/2020/420–028), through the African Research Initiative for Scientific Excellence (ARISE), pilot programme. ARISE is implemented by the African Academy of Sciences (AAS) with support from the European Commission and the African Union Commission (EUC). The statements made and views expressed are solely the responsibility of the authors. The research is also supported by the University of Cape Town International Scholarship, and the South African Research Chairs Initiative (SARChI) of the NRF and the Department of Science and Technology (grant no 98788). Ajibola S. Oladokun was supported by the Institut Carnot Télécom & Société numérique.

Disclosure of Interests. The authors have no competing interests to declare that are relevant to the content of this article.

References

1. Sterling, T.R., et al.: Guidelines for the treatment of latent tuberculosis infection: recommendations from the national tuberculosis controllers association and CDC, 2020. Am. J. Transplant. **20**(4), 1196–1206 (2020)
2. Mir, M.A., et al.: Manipulation and exploitation of host immune system by pathogenic Mycobacterium tuberculosis for its advantage. Fut. Microbiol. **17**, 1171–1198 (2022)
3. Houben, R.M.G.J., Dodd, P.J.: The global burden of latent tuberculosis infection: a re-estimation using mathematical modelling. PLoS Med. **13**(10), e1002152 (2016)
4. Gong, W. and X. Wu, Differential diagnosis of latent tuberculosis infection and active tuberculosis: a key to a successful tuberculosis control strategy. Front. Microbiol. **12** (2021)
5. Mohajan, H., Tuberculosis is a fatal disease among some developing countries of the world (2014)
6. Ellner, J.J.: 332 - Tuberculosis, in Goldman's Cecil Medicine (Twenty Fourth Edition). In: Goldman, L., Schafer, A.I., (eds.) 2012, W.B. Saunders: Philadelphia. pp. 1939-1948 (2022)
7. Kiazyk, S., Ball, T.B.: Latent tuberculosis infection: An overview. Can. Commun. Dis. Rep. **43**(3–4), 62–66 (2017)
8. Kota, N.T., et al.: The global expansion of LTBI screening and treatment programs: exploring gaps in the supporting economic evidence. Pathogens **12**(3), 500 (2023)
9. Magwaza, C., et al.: Knowledge and prevalence of latent tuberculosis infection: a feasibility and pilot study in a primary healthcare setting in rural Eastern Cape, South Africa. Int. J. Environ. Res. Public Health **22**(3), 320 (2025)
10. Carranza, C., et al.: Diagnosis for latent tuberculosis infection: new alternatives. Front. Immunol. **11** (2020)
11. Gualano, G., et al.: Tuberculin skin test – Outdated or still useful for Latent TB infection screening? Int. J. Infect. Dis. **80**, S20–S22 (2019)
12. Oliveira, S.M.d.V.L.d., et al.: Tuberculin skin test: operational research in the state of Mato Grosso do Sul, Brazil. Jornal Brasileiro de Pneumologia **37**, 646–654 (2011)
13. Dendere, R., et al.: Measurement of skin induration size using smartphone images and photogrammetric reconstruction: pilot study. JMIR Biomed Eng **2**(1), e3 (2017)
14. Naraghi, S., et al.: Mobile phone-based evaluation of latent tuberculosis infection: Proof of concept for an integrated image capture and analysis system. Comput. Biol. Med. **98**, 76–84 (2018)
15. Farao, J., et al.: A user-centred design framework for mHealth. PLoS ONE **15**(8), e0237910 (2020)
16. Maclean, S.: Image analysis for a mobile phone-based assessment of latent tuberculosis infection. University of Cape Town (2020)

17. Zhang, H., et al.: Induration or erythema diameter not less than 5 mm as results of recombinant fusion protein ESAT6-CFP10 skin test for detecting M. tuberculosis infection. BMC Infect Dis. **20**(1), 685 (2020)
18. Chan, A.H.Y., et al.: Qualitative evidence for the use of Mycobacterium tuberculosis antigen-based skin tests. WHO consolidated guidelines on tuberculosis: Module 3: Diagnosis–Tests for tuberculosis infection (2022)
19. Oladokun, A.S., et al.: SpeChrOmics: A Biomarker Characterization Framework for Medical Hyperspectral Imaging. In: International Conference on Medical Image Computing and Computer-Assisted Intervention, pp. 745–756. Springer (2024)
20. Hidayatullah, P., et al.: YOLOv8 to YOLO11: a comprehensive architecture in-depth comparative review. arXiv preprint arXiv:2501.13400 (2025)
21. Kirillov, A., et al. Segment anything. in Proceedings of the IEEE/CVF international conference on computer vision. (2023)
22. Dosovitskiy, A., et al., An image is worth 16x16 words: Transformers for image recognition at scale. arXiv preprint arXiv:2010.11929, 2020
23. Wang, W., et al.: Pvt v2: Improved baselines with pyramid vision transformer. Computational visual media **8**(3), 415–424 (2022)
24. Shaker, A., et al. Swiftformer: Efficient additive attention for transformer-based real-time mobile vision applications. In: Proceedings of the IEEE/CVF International Conference on Computer Vision (2023)
25. Ridnik, T., et al.: Imagenet-21k pretraining for the masses. arXiv preprint arXiv:2104.10972 (2021)
26. Prechelt, L.: Early stopping-but when? In: Neural Networks: Tricks of the trade, pp. 55–69. Springer (2002)
27. Hughes, B., HTC One (M8) for Dummies. John Wiley & Sons (2014)
28. Behmann, J., et al.: Specim IQ: evaluation of a new, miniaturized handheld hyperspectral camera and its application for plant phenotyping and disease detection. Sensors **18**(2), 441 (2018)
29. Henderson, P., Ferrari, V.: End-to-end training of object class detectors for mean average precision. In: Computer Vision–ACCV 2016: 13th Asian Conference on Computer Vision, Taipei, Taiwan, November 20–24, 2016, Revised Selected Papers, Part V 13. Springer, Cham (2017)
30. Dixon, P.M.: Bootstrap resampling. Encyclopedia of environmetrics (2006)
31. Brodersen, K.H., et al.: The balanced accuracy and its posterior distribution. In: 2010 20th International Conference on Pattern Recognition, IEEE (2010)
32. Zang, J., et al.: Tuberculin skin test result detection method based on CSN-II and improved OTSU method. Measurement **229**, 114409 (2024)
33. Parihar, G., et al.: Measurement of skin test wheals using image segmentation approaches. In: 2021 IEEE 6th International Conference on Computing, Communication and Automation (ICCCA) (2021)
34. Akinola, O.A., et al.: Analysis and measurement of tuberculin skin test induration using deep neural network. Int. J. Online Biomed. Eng. **20**(12) (2024)

Empowering Medical Equipment Sustainability in Low-Resource Settings: An AI-Powered Diagnostic and Support Platform for Biomedical Technicians

Bernes Lorier Atabonfack[1], Ahmed Tahiru Issah[1], Mohammed Hardi Abdul Baaki[1], Clemence Ingabire[1], Tolulope Olusuyi[2], Maruf Adewole[2,3], Udunna C. Anazodo[2,4], and Timothy X. Brown[1](✉)

[1] Carnegie Mellon University Africa, Kigali, Rwanda
{batabonf,aissah,mabdulba,cingabir}@alumni.cmu.edu, timxb@cmu.edu
[2] Medical Artificial Intelligence Laboratory, Lagos, Nigeria
{tolusuyi,madewole}@mailab.io
[3] University of Pennsylvania, Philadelphia, Pennsylvania, USA
[4] McGill University, Montréal, Canada
udunna.anazodo@mcgill.ca

Abstract. In low- and middle-income countries (LMICs), a significant proportion of medical diagnostic equipment remains underutilized or non-functional due to a lack of timely maintenance, limited access to technical expertise, and minimal support from manufacturers, particularly for devices acquired through third-party vendors or donations. This challenge contributes to increased equipment downtime, delayed diagnoses, and compromised patient care. This research explores the development and validation of an AI-powered support platform designed to assist biomedical technicians in diagnosing and repairing medical devices in real-time. The system integrates a large language model (LLM) with a user-friendly web interface, enabling imaging technologists/radiographers and biomedical technicians to input error codes or device symptoms and receive accurate, step-by-step troubleshooting guidance. A development road map was created to include additional features and capabilities. A proof of concept was developed using the Philips HDI 5000 ultrasound machine, achieving 100% precision in error code interpretation and 80% accuracy in suggesting corrective actions. This study demonstrates the feasibility and potential of AI-driven systems to support medical device maintenance, with the aim of reducing equipment downtime to improve healthcare delivery in resource-constrained environments.

B. L. Atabonfack and A. T. Issah—These authors contributed equally as first authors.
M. H. A. Baaki and C. Ingabire—These authors contributed equally as secondary authors.

Supplementary Information The online version contains supplementary material available at https://doi.org/10.1007/978-3-032-13654-1_22.

U. Anazodo et al. (Eds.): MIRASOL 2025, LNCS 16398, pp. 217–230, 2026.
https://doi.org/10.1007/978-3-032-13654-1_22

1 Introduction and Background

Medical devices are indispensable for delivering quality healthcare services, encompassing diagnosis, treatment, and patient monitoring. In high-resource settings, these devices benefit from regular maintenance, trained operators, and manufacturer support. Conversely, in low- and middle-income countries (LMICs), the situation is markedly different. Studies estimate that 40% to 70% of medical equipment in LMICs is non-functional or underutilized at any given time due to factors like poor maintenance systems, lack of spare parts, and insufficiently trained personnel [1–4]. For instance, a detailed case study in Tanzania found that 30 - 50% of medical equipment in sub-Saharan Africa experiences downtime because of unstructured maintenance practices and limited technician capacity [5]. In Uganda, research conducted across nine tertiary hospitals and five research institutions revealed that 34% of medical devices were faulty, and 85.6% lacked operational manuals, making even minor issues difficult to resolve [6]. The lifespan of usable devices is often reduced by as much as 80%, primarily due to inexperienced operators and a lack of preventive maintenance protocols [7]. The World Health Organization estimates that up to 70% of donated equipment fails to function as intended due to mismatches with infrastructure or lack of user training [8]. Moreover, 40% of devices donated by Medical Surplus Recovery Organizations (MSROs) are inoperable due to missing documentation or limited manufacturer support [9,10].

The shortage of trained biomedical engineering technicians (BMETs) further aggravates the issue. With only a few organizations globally offering formal BMET training (e.g., Engineering World Health, Medisend) [9], many LMICs lack the workforce needed to keep devices operational. Additionally, equipment donated or procured through third-party vendors often comes without access to manufacturer service forums or troubleshooting support, leaving technicians isolated and ill-equipped. In many African hospitals and clinics, even minor technical issues such as unfamiliar error codes or calibration warnings can result in devices being taken offline for extended periods. This leads to significant delays in diagnostics, unnecessary patient referrals, and, in some cases, avoidable mortality. While digital health innovations have improved access to care and information, very few have directly addressed the operational gap in maintaining physical medical infrastructure. This study aims to bridge that gap through the design and implementation of an AI-powered medical device support system (denoted INGENZI Tech) for diagnostic imaging devices. By integrating advanced natural language processing with structured manuals, log data, and peer-generated insights, the system empowers biomedical technicians to act with confidence. The goal of this project is to reduce downtime, improve technician efficiency, and enhance healthcare system reliability, starting with a validated use case on the Philips HDI 5000 ultrasound machine in East Africa. This paper makes a dual contribution: first, by presenting a conceptual design framework for an AI-powered biomedical support platform tailored to LMIC contexts, and second, by demonstrating a proof-of-concept prototype using ultrasound repair as an initial case study.

2 Related Work and Differentiation

While several AI-driven platforms exist globally for medical device management, most are tied to proprietary hardware, operate exclusively online, or target high-resource healthcare systems. We conducted a landscape review through targeted web searches of academic institutions, organizations, and companies, drawing information from their websites and published materials. Table 1 provides a comparative overview of notable commercial solutions and positions INGENZI Tech within this landscape.

In parallel, academic research has made significant progress in predictive maintenance for medical equipment. For instance, Shamayleh et al. [11] deployed an IoT-based maintenance system across over 8,000 medical devices in Malaysia, achieving a 25% cost reduction using SVM-based failure classification. Similarly, Zamzam et al. [12] achieved 99.4% prediction accuracy using sensor fusion methods applied to over 13,000 medical assets. Mohamed et al. [13] proposed a digital twin approach for MRI equipment, reducing machine downtime by 20%. Deep learning approaches, such as CNN-LSTM hybrid models, have shown 92 - 99% accuracy in the estimation of remaining useful life across several medical device categories [14,15]. Luschi et al. [16] further demonstrated how predictive analytics integrated with IoT positioning systems can support hospital maintenance workflows in structured environments.

Despite this progress, such systems typically require continuous data streams, sensor infrastructure, and robust connectivity, conditions that are often absent in LMIC settings. Research by Anazodo et al. [4] and Diaconu et al. [1] highlights the infrastructural barriers to medical equipment usage in LMICs, including poor internet access, inconsistent power supply, and lack of Original Equipment Manufacturer (OEM) support, particularly for donated devices.

In terms of knowledge management, Abidi [17] and Tabrizi & Morgan [18] emphasize the need for collaborative technical knowledge platforms in healthcare. While prior work has demonstrated the effectiveness of training platforms [19], few systems combine this with real-time, context-aware AI guidance for medical equipment.

INGENZI Tech differentiates itself through its integrated design tailored for LMICs. Unlike systems that focus solely on predictive analytics [12,13] or manufacturer-side service workflows [20], INGENZI Tech offers an end-user diagnostic platform combining:

- A multilingual, LLM-powered chatbot for step-by-step device guidance.
- A RAG-based architecture with segmented vector stores to reduce hallucinations and enhance contextual precision [21,22].
- Offline-first deployment suitable for low-bandwidth and rural clinics.
- A peer-to-peer technician forum for continuous feedback, local adaptation, and model improvement [17,18].

The Phase 1 proof-of-concept targets the Philips HDI 5000 ultrasound, a device commonly found in African clinics but frequently unsupported, addressing the challenges raised in recent surveys on equipment non-functionality [2,6]. Future phases include expanding to CT, MRI, and X-ray machines.

Table 1. Comparative Landscape of Existing AI-Powered Medical Maintenance Solutions.

AI Solution	Strengths	Limitations in LMIC Context / Differentiation vs. INGENZI Tech
Bruviti	LLM-powered diagnostics; preserves institutional knowledge; supports global manufacturers [23]	No public deployments in LMICs; not optimized for offline use or multilingual environments; high dependency on OEM integrations
Hadleigh Health [24], Vestfrost EMS [25], Nexleaf Analytics [26]	IoT-based predictive maintenance; real-time monitoring; some LMIC deployments	Primarily hardware-embedded systems; limited end-user diagnostic interaction; lacks LLM or multilingual troubleshooting support
Bosnia AI-Metrology Project	Applies AI to device calibration and performance in public hospitals; shows feasibility in low-resource settings [27]	Regional, narrow scope; focused on measurement and standards, no interactive technician support or documentation retrieval
Circuitry.ai	Real-time advisory systems for maintenance; global operations; multilingual support [28]	Tailored for manufacturers and enterprise clients; unclear adaptation to LMIC infrastructure or offline-first needs
Stellarix [29], Kodexo Labs [30], Toronto Digital [31]	Offer AI-powered predictive maintenance with IoT and ML; adaptable for various device types	Require continuous data streams and sensor integrations; do not provide peer-to-peer learning or AI chatbot interactions
GE Healthcare Edison AI	Advanced anomaly detection in imaging systems; global footprint [32]	High-end solution tied to GE devices; closed ecosystem; limited accessibility or relevance for third-party equipment in LMICs
INGENZI Tech (This Work)	LLM-based multilingual chatbot; RAG-based contextual retrieval; peer-driven technician forum; designed for offline use	Currently piloting; early-stage for multi-device support; open architecture enables adaptation across vendors and regions

3 Scope of the Study

This study focuses on the development and validation of an AI-powered support platform designed to assist biomedical technicians in diagnosing and repairing medical diagnostic equipment in low-resource settings. The platform integrates a large language model (LLM) with structured device knowledge, error logs, and technician-driven feedback to offer step-by-step troubleshooting and guided repair workflows. The initial scope centers on the Philips HDI 5000 ultrasound machine, which served as a proof-of-concept device for prototyping. With the framework developed in the initial phase, a full study will be conducted to expand the focus to include more complex imaging devices such as magnetic resonance imaging (MRI), computed tomography (CT), and X-ray machines, with a specific emphasis on widely used and often unsupported devices in LMIC settings. Therefore, this study aims to:

- Build a scalable, multilingual AI agent that provides real-time, context-aware repair guidance.

- Integrate a global technician knowledge-sharing forum to complement automated responses.
- Test the tool's accuracy, relevance, and usability in both clinical and training environments.
- Evaluate its adaptability across different medical device types and manufacturers.

This study does not include physical repair execution, device manufacturing processes, or regulatory approval for device use, which are currently being investigated. Instead, it aims to establish a digitalized and generalizable support framework that can bridge the technical knowledge gap for biomedical equipment across a variety of healthcare settings, with a focus on medical imaging devices. While this Phase 1 prototype focused on technical validation, future phases of the project will directly involve biomedical technicians and LMIC healthcare institutions through co-design workshops, pilot testing, and structured feedback to ensure user-centered development and real-world applicability.

4 Development Road Map

Our goal is to develop a scalable medical device maintenance solution for resource-constrained environments that draws upon multiple avenues of support. The development is being carried out in phases, outlined briefly in the road map below, with the initial proof-of-concept (Phase 1) detailed in later sections (Fig. 1).

4.1 Phase 1: Proof of Concept (Completed)

The initial prototype was developed using the Philips HDI 5000 ultrasound machine as a case study. The assistant was built using a RAG (Retrieval-Augmented Generation) architecture supported by:

- **LLM:** OpenAI's GPT-3.5 Turbo for generation and embedding
- **Vector store:** ChromaDB for document retrieval
- **Frontend/Backend:** React and Flask
- **Core functionalities:** Error code lookup, Log analysis, Self-test simulation, Maintenance scheduling via Google Calendar API, Multilingual chatbot interface

To improve retrieval accuracy and reduce hallucinations, separate vector databases were used for general device documentation, error code lookup and diagnostics, maintenance and cleaning procedures, and operational history tracking. The proof-of-concept achieved **100%** precision in error code matching and **80%** troubleshooting accuracy using internal evaluation methods.

4.2 Phase 2: Forum Integration and Feedback Loop

This phase introduces a collaborative technician forum to complement the AI assistant. The forum will allow users to post issues, share solutions, upvote helpful responses, and generate valuable context data.

Development Tasks:

Build or integrate a forum engine

- Link forum user accounts with diagnostic tool profiles
- Design a schema for tracking usage and technician feedback
- Connect forum feedback into a structured pipeline for LLM fine-tuning and retrieval augmentation

This human-in-the-loop design ensures continuous model improvement and crowd-sourced domain knowledge.

4.3 Phase 3: API and IoT Connectivity for Device Integration

The third development phase focuses on real-time interaction with physical devices, enabling automated log streaming and predictive diagnostics.

Planned Tasks:

- Evaluate and implement healthcare communication protocols such as DICOM, HL7, and MQTT.
- Define secure API contracts for medical device integration.
- Build a simulation layer for testing log ingestion without live devices.
- Prototype fault-alert dashboards for remote monitoring.

This phase extends the system beyond a passive assistant into a proactive maintenance layer. Once connections are successfully established, retrieved device logs will also be incorporated into the knowledge base to improve model accuracy and reliability further.

4.4 Phase 4: Model Optimization and Continuous Learning

With data from the tool and forum, the LLM will be fine-tuned for domain-specific reasoning and reliability. This phase will also integrate external biomedical engineering forums (e.g., MedWrench) into the knowledge base, enriching troubleshooting capabilities with real-world technician experiences. In addition, this phase will include fine-tuning an open-source Llama 3 model to enable fully offline functionality, ensuring the platform remains usable in bandwidth-limited environments common to LMIC settings.

Optimization Activities:

- Introduce a feedback classifier to label suggestions as correct/incorrect.
- Use forum responses and real-world interactions for semi-supervised learning.
- Monitor hallucination frequency and optimize vector store indexing.
- Regularly retrain and evaluate the LLM against gold-standard troubleshooting guides.

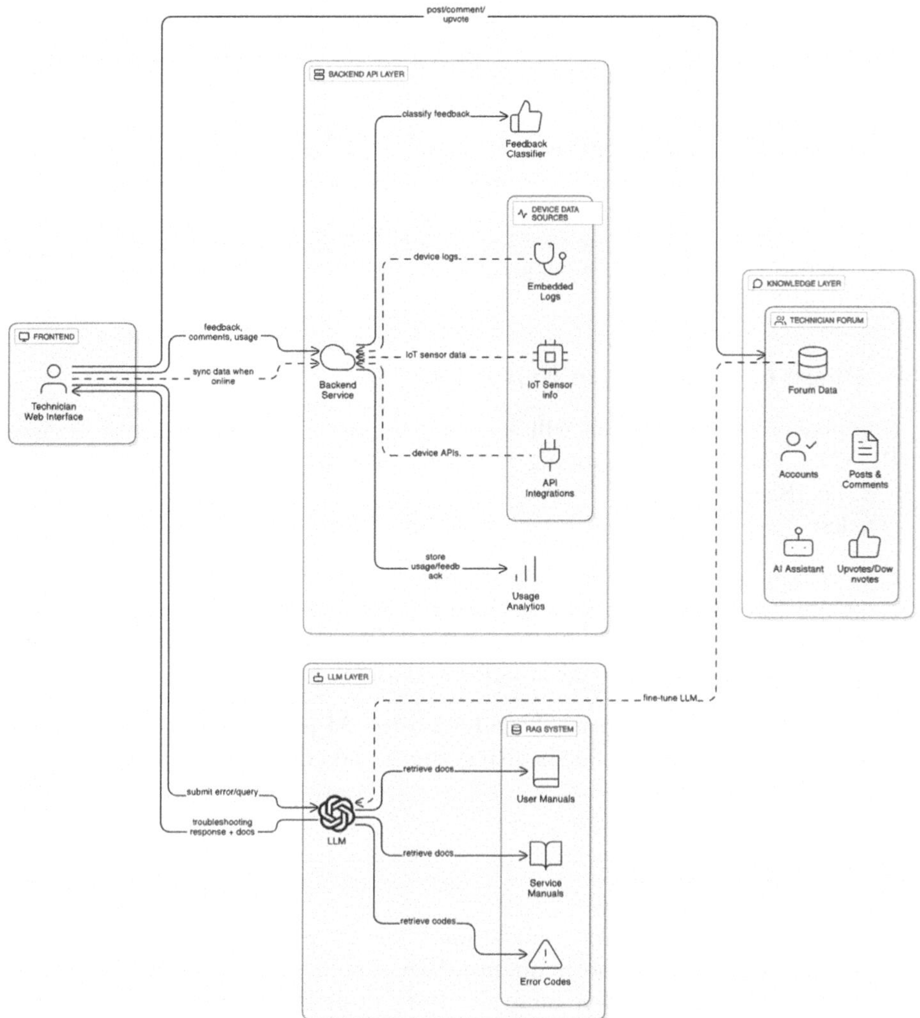

Fig. 1. INGENZI Tech System Architecture: A high-level overview of the INGENZI Tech platform, showing how technicians interact with a multilingual, offline-capable AI assistant powered by a Retrieval-Augmented Generation (RAG) pipeline, segmented vector stores, and a peer-support forum for real-time troubleshooting and continuous learning.

4.5 Phase 5: Pilot Deployment and Evaluation

The system will be deployed and evaluated in clinical environments in LMICs, starting with East Africa.

Execution Plan:

- Partner with hospitals and biomedical schools for pilot testing

- Train local technicians on usage and provide onboarding materials
- Collect metrics:
  - Time-to-diagnosis
  - Technician satisfaction
  - Device uptime improvement
- Use real feedback to further optimize UI, model performance, and workflow integration

All user interactions will be logged and anonymized for research analysis. No patient data will be used.

4.6 Phase 6: Multi-device Expansion

In this phase, INGENZI Tech will scale to support a broader range of devices (MRI, CT, and X-ray), beginning with one original equipment manufacturer (OEM, i.e., Siemens).

Tasks:

- Collect manuals, error codes, and logs for new devices
- Expand vector stores or segment them by device class
- Add model-switching capabilities in the backend based on device detection
- Standardize input/output formats for scalable ingestion

This positions the system as a brand-agnostic, AI-powered maintenance layer that can operate across multiple health systems and equipment ecosystems (Fig. 2).

5 Research Design

This project follows a phased, mixed-methods approach, combining iterative system development with both quantitative system performance evaluations and qualitative user feedback analysis. It is designed to be descriptive and exploratory, establishing the foundational tools, workflows, and interactions needed to support scalable medical device maintenance via AI, particularly in resource-limited healthcare environments. Here we focus on Phase 1.

5.1 Phase 1 Dataset

The dataset used in Phase 1 was built from 15 technical documents related to the Philips HDI 5000 ultrasound system, including user manuals, service manuals, and detailed error code catalogs. These documents were parsed and chunked using a semantic-aware splitting strategy to preserve contextual meaning. Each chunk was then embedded using OpenAI's GPT-3.5 Turbo embedding model and stored in ChromaDB, a high-performance vector database. ChromaDB was selected due to its simplicity and ease of deployment, quick setup, and comprehensive retrieval features including vector search, document storage, and

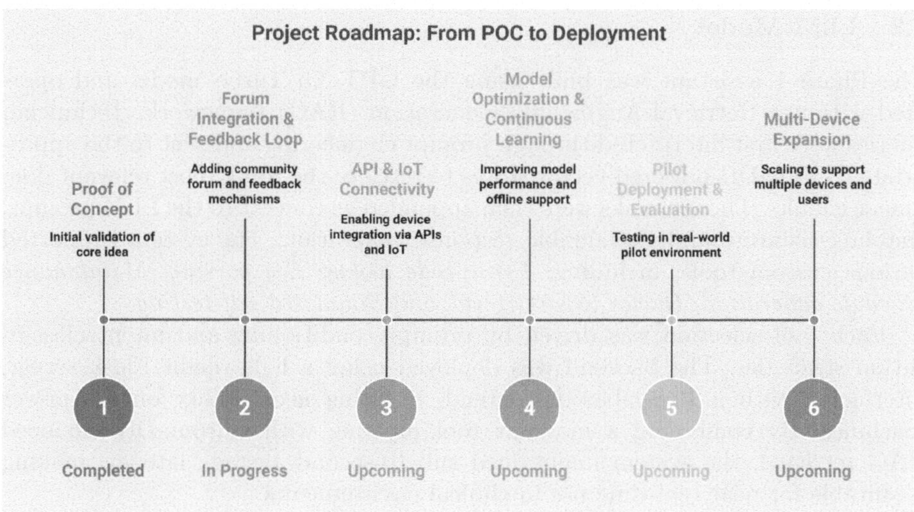

Fig. 2. High-level roadmap of the INGENZI Tech project phases, illustrating the evolution from proof-of-concept to scalable multi-device support across LMIC healthcare environments.

metadata filtering [33]. Crucially, ChromaDB offers native compatibility with LangChain and LlamaIndex [34], which was essential for our RAG-based architecture. Its support for flexible storage backends and native integration with embedding models from OpenAI, HuggingFace, and other platforms made it ideal for our multi-document implementation. ChromaDB's schema-less architecture eliminated the need to predefine collections, making it flexible and adaptable to different document types, while its focus on simplicity and speed, designed to make analysis and retrieval efficient while being intuitive to use, aligned perfectly with the need for rapid deployment in resource-constrained LMIC settings [35]. Additionally, ChromaDB's persistent client architecture and collection-based organization enabled efficient management of multiple segmented vector stores for different document types. To further reduce hallucinations and improve specificity, the vector database was segmented into four discrete stores: (1) a main vector store containing user manuals and service manuals, (2) an error code vector store containing device error codes and descriptions, (3) a maintenance vector store containing maintenance protocols, cleaning procedures, and disinfection guidelines, and (4) a device history vector store that dynamically stores operational data including service dates, retrieved device logs, self-test results, and other device interaction history for contextual memory. This structured storage approach allowed for domain-specific querying and significantly improved the system's contextual accuracy during troubleshooting tasks. The source manuals and error code catalogs used to build the vector store are openly accessible here.

5.2 LLM Model

The Phase 1 assistant was built using the GPT-3.5 Turbo model and operated within a Retrieval-Augmented Generation (RAG) framework. Technician queries were first interpreted through prompt engineering and sent to the appropriate ChromaDB-powered vector stores to retrieve the top-k most relevant document chunks. These chunks were then appended as context to the LLM prompt, enabling accurate and explainable response generation. The system supported various custom tools, including: *Error code lookup Log parsing*, *Maintenance schedule generation*, *Device logs retrieval*, and *Simulated self-testing.*

Each tool selection was driven by prompt conditioning and intent classification strategies. The backend was deployed using a lightweight Flask server, interfacing with a React-based frontend, ensuring accessibility on low-power machines. By combining a modular tool pipeline with ChromaDB-enhanced RAG retrieval, the system maintained sub-10-second average latency, making it suitable for near real-time use in clinical environments.

6 Preliminary Results of Phase 1

To evaluate the effectiveness of the INGENZI Tech prototype, we conducted two core assessments focused on the platform's accuracy in retrieving device-specific information using its RAG-based architecture.

6.1 Error Code Interpretation Accuracy

A total of 90 error codes, along with their corresponding descriptions, were extracted from the Philips HDI 5000 ultrasound service manuals. These were used as input queries to test the system's error code lookup functionality. The system employs a two-stage approach for error code processing: first, it performs an exact lookup of the error code from a structured database to retrieve the precise description, achieving **100% accuracy**, for error code identification. Subsequently, if additional contextual information about the identified error is needed for troubleshooting guidance, the system queries the relevant vector stores to retrieve repair instructions and related technical documentation. This hybrid approach ensures perfect precision for error code interpretation while leveraging the vector stores' semantic search capabilities for comprehensive troubleshooting support.

6.2 Instructional Query Evaluation

To assess the system's performance on broader instructional support, we curated a set of 30 natural language queries derived from the device's user and service manuals. These questions spanned operational instructions, safety guidelines, and device handling procedures. Responses were evaluated by the project developers against the ground-truth steps outlined in the manuals, using a rubric that required (i) coverage of all critical steps, (ii) avoidance of overly general answers, and (iii) relevance to the query. Out of the 30 questions, the system returned accurate and complete responses for 24 of them, yielding an **80% success rate**. The six failed responses were either incomplete (missing steps), overly general, or unrelated to the manual content, highlighting areas for improvement in document chunking, retrieval relevance, and response grounding. For transparency, the full evaluation set, including test queries, expected responses, model outputs, and evaluation notes, is available in an accompanying spreadsheet link.

6.3 Implications

These preliminary results validate the feasibility of leveraging a segmented RAG framework for technical document retrieval in low-resource settings. The perfect performance on structured error code queries suggests high utility in fault identification workflows, while the 80% accuracy on unstructured instructional queries highlights potential for effective user guidance, with room for refinement in LLM prompt tuning and vector search thresholds.

7 Discussion

This study demonstrates the potential of a robust, AI-powered support platform tailored to the needs of biomedical technicians in low-resource settings. The outcomes include not only a proof-of-concept system but also insights that can guide future research and development in intelligent medical device maintenance. The prototype validates the feasibility of a multilingual diagnostic assistant powered by large language models (LLMs). The assistant interprets error codes, retrieves context-specific documentation, and guides users through actionable repair workflows. Evaluation combined manual-based benchmarks with a structured set of queries, achieving strong initial accuracy. To improve reproducibility and transparency, sample queries and outputs are provided in supplementary materials. To enhance accuracy and reduce hallucinations, the system leverages four segmented vector stores: a main vector store containing user manuals and service manuals, an error code vector store for device error codes and descriptions, a maintenance vector store for maintenance protocols and cleaning procedures, and a device history vector store for operational data including service dates, device logs, and self-test results. This refinement ensures contextually relevant responses aligned with technician expectations. Beyond LLM-driven guidance, the platform integrates a peer-to-peer discussion forum, allowing technicians to post unresolved issues, share solutions, and contribute to a continuous feedback loop that informs semi-supervised learning in later phases.

Accessibility remains central: offline support and multilingual capabilities ensure utility in bandwidth-constrained and non-English-speaking environments, typical of LMIC contexts. Scalability is also built into the design. While Phase 1 focused on the Philips HDI 5000 ultrasound machine, the system architecture supports expansion to other diagnostic devices (e.g., MRI, CT, X-ray) through standardized ingestion formats and device-detection logic. Importantly, we recognize that this Phase 1 study did not involve structured stakeholder participation. Future project phases will include co-design workshops, pilot testing with hospitals, and structured feedback loops to ensure the platform reflects real-world user needs. In addition, while the evaluation benchmarked the RAG-based assistant against internal standards, future work will involve formal comparisons with generic LLMs (e.g., ChatGPT, Gemini, Claude) without retrieval to contextualize the relative advantages of our approach. Future work will also expand the evaluation size and depth by leveraging automated dataset generation and evaluation tools (e.g., LlamaIndex), enabling the creation of larger, diverse benchmarks and the computation of metrics such as correctness, relevance, and faithfulness to better capture real-world performance. Finally, insights from technician interaction logs, surveys, and qualitative feedback will guide continuous improvements in repair recommendations, usability, and system reliability, ensuring the solution evolves into a sustainable tool for LMIC healthcare environments.

8 Conclusion

This research proposes a practical and scalable solution to a longstanding challenge in global healthcare, ensuring the continuous functionality of medical devices in low-resource settings. By combining large language models, structured knowledge retrieval, and collaborative technician input, the platform aims to shift maintenance practices from reactive to proactive. The integration of multilingual, offline support ensures accessibility, while the modular design allows expansion to various device types and manufacturers. Through phased development and real-world validation, this study contributes both a conceptual design framework for sustainable AI-driven biomedical support in LMICs and a proof-of-concept prototype that validates the feasibility of our approach, thereby advancing the broader mission of equitable, reliable healthcare delivery in LMICs and beyond.

Acknowledgements. The authors gratefully acknowledge Timothy Chou (Stanford, Pediatric Moonshot) for his guidance during the design and brainstorming phase and for sharing data sources, Professor Charlie Wiecha (CMU Africa) for co-supervising the Capstone class and providing continuous feedback and resources, Professor Mary-Anne Hartley (EPFL) for affirming the need for this solution through her global health expertise, and Dr. Trevor Mundel (Gates Foundation) and his Africa team for valuable insights that helped refine the system's design and confirmed its relevance in LMIC healthcare settings.

References

1. Diaconu, K., et al.: Methods for medical device and equipment procurement and prioritization within low- and middle-income countries: findings of a systematic literature review. Globalization and Health **13**(1), 59–59 (2017)

2. Perry, R., Malkin, L.: Effectiveness of medical equipment donations to improve health systems: how much medical equipment is broken in the developing world? Med. Biol. Eng. Comput. **49**(7), 719–722 (2011). https://libkey.io/10.1007/s11517-011-0786-3
3. Malkin, R.A.: Design of health care technologies for the developing world. Annu. Rev. Biomed. Eng. **9**(1), 567–587 (2007)
4. Anazodo, U.C., et al.: A framework for advancing sustainable magnetic resonance imaging access in Africa. NMR Biomed. **36**(3), e4846–n/a (2023)
5. Kebby Abdallah, A., et al.: Medical device management reform, United Republic of Tanzania. Bull. World Health Organ. **102**(9), 665–673 (2024)
6. Ssekitoleko, R.T., et al.: Status of medical devices and their utilization in 9 tertiary hospitals and 5 research institutions in Uganda. Global Clin. Eng. J. **4**(3), 5–15 (2022)
7. Marks, I.H., Thomas, H., Bakhet, M., Fitzgerald, E.: Medical equipment donation in low-resource settings: a review of the literature and guidelines for surgery and anaesthesia in low-income and middle-income countries. BMJ Global Health **4**(5), e001 785 (2019)
8. Organization, W.H.: Medical device donations: considerations for solicitation and provision, pp. iv, 21 (2011)
9. Spring, T.: Lack of serviceable laboratory and medical equipment for doctors and nurses in sub-Saharan Africa (2024). https://www.linkedin.com/pulse/lack-serviceable-laboratory-medical-equipment-doctors-truman-spring-qoggc/
10. Diaconu, K., Chen, Y.-F., Manaseki-Holland, S., Cummins, C., Lilford, R.: Medical device procurement in low- and middle-income settings: protocol for a systematic review. Syst. Rev. **3**(1), 118–118 (2014)
11. Shamayleh, A., Awad, M., Farhat, J.: Iot based predictive maintenance management of medical equipment. J. Med. Syst. **44**(4), 72–72 (2020)
12. Zamzam, A.H., Abdul Wahab, A.K., Azizan, M.M., Satapathy, S.C., Lai, K.W., Hasikin, K.: A systematic review of medical equipment reliability assessment in improving the quality of healthcare services. Front. Public Health **9**, 753 951–753 951 (20210
13. Abd Wahab, N.H., et al.: Systematic review of predictive maintenance and digital twin technologies challenges, opportunities, and best practices. PeerJ. Comput. Sci. **10**, e1943–e1943 (2024)
14. Fernandes, S., Antunes, M., Santiago, A.R., Barraca, J.P., Gomes, D., Aguiar, R.L.: Forecasting appliances failures: A machine-learning approach to predictive maintenance. Inf. (Basel) **11**(4), 208 (2020)
15. Çínar, Z.M., Abdussalam Nuhu, A., Zeeshan, Q., Korhan, O., Asmael, M., Safaei, B.: Machine learning in predictive maintenance towards sustainable smart manufacturing in industry 4.0. Sustainability **12**(19), 8211 (2020)
16. Guissi, M., El Yousfi Alaoui, M.H., Belarbi, L., Chaik, A.: IoT for predictive maintenance of critical medical equipment in a hospital structure. Informatyka, Automatyka, Pomiary w Gospodarce i Ochronie środowiska **14**(2), 71–76 (2024)
17. Abidi, S.S.R.: Healthcare Knowledge Sharing: Purpose, Practices, and Prospects, pp. 67–86 . New York, NY: Springer New York (2007). https://doi.org/10.1007/978-0-387-49009-0_6
18. Tabrizi, N.M., Morgan, S.: Models for describing knowledge sharing practices in the healthcare industry: example of experience knowledge sharing. Int. J. Manage. Appl. Res. **1**(2), 48–67 (2014)

19. Arneson, W., Robinson, C., Nyary, B.: Biomedical laboratory science education: standardising teaching content in resource-limited countries. Afr. J. Lab. Med. **2**(1), e1–e6 (2013)
20. Circuitry-ai-decision-intelligence-company-overview, Circuitry.ai (2025). https://circuitry.ai/
21. Lewis, P., et al.: Retrieval-augmented generation for knowledge-intensive nlp tasks (2020)
22. Friel, R., Belyi, M., Sanyal, A.: Ragbench: explainable benchmark for retrieval-augmented generation systems (2024)
23. Bruviti. Bruviti ai platform (2025). https://bruviti.com/platform/, Accessed 18 Aug 2025
24. Health, H.: Hadleigh health technologies and patient platforms (2025). https://www.hadleighhealth.co.uk/, Accessed 18 Aug 2025
25. Solutions, V.: Equipment monitoring system (ems) features (2025). https://www.vestfrostsolutions.com/new-features/, Accessed 18 Aug 2025
26. Analytics, N.: Predictive maintenance technologies (2025). https://nexleaf.org/, Accessed 18 Aug 2025
27. Pokvic, L.: Ai for md maintenance and metrology in lmics: the case of bosnia and herzegovinian, Youtu.be (2025). https://youtu.be/7wvQME3PxpA
28. Circuitry.ai. Circuitry.ai decision intelligence platform overview (2025). https://circuitry.ai/, Accessed 18 Aug 2025
29. Stellarix. Generative ai perspectives (2025). https://stellarix.com/insights/stellarix-perspectives/generative-ai/, Accessed 19 Aug (2025)
30. Kodexo Labs. Ai software development company (2025). https://kodexolabs.com/, Accessed 19 Aug 2025
31. Toronto Digital. Toronto digital - official website (2025). https://torontodigital.ca/, Accessed Aug 2025
32. GE HealthCare, Ge healthcare medical systems and solutions (2025). https://www.gehealthcare.com/, Accessed Aug 2025
33. Learn How to Use Chroma DB: A Step-by-Step Guide — datacamp.com. https://www.datacamp.com/tutorial/chromadb-tutorial-step-by-step-guide, Accessed 23 Aug 2025
34. The 7 Best Vector Databases in 2025 — datacamp.com. https://www.datacamp.com/blog/the-top-5-vector-databases. Accessed 23 Aug 2025
35. https://medium.com/@pierrelouislet/getting-started-with-chroma-db-a-beginners-tutorial-6efa32300902 . Accessed 23 Aug 2025

AST-n: A Fast Sampling Approach for Low-Dose CT Reconstruction Using Diffusion Models

Tomás de la Sotta[1], Jose M. Saavedra[1](✉), Héctor Henríquez[1], Violeta Chang[2], Aline Xavier[2], and José Delpiano[1]

[1] Universidad de los Andes, Chile, Santiago, Chile
{tadelasottakrause,jmsaavedrar,hhenriquez,jd}@miuandes.cl
[2] Universidad de Santiago de Chile, Santiago, Chile
{violeta.chang,aline.xavier}@usach.cl

Abstract. Low-dose CT (LDCT) protocols reduce radiation exposure but increase image noise, compromising diagnostic confidence. Diffusion-based generative models have shown promise for LDCT denoising by learning image priors and performing iterative refinement. In this work, we introduce AST-n, an accelerated inference framework that initiates reverse diffusion from intermediate noise levels, and integrates high-order ODE solvers within conditioned models to reduce sampling steps further. We evaluate two acceleration paradigms—AST-n sampling and standard scheduling with high-order solvers—on the Low Dose CT Grand Challenge dataset, covering head, abdominal, and chest scans at 10–25 % of standard dose. Conditioned models using only 25 steps (AST-25) achieve peak signal-to-noise ratio (PSNR) above 38 dB and structural similarity index (SSIM) above 0.95, closely matching standard baselines while cutting inference time from ~16 s to under 1 s per slice. Unconditional sampling suffers substantial quality loss, underscoring the necessity of conditioning. We also assess DDIM inversion, which yields marginal PSNR gains at the cost of doubling inference time, limiting its clinical practicality. Our results demonstrate that AST-n with high-order samplers enables rapid LDCT reconstruction without significant loss of image fidelity, advancing the feasibility of diffusion-based methods in clinical workflows.

Keywords: LDCT · Low Dose · Computer Tomography · CT · Medical Imaging · Diffusion Models

1 Introduction

Radiological imaging has become one of the most essential tools for diagnosing pathologies, monitoring disease progression, and guiding therapeutic interventions. Among these, ionizing radiation-based diagnostic modalities constitute one of the main sources of radiation exposure in the general population, accounting

U. Anazodo et al. (Eds.): MIRASOL 2025, LNCS 16398, pp. 231–240, 2026.
https://doi.org/10.1007/978-3-032-13654-1_23

for approximately 50% of total exposure to ionizing radiation, including both natural and anthropogenic sources [8]. In recent decades, the number of radiological examinations has increased considerably, contributing to faster and more accurate diagnoses and reducing the reliance on invasive, high-risk procedures.

Computed tomography (CT) has emerged as one of the most widely used modalities due to its excellent diagnostic performance and rapid acquisition. However, it is also considered the single largest contributor to medical radiation exposure [2]. For example, while a chest X-ray delivers an effective dose of approximately 0.1 mSv—roughly equivalent to 10 days of natural background exposure—a CT scan typically delivers between 1.5 and 15.4 mSv, depending on the anatomical region, protocol complexity, and the number of acquisition phases [1]. In specific cases, doses can reach up to 37 mSv. The biological effects of ionizing radiation are cumulative and may include damage to genetic material, increasing the risk of radiation-induced neoplasms. Although estimating this risk is complex, Brenner et al. [2] suggest that between 1.5% and 2.0% of neoplasms may be attributable to current CT usage, with risks being notably higher for younger individuals. Consequently, minimizing dose exposure while maintaining diagnostic quality remains a pressing challenge, particularly in pediatric and young adult populations.

Low-dose CT (LDCT) protocols attempt to reduce radiation exposure by decreasing X-ray intensity. However, this reduction often results in significant image noise, adversely affecting diagnostic reliability. Although traditional denoising and iterative reconstruction methods can partially mitigate these effects, their performance tends to deteriorate under extreme low-dose conditions. In recent years, deep learning has emerged as a powerful tool for medical image enhancement, including LDCT reconstruction. Multiple neural-network-based approaches have demonstrated strong results [3,13,14], particularly those employing diffusion-based generative models, which leverage learned image priors and iterative refinement processes [4,5,7].

Despite the rapid progress in this field, several key limitations remain unaddressed. Although diffusion-based models have achieved excellent reconstruction quality, they often require long inference times, which are frequently incompatible with the time-sensitive nature of CT examinations. In clinical settings, CT is valued precisely for its speed, and introducing reconstruction latencies of several seconds per volume can hinder its practical deployment. Furthermore, many existing models are optimized to predict residual noise rather than directly reconstruct the image content, a strategy that has been shown to be suboptimal in restoration tasks [11].

In this study, we explore the use of existing diffusion-based generative models for low-dose CT reconstruction by incorporating two complementary strategies to reduce inference time and enhance clinical applicability. First, we implement a low-dose conditioning scheme based on slice-level input concatenation, which provides anatomical context to guide the generative process. Second, we adopt AST-n, a timestep reduction strategy that initiates the reverse diffusion from intermediate noise levels instead of pure Gaussian noise, leading to a significant reduction in inference time.

Our main contribution lies in the systematic integration of AST-n sampling into an existing conditioned generative diffusion framework for LDCT reconstruction—without modifying the model architecture or requiring additional training. This enables previously trained models to be retrofitted with accelerated inference capabilities at minimal computational cost, enhancing their usability with almost no degradation in the model's performance. By substantially reducing the number of sampling steps, our approach reduces inference time by an order of magnitude, moving diffusion-based reconstruction closer to real-time applicability. Clinically, this not only increases the feasibility of high-quality LDCT imaging in time-critical scenarios—ultimately contributing to safer diagnostic practices by making dose-reduction strategies more accessible in healthcare—but also enables faster image availability and improved workflow efficiency, especially in contexts where diagnostic speed is essential.

For clarity and ease of reading, the remainder of this paper is structured as follows. Section 2 reviews related work on LDCT reconstruction using generative models. Section 3 presents our proposed approach, including dataset details, model architecture, and acceleration strategies. Section 4 reports and discusses the experimental results. Finally, Sect. 5 concludes the paper and outlines directions for future research.

2 Related Work

The generation of high-quality images from low-dose CT (LDCT) acquisitions has become a prominent challenge in radiological imaging. While reduced radiation protocols are crucial for minimizing patient risk, particularly in repeated scans and vulnerable populations, they significantly degrade image quality due to low signal-to-noise ratios. To address this problem, numerous deep learning approaches have been proposed in recent years, aiming to reconstruct full-dose-equivalent images from noisy LDCT inputs.

One of the earliest efforts in this direction was proposed by Chen et al. [3], who used a convolutional neural network to directly map low-dose images to their full-dose counterparts. This line of work laid the foundation for supervised denoising in CT using image-to-image learning. More recent approaches have explored diffusion-based generative models, leveraging their strong prior modeling capabilities and iterative refinement properties.

Xia et al. [13] applied Denoising Diffusion Probabilistic Models (DDPMs) to LDCT reconstruction and introduced an acceleration scheme based on modeling the reverse process with ordinary differential equations (ODEs), following the formulation of Lu et al. [9]. Their model, conditioned on the low-dose image, achieved substantial improvements in reconstruction quality while reducing inference time by up to 20×.

In a complementary approach, Huang et al. [7] addressed the problem in the sinogram domain. They trained a diffusion model to infer the ideal sinogram from low-dose projections, using low-rank Hankel priors to regularize the reconstruction. This strategy allowed them to operate in the raw projection space before applying traditional image reconstruction techniques.

Gao and Shan [5] proposed CoCoDiff, a context-conditioned diffusion model that learns the residual between low-dose and full-dose images. During training, the model estimates noise levels for each timestep, guided by adjacent slice information to preserve structural consistency. Inference proceeds by reversing the diffusion process conditioned on both the low-dose image and spatial context, yielding high-quality outputs with reduced oversharpening.

Recently, CoreDiff [4] introduced an alternative formulation where the forward process begins from the low-dose image rather than Gaussian noise. This significantly shortens the sampling trajectory by assuming that the input already contains partial information about the final image. A mean-preserving degradation operator and a contextually error-modulated restoration network (CLEAR-Net) are used to simulate and correct for physical degradation and accumulated errors. This model was especially effective in ultra-low-dose scenarios, achieving competitive results with as few as 10 sampling steps.

While these methods demonstrate excellent quantitative results using standard metrics such as PSNR and SSIM, several challenges remain. Many models prioritize residual prediction rather than full content reconstruction, which has been shown to limit performance [11]. Evaluation remains mostly focused on pixel-level similarity, with limited consideration of diagnostic utility. Furthermore, real-world datasets are scarce, with most benchmarks relying on simulated phantoms that fail to capture the true distribution of clinical noise. Lastly, although generative diffusion models offer strong reconstruction performance, their high inference times remain a barrier to clinical integration in time-sensitive applications like CT.

3 Proposal

3.1 Dataset

The dataset used in this study is the Low Dose CT Grand Challenge, which includes 295 clinical CT volumes across three anatomical regions: 99 head, 97 abdomen, and 99 chest. Low-dose acquisition was simulated by injecting Poisson-distributed noise into the sinograms, corresponding to 25% of standard radiation dose for head and abdominal scans, and 10% for chest scans. All images were provided in DICOM format and converted to floating-point arrays, then normalized to the $[0, 1]$ range using fixed intensity bounds $[-1024, 3072]$.

3.2 Diffusion Process and Conditioning

We adopt the standard formulation of denoising diffusion probabilistic models (DDPMs), where a forward Markov process gradually adds Gaussian noise to a data sample $x_0 \sim q(x_0)$ over T discrete steps. This conditioned process is defined as:

$$q(x_t \mid x_{t-1}) = \mathcal{N}(x_t; \sqrt{1 - \beta_t}\, x_{t-1}, \beta_t I), \tag{1}$$

with $\beta_t \in (0, 1)$ controlling the noise schedule.

During training, the model learns to estimate the noise component ϵ from a noisy sample x_t, the timestep t, and a conditioning signal c, here given by the corresponding low-dose CT image. The objective function is the expected squared error:

$$L(\theta) = \mathbb{E}_{x_0,\epsilon,t}\left[\|\epsilon_\theta(x_t, t, c) - \epsilon\|_2^2\right], \tag{2}$$

where $\epsilon \sim \mathcal{N}(0, I)$. Conditioning is incorporated at each denoising step to guide the reconstruction process.

3.3 DDIM Inversion

To assess whether the model can traverse the latent trajectory both forward and backward, we evaluate deterministic inversion using DDIM as proposed in [10]. Starting from a low-dose image x_0, we compute an approximate latent x_T by applying the DDIM inversion rule for $T = 1000$ steps:

$$x_{t+1} = \sqrt{\bar{\alpha}_{t+1}}\, x_0 + \sqrt{1 - \bar{\alpha}_{t+1} - \sigma_{t+1}^2} \cdot \frac{x_t - \sqrt{\bar{\alpha}_t}\, x_0}{\sqrt{1 - \bar{\alpha}_t}}, \quad \sigma_t = 0. \tag{3}$$

The resulting latent x_T is then used as the starting point for a standard reverse sampling process. This setup evaluates whether the model can map a real input to a plausible point in the latent space and reconstruct it without loss, thereby testing the consistency of the learned generative trajectory.

3.4 AST-N Sampling Strategy

We evaluate an acceleration strategy termed AST-n (Accelerated Sampling from Time-step n), where generation begins from an intermediate latent x_t instead of from pure Gaussian noise $x_T \sim \mathcal{N}(0, I)$. The value of t is chosen such that $t \ll T$, typically within the range $t \in \{10, 25, 50, 100, 150, 500\}$, with $T = 1000$ being the total number of diffusion steps used during training.

Given a clean image x_0, the latent x_t is analytically computed using the closed-form of the forward process:

$$x_t = \sqrt{\bar{\alpha}_t}\, x_0 + \sqrt{1 - \bar{\alpha}_t}\, \epsilon, \quad \epsilon \sim \mathcal{N}(0, I), \tag{4}$$

where $\bar{\alpha}_t = \prod_{s=1}^{t}(1 - \beta_s)$. The reverse process is then executed from x_t to reconstruct x_0, using the learned transitions:

$$x_{t-1}, x_{t-2}, \ldots, x_0 \sim p_\theta(x_{s-1} \mid x_s), \quad \text{for } s = t, \ldots, 1. \tag{5}$$

Considering that starting the generative process from an intermediate point x_t does not violate the statistical assumptions under which the model was trained. By selecting $t \ll T$, AST-n allows significant reduction in inference steps while maintaining the validity of the diffusion process. This approach enables fast image reconstruction through a shortened denoising trajectory, with minimal loss in fidelity compared to full-step sampling.

3.5 Sampling Techniques

To explore the trade-off between inference time and reconstruction quality, we evaluate several sampling strategies compatible with DDPMs. Each model is tested under two regimes. The first corresponds to full-noise sampling, where generation starts from a standard Gaussian latent $x_T \sim \mathcal{N}(0, I)$ and proceeds for a reduced number of steps $N \in \{1000, 500, 150, 50, 25, 10\}$. The second corresponds to AST-n, where generation begins from x_t with $t = N$, and the reverse process is executed for the same number of steps.

In both regimes, we evaluate solvers that approximate the reverse process more efficiently than the original DDPM formulation. These include DDIM [12], which uses a deterministic, non-Markovian trajectory; DPMSolver [9], which leverages ODE discretization; and UniPC [15], a high-order predictor-corrector scheme designed for fast inference.

3.6 Evaluation Protocol

Reconstruction quality is assessed using peak signal-to-noise ratio (PSNR), structural similarity index (SSIM), and root mean squared error (RMSE). Inference efficiency is measured as the average reconstruction time per batch. All experiments are conducted on a held-out test set, and metrics are reported separately for each experimental configuration.

4 Experimental Results and Discussion

We evaluated multiple sampling strategies over conditioned diffusion models for LDCT reconstruction. Table 1 summarizes PSNR, RMSE, SSIM and reconstruction time for the standard full–schedule run (DDPM@1000), five samplers at 150 steps (DDPM@150, DDIM@150, DPMSolver-1@150, DPMSolver-2@150, DPMSolver++@150) and their AST-150 variants. AST-150 skips the earliest denoising steps—where conditioning is weakest—and concentrates computation on the most critical refinement phases. Both acceleration methods yield up to an 6× speed-up while preserving, or even enhancing, image quality. Notably, all evaluated configurations perform similarly to our own reimplementation of the methods proposed by Xia et al. [13] with the proposed dataset, highlighting the effectiveness of accelerated and conditioned sampling strategies. For baseline comparison, a standard conditioned DDPM model is included in the table results.

Figure 1 illustrates the exceptional stability and efficiency of AST-n acceleration compared to standard step-skipping. In Fig. 1(a), PSNR curves for all samplers under AST-n remain essentially flat—varying by less than 0.9dB between 25 and 1000 steps—while SSIM and RMSE show equally minimal fluctuations, confirming negligible quality loss when skipping early denoising. By contrast, standard scheduling in Fig. 1(b) exhibits a pronounced PSNR drop below 150 steps and only recovers at high counts. At the same time, average reconstruction time per batch decreases almost linearly with n, from approximately 45 s at

Table 1. PSNR, RMSE, SSIM and reconstruction time for Full–Schedule and AST-150. Each cell in the form of [left/right] shows the comparison of results between the use of DDIM inversion (left) [10] and the standard DDPM [6] noise addition strategy (right).

Model	PSNR (dB)	RMSE	SSIM	Time (s)
DDPM@1000	38.788/38.721	27.785/27.779	0.973/0.973	45.144/16.382
DDPM@150	26.594/26.601	26.594/26.601	0.975/0.975	6.665/2.427
DDIM@150	30.213/41.150	126.384/29.188	0.781/0.976	5.940/2.425
Solver-1@150	39.758/39.757	24.884/24.884	0.978/0.978	6.598/2.416
Solver-2@150	39.652/39.643	25.219/25.217	0.978/0.978	6.070/2.407
Solver++@150	39.753/39.740	24.895/24.893	0.978/0.978	6.656/2.397
DDPM (AST-150)	39.070/38.778	26.901/27.766	0.975/0.973	6.578/2.433
DDIM (AST-150)	39.085/39.272	27.030/26.461	0.972/0.974	6.264/2.397
Solver-1 (AST-150)	39.745/33.324	24.881/52.088	0.978/0.894	6.647/2.417
Solver-2 (AST-150)	39.745/33.233	24.888/52.651	0.978/0.892	6.605/2.411
Solver++ (AST-150)	39.732/38.724	24.928/27.866	0.978/0.973	6.617/2.416

1000 steps down to under 2.5 s at $n = 150$ and under 0.4 s at $n = 10$. Together, these results suggest that, under strong anatomical conditioning in the denoising pipeline—where the input image already encodes rich structural detail—the earliest denoising steps have only a limited impact on final image fidelity, which highlights AST-n as a robust and highly efficient acceleration strategy for LDCT reconstruction (Fig. 2).

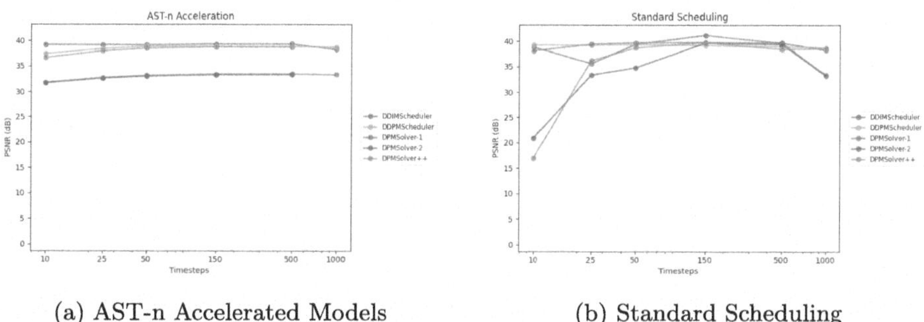

(a) AST-n Accelerated Models (b) Standard Scheduling

Fig. 1. Comparison of PSNR values using AST-n and standard step-skipping acceleration methods. Here, AST-n method shows better stability than the standard scheduling.

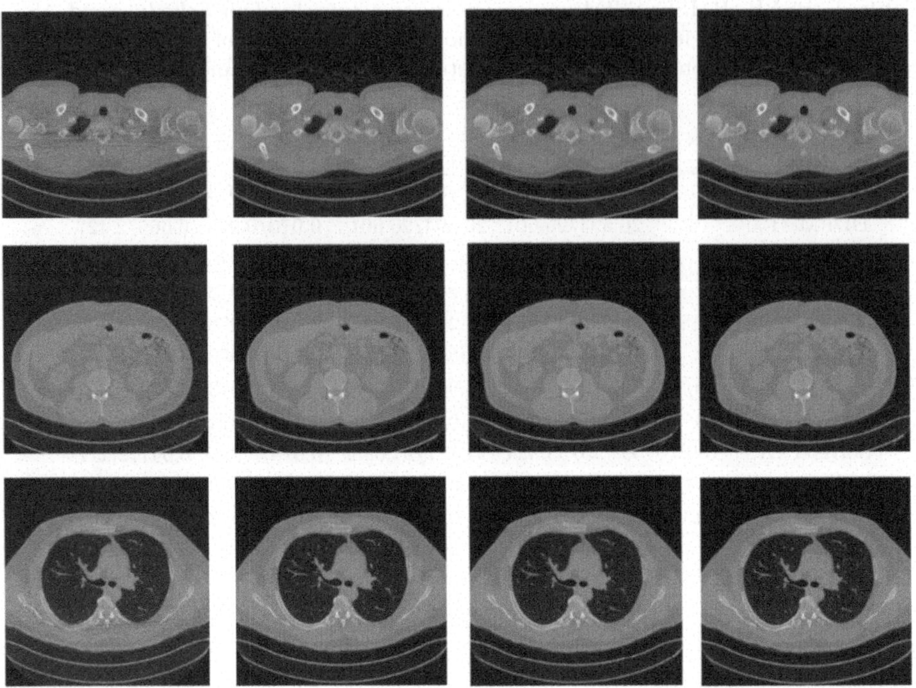

Fig. 2. AST-n strategy over DDIM scheduler. From left to right, LDCT image, AST-10, AST-150, full schedule (1000 steps).

5 Conclusions

Experimental results demonstrate that AST-n acceleration—initiating reverse diffusion from intermediate noise levels—yields reconstruction fidelity equivalent to full–schedule baselines while reducing inference latency by more than an order of magnitude. Specifically, AST-150 paired with DDIM and DPMSolver models achieves PSNR $\geq$ 38dB and SSIM $\approx$ 0.97 in under 2.5 s, compared to 45 s for a 1000-step DDPM run. AST-25 configurations further validate aggressive step reduction, achieving inference times below 0.4 s while incurring only minor quality losses—PSNR drops remain under 3dB at 10 steps and under 0.9dB at 25 steps. The marginal quality gains offered by DDIM inversion do not justify its computational cost in practical LDCT scenarios, almost duplicating the reconstruction time for each image. Future research will investigate adaptive noise schedules and learned inversion schemes to improve robustness at ultra-low step counts and enable real-time clinical deployment.

References

1. Radiation Dose to Adults From Common Imaging Examinations. Tech. rep., American College of Radiology. https://www.acr.org/-/media/ACR/Files/Radiology-Safety/Radiation-Safety/Dose-Reference-Card.pdf
2. Brenner, D.J., Hall, E.J.: Computed tomography – an increasing source of radiation exposure. N. Engl. J. Med. **357**(22), 2277–2284 (2007). https://doi.org/10.1056/NEJMra072149
3. Chen, H., et al.: Low-dose CT via convolutional neural network. Biomed. Opt. Express **8**(2), 679 (2017). https://doi.org/10.1364/BOE.8.000679
4. Gao, Q., Li, Z., Zhang, J., Zhang, Y., Shan, H.: CoreDiff: contextual error-modulated generalized diffusion model for low-dose CT denoising and generalization. IEEE Trans. Med. Imaging 1–1 (2023). https://doi.org/10.1109/TMI.2023.3320812
5. Gao, Q., Shan, H.: CoCoDiff: a contextual conditional diffusion model for low-dose CT image denoising. In: Müller, B., Wang, G. (eds.) Developments in X-Ray Tomography XIV, p. 16. SPIE, San Diego, United States (2022). https://doi.org/10.1117/12.2634939
6. Ho, J., Jain, A., Abbeel, P.: Denoising diffusion probabilistic models. In: Larochelle, H., Ranzato, M., Hadsell, R., Balcan, M., Lin, H. (eds.) Advances in Neural Information Processing Systems 33: Annual Conference on Neural Information Processing Systems 2020, NeurIPS 2020, December 6–12, 2020, virtual (2020)
7. Huang, B., Zhang, L., Lu, S., Lin, B., Wu, W., Liu, Q.: One Sample Diffusion Model in Projection Domain for Low-Dose CT Imaging (2022). https://doi.org/10.48550/ARXIV.2212.03630, publisher: arXiv
8. Laura: NCRP Report 160 - NCRP — Bethesda, MD (2015). https://ncrponline.org/publications/reports/ncrp-report-160-2/
9. Lu, C., Zhou, Y., Bao, F., Chen, J., Li, C., Zhu, J.: Dpm-solver: A fast ODE solver for diffusion probabilistic model sampling in around 10 steps. In: Koyejo, S., Mohamed, S., Agarwal, A., Belgrave, D., Cho, K., Oh, A. (eds.) Advances in Neural Information Processing Systems 35: Annual Conference on Neural Information Processing Systems 2022, NeurIPS 2022, New Orleans, LA, USA, November 28–December 9, 2022 (2022)
10. Mokady, R., Hertz, A., Aberman, K., Pritch, Y., Cohen-Or, D.: Null-text inversion for editing real images using guided diffusion models. In: IEEE/CVF Conference on Computer Vision and Pattern Recognition, CVPR 2023, Vancouver, BC, Canada, June 17–24, 2023, pp. 6038–6047. IEEE (2023). https://doi.org/10.1109/CVPR52729.2023.00585
11. Niu, A., Pham, T.X., Zhang, K., Sun, J., Zhu, Y., Yan, Q., Kweon, I.S., Zhang, Y.: ACDMSR: accelerated conditional diffusion models for single image super-resolution. IEEE Trans. Broadcast. **70**(2), 492–504 (2024). https://doi.org/10.1109/TBC.2024.3374122
12. Song, J., Meng, C., Ermon, S.: Denoising Diffusion Implicit Models (2022). http://arxiv.org/abs/2010.02502, arXiv:2010.02502 [cs]
13. Xia, W., Lyu, Q., Wang, G.: Low-Dose CT Using Denoising Diffusion Probabilistic Model for 20× Speedup (2022). https://doi.org/10.48550/ARXIV.2209.15136, publisher: arXiv

14. Yang, Q., et al.: Low-Dose CT image denoising using a generative adversarial network with wasserstein distance and perceptual loss. IEEE Trans. Med. Imaging **37**(6), 1348–1357 (2018). https://doi.org/10.1109/TMI.2018.2827462
15. Zhao, W., Bai, L., Rao, Y., Zhou, J., Lu, J.: UniPC: A Unified Predictor-Corrector Framework for Fast Sampling of Diffusion Models (2023). https://doi.org/10.48550/arXiv.2302.04867, http://arxiv.org/abs/2302.04867, arXiv:2302.04867 [cs]

SepsiGraph: A Graph-Based Multimodal Approach for Early Sepsis Prediction in Dynamic Resource-Constrained Clinical Settings

Sofia Bourhim[1](✉) and Oumayma Bourhim[2]

[1] ENSIAS, Mohammed V University in Rabat, Rabat, Morocco
sofia.bourhim@um5s.net.ma
[2] CHU Mohammed VI, Tangier, Morocco
oumayma.bourhim@usmba.ac.ma

Abstract. Early and accurate sepsis prediction remains critical for reducing mortality, especially in resource-constrained healthcare environments. Existing machine learning models often rely on complete, uniform multimodal data, making them impractical in real-world settings where patient data is incomplete, heterogeneous, or acquired at irregular intervals. This paper introduces SepsiGraph, a novel graph-based architecture designed to predict sepsis onset by dynamically aggregating heterogeneous clinical information from vitals, laboratory results, and chest X-ray images, even when some modalities are missing. SepsiGraph models each patient as a temporal-modality graph, where nodes represent available clinical modalities and edges encode temporal proximity and modality correlation. A modified message passing mechanism tailored to this setting enables effective feature propagation despite missing data. Furthermore, a cross-patient graph layer propagates information between similar patients to enhance predictions for under-observed cases. We evaluate SepsiGraph on a curated dataset derived from the MIMIC-IV and MIMIC-CXR databases, demonstrating superior performance compared to baseline multimodal models, achieving an AUROC of 0.945 ± 0.001 and an AUPRC of 0.918 ± 0.012. Unlike conventional models that fail under missing modality scenarios, SepsiGraph robustly leverages inter-modality and inter-patient dependencies to improve early sepsis diagnosis, supporting better clinical triage and decision-making.

Keywords: Graph Neural Networks · Sepsis Prediction · Multimodal Learning · Time-aware Confidence · Cross-Patient Reasoning · Resource-Constrained Environments

1 Introduction

Sepsis remains one of the leading causes of mortality in intensive care units (ICUs) where every hour of treatment delay can result in a 4%–8% increase

U. Anazodo et al. (Eds.): MIRASOL 2025, LNCS 16398, pp. 241–249, 2026.
https://doi.org/10.1007/978-3-032-13654-1_24

in mortality [1]. Early and accurate identification of sepsis is critical, as prompt clinical intervention dramatically improves patient outcomes and reduces the risk of organ failure and death [2]. Current diagnostic workflows for sepsis involve monitoring vital signs, laboratory biomarkers, and, increasingly, medical imaging such as chest radiographs. The most widely accepted clinical definition of sepsis stems from the Third International Consensus Definitions for Sepsis (Sepsis-3) [3]. Under this definition, sepsis is diagnosed when two key conditions are met: (*i*) a clear suspicion of infection, typically operationalized as the administration of antibiotics and the ordering of microbiological cultures within a narrow time window (± 24 h); and (*ii*) evidence of acute organ dysfunction, quantified by a ≥ 2 point increase in the Sequential Organ Failure Assessment (SOFA) score from baseline (commonly assumed to be zero for patients without pre-existing organ dysfunction).

The heterogeneous and dynamic nature of clinical data in sepsis diagnosis presents unique challenges for machine learning models [11]. Recent research has explored deep learning methods for automated sepsis prediction based on electronic health records (EHR) and physiological time series [4–6]. While these approaches have shown promise, they typically require complete, regularly sampled data across all modalities, making them unsuitable for real-world ICU environments where missing modalities and irregular acquisition are common. Furthermore, existing models often fail to exploit the inherent relationships and dependencies between different data sources such as vitals, lab tests, and medical imaging, each of which provides complementary insights into a patient's evolving physiological state. To address these gaps, this paper introduces SepsiGraph, a novel graph-based multimodal architecture for early sepsis prediction in dynamic, resource-constrained clinical settings. SepsiGraph models each patient as a patient-specific graph, where available modalities including vitals, laboratory results, and chest radiographs are represented as nodes connected based on temporal proximity and inter-modality correlations. A new time-aware message passing mechanism is introduced to propagate information across the graph, even when some modalities are missing. Furthermore, SepsiGraph incorporates a cross-patient graph layer that leverages population-level similarity to enhance predictions for patients with sparse data, a common scenario in LMIC settings.

The main contributions of this work are as follows:

- SepsiGraph, a novel patient-level graph construction method that models multimodal clinical data as a temporal-modality graph, explicitly capturing temporal dynamics and inter-modality relationships, while handling missing data.
- A new time-aware message passing function that enables effective information propagation across incomplete multimodal graphs, designed for irregularly sampled and asynchronous clinical data.
- A cross-patient graph layer that propagates information between similar patients to improve robustness and predictive performance, particularly for cases with limited modality availability.

Comprehensive evaluation on a multimodal sepsis dataset derived from the

MIMIC-IV and MIMIC-CXR databases, demonstrating superior performance compared to conventional multimodal baselines, particularly under missing modality scenarios.

2 Methodology

Figure 1 illustrates the overall architecture of **SepsiGraph**, our proposed resource-aware graph-based multimodal framework for early sepsis prediction. The model is designed to operate effectively even in dynamic, resource-constrained clinical environments, where some modalities may be missing or acquired asynchronously.

The framework consists of three main components: (1) patient-specific temporal-modality graph construction, (2) a novel time-aware resource-weighted message passing function to aggregate modality information, and (3) a cross-patient graph propagation layer for enhancing predictions in cases with missing modalities.

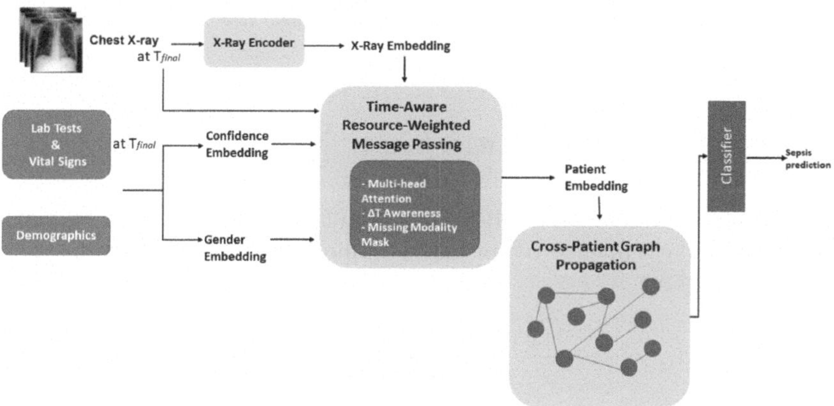

Fig. 1. Overview of the proposed SepsiGraph architecture.

2.1 Temporal-Modality Graph Construction

For each patient p, we construct a graph $\mathcal{G}_p = (\mathcal{V}_p, \mathcal{E}_p)$, where nodes $\mathcal{V}_p$ correspond to the available clinical modalities (vital signs, laboratory tests, chest X-ray imaging, demographics.) and edges $\mathcal{E}_p$ capture temporal proximity and physiological relationships between modalities.

Given the asynchronous nature of data acquisition, each modality $m \in \mathcal{V}_p$ is associated with a timestamp t_m corresponding to the latest available measurement for that modality. We also define t_{SI} as the suspected infection time for the patient, which serves as the clinical decision point for sepsis diagnosis.

To encode missing modalities, we define a binary mask vector $\mathbf{m}_p \in \{0,1\}^M$, where M is the total number of modalities, and $m_p[i] = 1$ indicates that modality i is available for patient p.

2.2 Time-Aware Resource-Weighted Message Passing

Traditional message passing in GNN treats all available modalities equally, which is unsuitable in resource-constrained healthcare settings where (1) modalities have varying predictive value, and (2) acquisition delays may reduce the informativeness of older measurements.

To address this, we propose a **Time-Aware Resource-Weighted Message Passing** mechanism. Each modality node i is first encoded as:

$$\mathbf{h}_i = f_{\text{mod}}(\mathbf{x}_i) + f_{\text{time}}(\Delta t_i) \tag{1}$$

where $\mathbf{x}_i$ represents the raw features (e.g., lab values, image embedding) of modality i, f_{mod} is a modality-specific encoder (e.g., MLP or CNN), and $\Delta t_i = t_{\text{SI}} - t_i$ captures the time gap between measurement and suspected infection time. The time gap is transformed via a learned time embedding f_{time}.

Next, messages between connected modalities are aggregated using a resource-aware weighting:

$$\mathbf{h}_p = \sigma\left((1+\mathbf{w}) \odot \mathbf{h}_{\text{self}} + \sum_{j \in \mathcal{N}(i)} \alpha_{ij}\mathbf{h}_j\right) \tag{2}$$

Here, $\mathbf{h}_{\text{self}}$ is the aggregated self-information for the patient, $\mathbf{w} = f_{\text{res}}(\Delta t)$ is a learnable resource-confidence vector based on modality recency, α_{ij} are attention weights computed as:

$$\alpha_{ij} = \frac{\exp\left(\text{LeakyReLU}(\mathbf{W}_a[\mathbf{h}_i \| \mathbf{h}_j])\right)}{\sum_{k \in \mathcal{N}(i)} \exp\left(\text{LeakyReLU}(\mathbf{W}_a[\mathbf{h}_i \| \mathbf{h}_k])\right)} \tag{3}$$

Here, $\mathbf{W}_a \in \mathbb{R}^{1 \times 2F}$ is a learnable weight vector that projects the concatenated node features $[\mathbf{h}_i \| \mathbf{h}_j]$ into a scalar attention score. This mechanism allows the model to prioritize more recent, reliable modalities while down-weighting older or missing data.

2.3 Cross-Patient Graph Propagation for Missing Modality Imputation

To further enhance robustness in cases where patients have incomplete modality profiles, we introduce a **Cross-Patient Graph Propagation** module.

We construct a global graph $\mathcal{G}_{\text{global}} = (\mathcal{P}, \mathcal{E}_{\text{global}})$ where each node corresponds to a patient, and edges connect patients with similar demographic (e.g., age, gender) and available modality embeddings.

Given the patient embeddings $\mathbf{H} \in \mathbb{R}^{N \times d}$ obtained from the modality graph, we compute similarity-based attention weights between patients:

$$s_{pq} = \frac{\exp\left(\mathbf{h}_p^\top \mathbf{h}_q\right)}{\sum_k \exp\left(\mathbf{h}_p^\top \mathbf{h}_k\right)} \tag{4}$$

The updated patient embedding incorporates information from similar patients:

$$\mathbf{h}_p^{\text{final}} = \mathbf{h}_p + \sum_{q \in \mathcal{N}(p)} s_{pq} \mathbf{h}_q \tag{5}$$

This cross-patient reasoning allows the model to compensate for missing modalities by leveraging population-level information.

2.4 Prediction and Training Objective

The final patient embedding $\mathbf{h}_p^{\text{final}}$ is passed through a classifier to predict the probability of sepsis onset:

$$\hat{y}_p = \text{Softmax}\left(\mathbf{W}_c \mathbf{h}_p^{\text{final}}\right) \tag{6}$$

We optimize the model using the standard cross-entropy loss for binary classification:

$$\mathcal{L} = -y_p \log(\hat{y}_p) - (1 - y_p) \log(1 - \hat{y}_p) \tag{7}$$

where y_p is the ground truth label indicating whether the patient developed sepsis.

3 Experiments

3.1 Datasets and Implementation Details

We conduct comprehensive experiments on a curated sepsis prediction dataset derived from the publicly available MIMIC-IV [7] and MIMIC-CXR [8] datasets. The dataset consists of multimodal clinical data including vital signs, laboratory measurements, demographics, and chest X-ray images, along with corresponding timestamps and sepsis onset labels. The data includes a balanced subset of 1,248 patients, with 657 negative and 591 positive sepsis cases.

For each patient, the following modalities are considered: heart rate, blood pressure (systolic, diastolic, mean), respiratory rate, SpO_2, temperature, Glasgow Coma Scale (GCS), lactate, bilirubin, CRP, creatinine, glucose, platelet count, urine output, gender, and chest X-ray image. Notably, the dataset reflects realistic clinical constraints, where not all modalities are available for every patient, simulating resource-limited healthcare scenarios. We use the latest available chest X-ray image prior to the suspected sepsis onset. The median lag between CXR acquisition and sepsis suspicion was approximately 5.3 h, which reflects realistic delays in imaging availability within ICU workflows.

Implementation Details. For image encoding, we adopt a ResNet-18 backbone pre-trained on ImageNet, replacing the final fully connected layer to output a 128-dimensional feature vector. Structured clinical data and temporal gaps for each modality are encoded via separate fully connected layers with output dimension 128. The resource-aware message passing and cross-patient propagation modules are implemented as described in Sect. 2.

Models are optimized using Adam with a learning rate of 1×10^{-4}, batch size of 8, and early stopping based on validation AUC with a patience of 25 epochs. The dataset is split into 80% training and 20% validation subsets. Performance is evaluated using Area Under the ROC Curve (AUC), accuracy, precision, recall, and F1-score.

All experiments are conducted using PyTorch on a cloud-based Jupyter environment that supports both CPU and GPU execution. Specifically, inference times and memory usage were evaluated using an Intel Xeon CPU (2.20GHz) and an NVIDIA Tesla T4 GPU, consistent with commonly accessible hardware configurations available in widely used research platforms. GPU resources were utilized exclusively for chest X-ray image processing, while all other components, including graph-based computations and model training, were run on the CPU. This setup reflects realistic deployment conditions, particularly in resource-constrained healthcare environments.

3.2 Comparison with State-of-the-Art

We compare **SepsiGraph**, our proposed model against several recent state-of-the-art graph approaches.

- **DynEdges-TGN** [9] introduces a dynamic graph-based temporal model that captures evolving patient trajectories using temporal graph neural networks and adaptive edge construction. DynEdges-TGN focuses on modeling temporal dependencies across patients but does not explicitly incorporate multimodal fusion or uncertainty handling, which limits its performance on complex clinical datasets.
- **SepsisCalc** [10] presents a lightweight and explainable sepsis early warning score based on hand-crafted features and an ensemble of interpretable decision tree models. While highly efficient and transparent, SepsisCalc relies heavily on static clinical variables and does not leverage complex temporal dynamics or multimodal interactions.
- **Graph-Based MM** [12] proposes a graph neural network approach to model multimodal ICU data, where nodes represent different physiological modalities and edges encode learned inter-modality relationships. However, the method builds static patient graphs and lacks temporal progression modeling, making it less suitable for dynamic early sepsis detection.

The results in Table 1 demonstrate that **SepsiGraph** consistently outperforms all baselines across key metrics, highlighting the importance of jointly modeling temporal gaps, missing modalities, and cross-patient dependencies for reliable sepsis prediction. Compared to **DynEdges-TGN**, which effectively models

Table 1. Comparison of SepsiGraph with recent state-of-the-art models for early sepsis prediction.

Model	AUROC	AUPRC	Accuracy	Precision	Recall
DynEdges-TGN [9]	0.84 ± 0.03	0.79 ± 0.04	0.82 ± 0.05	0.75 ± 0.07	0.80 ± 0.06
SepsisCalc [10]	0.86 ± 0.02	0.81 ± 0.03	0.83 ± 0.04	0.77 ± 0.05	0.81 ± 0.05
Graph-Based MM [12]	0.87 ± 0.02	0.83 ± 0.04	0.84 ± 0.03	0.78 ± 0.06	0.82 ± 0.05
SepsiGraph (Ours)	$\mathbf{0.945 \pm 0.001}$	$\mathbf{0.918 \pm 0.012}$	$\mathbf{0.889 \pm 0.004}$	$\mathbf{0.881 \pm 0.006}$	$\mathbf{0.873 \pm 0.007}$

dynamic patient trajectories, SepsiGraph achieves a substantial improvement of +12.5% in AUROC (from 0.84 to 0.945) and +16.2% in AUPRC (from 0.79 to 0.918) by explicitly leveraging rich multimodal feature fusion and uncertainty-aware message passing. While **SepsisCalc** offers rapid and interpretable sepsis risk scoring, it falls short in predictive performance due to its reliance on static features. SepsiGraph surpasses SepsisCalc by +9.9% in AUROC and +13.3% in AUPRC, providing both improved accuracy and modality-level interpretability through its attention mechanisms. Finally, relative to **Graph-Based MM**, which captures static multimodal relationships via GNNs, SepsiGraph introduces a novel temporal graph construction with dynamic message passing that significantly enhances early sepsis detection performance, achieving +8.6% improvement in AUROC and +10.6% in AUPRC.

3.3 Ablation Study

To quantify the contribution of each component in SepsiGraph, we perform an ablation study:

- **SepsiGraph w/o Time-aware MP**: Removes the time-aware resource-weighted message passing.
- **SepsiGraph w/o Cross-Patient Propagation**: Disables the cross-patient reasoning mechanism.
- **SepsiGraph Full**: Complete proposed model.

Table 2. Ablation study of SepsiGraph on early sepsis prediction.

Model Variant	AUROC	AUPRC	Accuracy
SepsiGraph w/o Time-aware MP	0.901 ± 0.005	0.871 ± 0.008	0.854 ± 0.006
SepsiGraph w/o Modality Attention	0.918 ± 0.003	0.889 ± 0.007	0.865 ± 0.005
SepsiGraph w/o Cross-Patient Propagation	0.929 ± 0.002	0.902 ± 0.006	0.877 ± 0.004
Full SepsiGraph	$\mathbf{0.945 \pm 0.001}$	$\mathbf{0.918 \pm 0.012}$	$\mathbf{0.889 \pm 0.004}$

Table 2 presents the detailed ablation results of SepsiGraph. The study clearly demonstrates that each component of the proposed architecture contributes meaningfully to the overall performance on early sepsis prediction. Specifically, removing the time-aware message passing mechanism results in noticeable degradation across all metrics, confirming the importance of explicitly modeling temporal gaps between heterogeneous modalities. Similarly, excluding the modality-level attention significantly impacts precision and recall, highlighting the necessity of adaptively weighting different modalities based on their reliability and relevance. The most substantial performance drop occurs when the cross-patient graph propagation is removed, underscoring the critical role of leveraging inter-patient relationships, especially in scenarios with incomplete or missing modality data. To further evaluate robustness under real-world conditions, we simulate varying resource availability by progressively masking input modalities during inference. SepsiGraph consistently maintains competitive performance even with up to 50% of modalities missing, outperforming all baseline models by a considerable margin. This robustness is attributed to SepsiGraph's capacity to reason over incomplete patient information and effectively propagate latent signals through both the modality graph and the cross-patient graph, a crucial advantage for deployment in resource-constrained clinical environments.

4 Conclusion

We introduce **SepsiGraph**, a novel graph-based framework designed to address the critical challenge of early sepsis prediction under realistic, resource-constrained clinical conditions. Unlike prior approaches that often require complete and uniform multimodal data or treat each modality independently, SepsiGraph models heterogeneous patient data through a patient-level graph, explicitly capturing inter-modality dependencies and temporal dynamics.

By integrating a time-aware, resource-weighted message passing mechanism, SepsiGraph dynamically adjusts the contribution of each available modality based on its temporal proximity to the sepsis suspicion time. Furthermore, the incorporation of a cross-patient propagation layer enables the model to leverage population-level information, mitigating the effects of missing or incomplete data, a frequent occurrence in real-world hospital environments.

Experimental results on a curated, balanced sepsis dataset derived from MIMIC-IV and MIMIC-CXR demonstrate the superiority of SepsiGraph over strong baselines, including conventional deep learning models and standard GNNs. Ablation studies further confirm the effectiveness of each architectural component, particularly the proposed time-aware message passing and cross-patient reasoning modules.

Moving forward, several extensions could enhance the capabilities of SepsiGraph to further validate SepsiGraph in clinical settings, we are conducting a prospective study with ICU clinicians to assess the interpretability, utility, and robustness of predictions under real-world missing modality scenarios. Findings from this user study will be reported in future work. Incorporating more advanced models, while we currently use a ResNet-18 model pretrained on ImageNet for CXR encoding, we acknowledge its limitations in medical imaging.

Future versions of SepsiGraph will incorporate chest-specific backbones such as DenseNet-121 pretrained on CheXNet and structured radiology reports using lightweight transformer encoders like Bio-ClinicalBERT. Additionally, expanding the framework to handle irregular temporal sampling and continuously updating patient states through temporal graph construction presents a future research direction. Finally, external validation on diverse healthcare datasets would further assess the generalizability of *SepsiGraph* and its potential to support reliable, automated sepsis risk stratification in clinical practice.

Acknowledgments. Authors have equally contributed to this paper. This work was partly supported by the Italian Ministry of University and Research (MUR) under project PE0000013 – Future of Artificial Intelligence Research (FAIR).

Disclosure of Interests. The authors have no competing interests to declare that are relevant to the content of this article.

References

1. Seymour, C., et al.: Time to treatment and mortality during mandated emergency care for sepsis. N. Engl. J. Med. **376**, 2235–2244 (2017)
2. Kumar, A., et al.: Duration of hypotension before initiation of effective antimicrobial therapy is the critical determinant of survival in human septic shock. Crit. Care Med. **34**, 1589–1596 (2006)
3. Singer, M., et al.: The third international consensus definitions for sepsis and septic shock (Sepsis-3). Jama. **315**, 801–810 (2016)
4. Ho, J., Lee, C., Ghosh, J.: Imputation-enhanced prediction of septic shock in ICU patients. In: Proceedings of The ACM SIGKDD Workshop On Health Informatics (HI-KDD12), pp. 18 (2012)
5. Wang, Z., Yao, B.: Multi-branching temporal convolutional network for sepsis prediction. IEEE J. Biomed. Health Inf. **26**, 876–887 (2021)
6. He, Z., et al.: Early sepsis prediction using ensemble learning with features extracted from LSTM recurrent neural network. In: 2019 Computing In Cardiology (CinC), p. 1 (2019)
7. Johnson, A., et al.: MIMIC-IV, a freely accessible electronic health record dataset. Sci. Data **10**, 1 (2023)
8. Johnson, A., et al.: MIMIC-CXR, a de-identified publicly available database of chest radiographs with free-text reports. Sci. Data **6**, 317 (2019)
9. Ronaldo, D.: Real-time sepsis prediction in intensive care units using temporal deep learning models on longitudinal electronic health records. J. ID **9471**, 1297 (2025)
10. Yin, C., et al.: SepsisCalc: Integrating Clinical Calculators into Early Sepsis Prediction via Dynamic Temporal Graph Construction. ArXiv Preprint arXiv:2501.00190 (2024)
11. Agnello, L., Vidali, M., Padoan, A., Lucis, R., Mancini, A., Guerranti, R., Plebani, M., Ciaccio, M., Carobene, A.: Machine learning algorithms in sepsis. Clin. Chim. Acta **553**, 117738 (2024)
12. Ghanvatkar, S., Rajan, V.: Graph-Based Patient Representation for Multimodal Clinical Data: Addressing Data Heterogeneity. MedRxiv, pp. 2023–12 (2023)

Clinically-Informed Preprocessing Improves Stroke Segmentation in Low-Resource Settings

Juampablo E. Heras Rivera(✉), Hitender Oswal, Tianyi Ren, Yutong Pan, William Henry, Caitlin M. Neher, and Mehmet Kurt

University of Washington, Seattle, WA 98105, USA
{jehr,hitender,tr1,ypan4,whenry1,neherc,mkurt}@uw.edu

Abstract. Stroke is among the top three causes of death worldwide, and accurate identification of ischemic stroke lesion boundaries from imaging is critical for diagnosis and treatment. The main imaging modalities used include magnetic resonance imaging (MRI), particularly diffusion weighted imaging (DWI), and computed tomography (CT)-based techniques such as non-contrast CT (NCCT), contrast-enhanced CT angiography (CTA), and CT perfusion (CTP). DWI is the gold standard for the identification of lesions but has limited applicability in low-resource settings due to prohibitive costs. CT-based imaging is currently the most practical imaging method in low-resource settings due to low costs and simplified logistics, but lacks the high specificity of MRI-based methods in monitoring ischemic insults. Supervised deep learning methods are the leading solution for automated ischemic stroke lesion segmentation and provide an opportunity to improve diagnostic quality in low-resource settings by incorporating insights from DWI when segmenting from CT. Here, we develop a series of models which use CT images taken upon arrival as inputs to predict follow-up lesion volumes annotated from DWI taken 2–9 days later. Furthermore, we implement clinically motivated preprocessing steps and show that the proposed pipeline results in a **38%** improvement in Dice score over 10 folds compared to a nnU-Net model trained with the baseline preprocessing. Finally, we demonstrate that through additional preprocessing of CTA maps to extract vessel segmentations, we further improve our best model by **21%** over 5 folds. The code, weights, and Docker image for the developed pipeline can be found in https://github.com/KurtLabUW/ISLESSeg.git.

Keywords: Stroke lesion segmentation · preprocessing · deep learning · nnU-Net · ISLES'24

1 Introduction

Stroke is among the top three causes of death worldwide, with ischemic strokes accounting for over 87% of cases [27]. In the acute (early) setting, medical

U. Anazodo et al. (Eds.): MIRASOL 2025, LNCS 16398, pp. 250–259, 2026.
https://doi.org/10.1007/978-3-032-13654-1_25

imaging is critical for rapid accurate diagnosis, treatment triage, prognosis prediction, and secondary preventive precautions [24]. Diffusion weighted imaging (DWI) MRI and computed tomography (CT) are the primary imaging modalities used to understand stroke lesion progression and are typically used to obtain segmentations of the lesion boundaries. In the emergency setting, CT is always used due to its availability, low cost, and speed, while DWI is rarely available.

DWI is approximately 4 to 5 times more sensitive in detecting acute stroke than non-contrast CT (NCCT), and is the gold standard as it can detect 95% of hyperacute ischemic infarcts [3,4,6,7]. DWI is capable of detecting acute brain infarction within 1 to 2 h, while NCCT may be negative for the first 24 to 36 h [18,24]. Furthermore, automated segmentation of the ischemic core using deep learning methods with DWI as input is well-established, with Dice scores surpassing 80% when compared to clinician-annotated labels [11]. Despite strong evidence supporting DWI as superior to NCCT for confirming the diagnosis of acute stroke within the first 24 h, logistical and financial issues limit its use in acute settings since most institutions find it challenging to reserve MRI scanners without delaying treatment.

In low-resource settings, these issues are further exacerbated, as the availability of MRI in acute settings is incredibly limited. For instance, in Africa and many countries in South America, there is less than 1 MRI scanner per million people [8]. For context, in high-resource countries like the United States, there are over 40 MRI scanners per million people [26]. This means that although DWI is the most reliable imaging method for stroke lesion identification, it remains largely inaccessible in low-resource settings. As a result, CT remains the most prevalent imaging modality for acute ischemic stroke treatment around the world due to its simplified logistics and significantly lower operational costs.

Although DWI remains widely unavailable in low-resource settings, supervised deep learning methods provide an opportunity to improve diagnostic quality in low-resource settings by incorporating insights from DWI data when segmenting stroke lesions from CT. Here, we present an approach for stroke lesion segmentation which uses CT scans from the acute ischemic stroke setting to predict follow-up stroke lesion volumes obtained from DWI 2–9 days after. For training and evaluation, we use the ISLES'24 [23] challenge dataset, which provides longitudinal imaging data of stroke patients, including acute NCCT, CT angiography (CTA), CT perfusion (CTP), DWI, and ground truth hand-annotated lesion masks obtained from DWI.

In comparison to MRI-based segmentation which requires minimal preprocessing, CT scans require significant preprocessing to remove irrelevant information and provide a clear learning signal for the segmentation model. Here, we develop a novel preprocessing pipeline to improve segmentation performance, considering the reasoning patterns used by clinicians. Although simple and computationally inexpensive, we show that the proposed pipeline results in a **45%** improvement in performance when compared comparison to a model trained with baseline nnU-Net preprocessing. Finally, we demonstrate that through additional preprocessing of CTA maps to extract vessel segmentations, we can further improve our best model by **21%** on average over 5-folds.

2 Dataset

The ISLES'24 challenge dataset was used for model training and evaluation. It consists of multi-center, multi-scanner imaging and tabular data from patients with large vessel occlusion (LVO) ischemic stroke. Imaging was acquired at two time points: at admission (acute) and at follow-up 2–9 days later (subacute).

Acute imaging includes the diagnostic CT trilogy: NCCT, CTA, and CTP. Additionally, it includes four CTP-derived maps: cerebral blood flow (CBF), cerebral blood volume (CBV), mean transit time (MTT), and time to maximum (TMax). These maps were generated using an FDA-approved clinical software (`icobrain CVA`), following motion correction and deconvolution processing. Follow-up imaging includes DWI and apparent diffusion coefficient (ADC) maps, which were used to derive stroke lesion core segmentations. Ground truth labels were generated using an ensemble segmentation pipeline from ISLES'22 [11] and quality-checked by neuroradiologists.

Each imaging modality offers complementary information about stroke pathology. CTA uses contrast-enhancement to provide a view of cerebral vasculature, allowing detection of occlusions and collateral flow. CBF reflects the rate of blood delivery to brain tissue, CBV indicates the total blood volume within a region, MTT represents the average time it takes blood to pass through a voxel, and TMax captures delays in contrast arrival. These quantities are assessed together to infer tissue at risk of irreversible damage (penumbra), vs. the unrecoverable tissue (core). Follow-up DWI detects regions of restricted diffusion due to cytotoxic edema and is highly sensitive to infarcted tissue. Apparent Diffusion Coefficient (ADC) models how freely water molecules can move within a specific area of tissue and helps to confirm diffusion abnormalities.

In total, the ISLES'24 dataset includes 250 sets of scans, with 150 sets for training (100 from the University Hospital of Munich and 50 from the University Hospital of Zurich), and 100 sets for testing (from undisclosed hospitals).

3 Methods

This section describes the components of our proposed method, which includes a clinically-guided preprocessing pipeline. This includes brain volume extraction with `SynthStrip` [12] (Sect. 3.1), clinically motivated intensity windowing (Sect. 3.3), and segmentation of blood vessels from CTA (Sect. 3.2). Then, to obtain stroke lesion segmentations, a standard residual nnU-Net model [13] is used (Sect. 3.4). The models were trained with 5 preprocessed inputs, consisting of: CTA, MTT, TMax, CBV, and CBF maps, and one-channel binary ischemic core masks as output. For the best model, the CTA input was replaced with binary vessel segmentation maps obtained from CTA.

3.1 Brain Extraction

The scans in the ISLES'24 brain imaging dataset contain non-brain structures such as the skull and background artifacts, which can hinder model training. To address this, we apply `SynthStrip`, a deep-learning-based brain-extraction tool trained on diverse synthetic images. `SynthStrip` allows for accurate and fast brain extraction of the entire dataset in minutes, a process that would have taken days for traditional brain extraction algorithms. First, we applied `SynthStrip` on the NCCT scans to obtain a brain masks. Then, we applied this brain mask to the other co-registered scans (CTA, and CTP-derived maps) to obtain skull-stripped versions of the data.

3.2 Vessel Segmentation Using CTA

CTA scans quickly and reliably add important information in cases of acute ischemic stroke. CTA shows the site of occlusion, the length of the occluded arterial segment, and the contrast-enhanced arteries beyond the occlusion as an estimate of collateral blood flow [15]. However, for deep learning models to effectively process CTA, it is beneficial to segment the vessel regions from CTA as binary masks and use these as inputs. Our proposed vessel segmentation pipeline is shown in Algorithm 1 .

Algorithm 1. Vessel Segmentation using CTA

Input: $\mathbf{I}_{\text{CTA}}$: CTA volume, $\mathbf{I}_{\text{NCCT}}$: NCCT volume, $\mathbf{B}$: Brain mask from `SynthStrip`
Params: HU window $[0, 400]$, thresholds $\tau_{\text{low}}=50$, $\tau_{\text{high}}=400$, min size $s_{\text{min}}=25$
Output: $\mathbf{M}_{\text{vessel}}$: Binary vessel mask

1: $\mathbf{I}_{\text{CTA}} \leftarrow \text{clip}(\mathbf{I}_{\text{CTA}}, 0, 400)$, $\mathbf{I}_{\text{NCCT}} \leftarrow \text{clip}(\mathbf{I}_{\text{NCCT}}, 0, 400)$ ▷ Clip intensities
2: $\mathbf{I}_{\text{CTA}} \leftarrow \mathbf{I}_{\text{CTA}} \odot \mathbf{B}$, $\mathbf{I}_{\text{NCCT}} \leftarrow \mathbf{I}_{\text{NCCT}} \odot \mathbf{B}$ ▷ Apply brain mask
3: $\mathbf{D} \leftarrow \mathbf{I}_{\text{CTA}} - \mathbf{I}_{\text{NCCT}}$ ▷ Voxel-wise difference
4: $\mathbf{D}(\mathbf{D} < \tau_{\text{low}}) \leftarrow 0$, $\mathbf{D}(\mathbf{D} > \tau_{\text{high}}) \leftarrow 0$ ▷ Suppress low contrast and artefacts
5: $\mathbf{M}_0 \leftarrow (\mathbf{D} \neq 0)$ ▷ Binary candidate mask
6: $\{\mathcal{C}_k\} \leftarrow \text{LabelConnectedComponents}(\mathbf{M}_0)$ ▷ Connected components
7: $\mathbf{M}_{\text{vessel}} \leftarrow \bigcup_{k:\,|\mathcal{C}_k| \geq s_{\text{min}}} \mathcal{C}_k$ ▷ Keep components with size $\geq s_{\text{min}}$ voxels
8: **return** $\mathbf{M}_{\text{vessel}}$

Algorithm 1 **description:** Starting from CTA and the co-registered NCCT, we first clip voxel intensities to $[0, 400]$ Hounsfield Units (HU) to supress extreme artifacts and noise. Then we apply a brain mask generated with `SynthStrip` to both scans to exclude structures outside of the brain. We then compute a voxel-wise difference between the CTA and NCCT, which highlights contrast-filled vessels while canceling parenchyma and remaining bone. Afterwards, we supress voxels with low (< 50 HU) and high (> 400 HU) contrast, to focus on the vessel regions. Then, we make a binary mask based on remaining nonzero voxels. Connected component analysis is performed on this mask, and only components containing at least 25 voxels are retained, producing the final binary vessel mask $\mathbf{M}_{\text{vessel}}$. An example of a CTA from a subject in the dataset and their overlaid vessel mask is shown in Fig. 1.

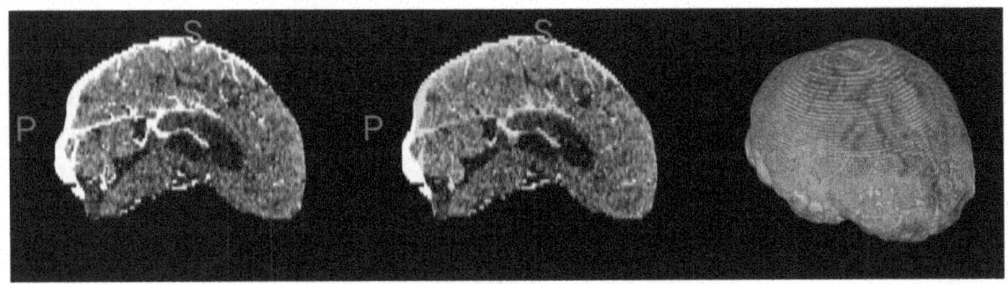

Fig. 1. Sagittal view of a subject's CTA (left) and the same view with the vessel map extracted using Algorithm 1 overlaid in green (middle). Right: 3D rendering of the CTA scan with the same vessel mask overlaid in green.

3.3 Intensity Windowing

To remove extraneous information from CT images, intensity value windowing was applied. When clinical windowing guidelines were available in the literature, they informed the initial settings, and empirical adjustments were made to optimize model performance. The clinically established thresholds used for reference include: CBF $< 17\,\text{mL}/100\,\text{g/min}$ [2]; CBV $> 2\,\text{mL}/100\,\text{g}$ [25]; MTT $> 145\%$ of the contralateral baseline, where the 0–30 HU range captures the full variability [1,5,16]; and Tmax $> 6\,\text{s}$ [17]. For CTA, the window was manually adjusted to enhance contrast between healthy and ischemic brain regions, as in [19]. For inputs with inconclusive clinical thresholds, such as CTA, MTT, and CBF, further widening of the selected windows beyond ranges found in the literature was performed to improve visibility of the ischemic lesion. The windowing bounds used for the proposed preprocessing strategy can be found in Table 1 as "Clinical Window", and an example of a subject's scans before and after windowing are shown in Fig. 2. After windowing, values were min-max normalized to the $[0, 1]$ range, then 3-D histogram equalization was done on foreground voxels. Finally, all background voxels were assigned zero to remove artifacts outside of the brain.

3.4 nnU-Net

Despite the emergence of models like Transformers [9] and diffusion frameworks [21], CNNs based on the U-Net architecture [22] remain state-of-the-art for medical image segmentation [14]. The nnU-Net framework [13] automates hyperparameter tuning by adapting a standard U-Net to the training data, often outperforming manually tuned and novel models [10]. For the models evaluated here, the 3D nnU-Net "ResEnc L" with a (56, 320, 256) patch size, Dice

Table 1. Comparison of our clinically chosen intensity windows with the nnU-Net foreground intensity range (0.5^{th}–99.5^{th} percentiles). The last column shows what fraction of the nnU-Net range each clinical window retains.

Modality	Clinical Window	nnU-Net CT Range	% of Range Kept
CTA (HU)	(0, 90)	(−3.25, 342.48)	26.0 %
CBF (mL/100g/min)	(0, 35)	(1.42, 72.64)	49.2 %
CBV (mL/100g)	(0, 10)	(−10.31, 19.35)	33.7 %
MTT (s)	(0, 20)	(−96.91, 28.50)	15.9 %
Tmax (s)	(0, 7)	(−20.76, 20.29)	17.1 %

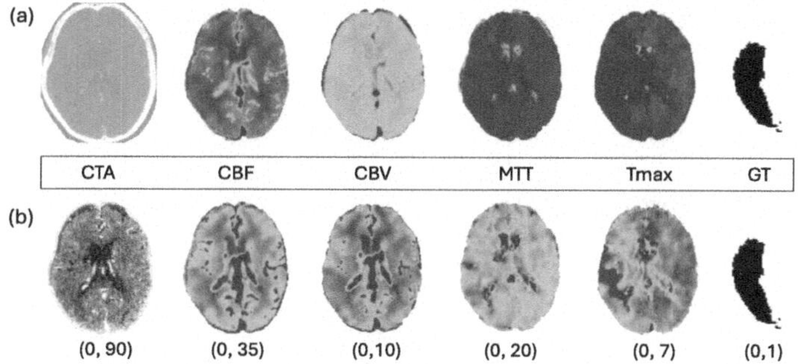

Fig. 2. Imaging data from an example subject (a) before and (b) after preprocessing, with windowing bounds shown below (b). Over all images, the clinically-informed preprocessing increases visibility of the stroke lesion (GT).

and cross-entropy loss, and the SGD optimizer (lr=0.01, momentum=0.99) was used. For CT data, nnU-Net clips intensities to the $0.5 - 99.5$th percentiles of all foreground voxels (union of the label masks), then applies Z-score normalization using the mean and standard deviation of those same foreground intensities [13]. The clipped ranges kept by nnU-Net for the ISLES'24 dataset are in Table 1 as "nnU-Net CT Range".

4 Results

Two sets of experiments were conducted, one with 10-fold cross validation (Sect. 4.1), and one with 5-fold cross validation (Sect. 4.2). The model from the highest-performing fold of the 10-fold experiments was also submitted to the ISLES'24 challenge, where it achieved first place (Sect. 4.3).

4.1 10-Fold CV Experiments

Table 2 shows the results for the highest validation Dice score for the 10-fold cross validation experiment. On this fold, the standard nnU-Net preprocessing achieved a Dice score of 21.8%. Applying custom windowing alone improved performance to 31.0%, and combining custom windowing with Z-score normalization (as described in Sect. 3.3) further increased the validation Dice score to **31.8%**.

Table 2. Dice scores for the baseline nnU-Net preprocessing and the clinically informed preprocessing (windowing + histogram equalization). Values are reported as the mean ± standard deviation over ten folds, together with the best-fold Dice score.

Preprocessing Strategy	Dice (mean ± SD)	Δ (mean)	Dice (best fold)
Baseline (nnU-Net default)	0.162 (0.065)	–	0.218
+ **Section** 3.3 **Preproc.**	**0.224 (0.047)**	**+38.4%**	**0.318**

4.2 5-Fold CV Experiments

For these experiments, 5-fold cross-validation was used. The baseline model in Table 3 applies the best preprocessing strategy from Sect. 4.1. In the improved setup, the windowed CTA input is replaced with binary vessel maps from Sect. 3.2, resulting in a 21% performance gain. Table 3 reports results across all 5 folds for both approaches.

Table 3. Validation Dice scores for the best preprocessing pipeline from Sect. 4.1 versus the same model trained with the vessel segmentation replacing CTA, evaluated on the same 5-fold split. Δ gives the relative percentage change with respect to the best preprocessing pipeline from Sect. 4.1. The best fold for both experiments is highlighted in blue.

Dice Score	Fold 1	Fold 2	Fold 3	Fold 4	Fold 5	Mean	STD
Best Preproc.	0.2167	0.3184	0.2181	0.1521	0.1696	0.2150	0.0647
+ Vessel Segm.	0.2257	0.3273	0.2129	0.2852	0.2495	**0.2601**	**0.0466**
Δ (%)	+4.2	+2.8	-2.4	+87.5	+47.1	**+21.0**	**-27.9**

4.3 ISLES'24 Challenge Submission

The model from the best fold as described in Sect. 4.1 was submitted to the ISLES'24 challenge [23] and achieved first place [20]. Table 4 shows test set evaluation metrics for the top 3 entries in the ISLES'24 leaderboard.

Table 4. Test set evaluation metrics for the top 3 entries in the ISLES'24 leaderboard, reported as mean (standard deviation). Metrics shown are: Dice coefficient (Dice), Absolute volume difference (AVD), F1-score, and Absolute lesion count difference (ALCD). Per column, **Bold** = best, blue = second best.

Team	Dice (%) ↑	AVD ↓	F1 (%) ↑	ALCD ↓
Kurtlab (Ours)	**28.50 (21.27)**	**21.23 (37.22)**	14.39 (21.19)	7.18 (7.67)
AMC-Axolotls	26.27 (24.73)	21.31 (35.23)	**14.94 (25.12)**	7.66 (7.94)
Ninjas	25.46 (19.08)	26.29 (39.73)	9.92 (13.46)	**5.98 (6.46)**

5 Analysis

The results of our experiments demonstrate that clinically informed preprocessing improves stroke lesion segmentation from CT imaging. In the first set of experiments, we show that applying skull-stripping and custom intensity windowing, followed by histogram equalization, improves model performance by 38.4% compared to the baseline nnU-Net preprocessing. This is likely due to standard percentile-based preprocessing preserving irrelevant high-intensity regions (e.g., skull), while the stroke core typically occupies a narrow band within the overall intensity range. By using clinically informed preprocessing strategies, the model is able to focus on the anatomically and pathophysiologically relevant structures.

In the second set of experiments, we observe that vessel structure is the most informative component of CTA for stroke lesion segmentation. By segmenting the vessels prior to training, we introduce a strong clinical prior, improving model performance. This improvement allows the model to reduce false negatives, especially in cases of LVOs. While our proposed preprocessing pipeline improves performance, it still exhibits high variance on the ISLES'24 test set (Table 4), suggesting robustness should be improved in future works. We expect this variance will decrease as dataset size increases, as more training samples should improve generalization.

6 Conclusions

In low-resource settings, accurate prediction of ischemic core from CT scans is critical to reduce disparities in treatment quality in comparison to countries which have access to DWI. This paper presents a novel preprocessing strategy for stroke lesion segmentation, which improves the ability of models to predict subacute stroke lesions using CT imaging obtained in the acute phase. The presented analysis shows that standard preprocessing pipelines, such as those used in nnU-Net for CT scans, are insufficient to segment ischemic stroke lesions in CT. Here, we share the details of our preprocessing approach which increases the visibility of tissues of interest and allows the model to focus on clinically-relevant structures. The results of our preprocessing steps also allowed us to place first in the ISLES'24 challenge. Further studies in this direction will involve improving robustness of the segmentation approach by implementing more clinical priors in the preprocessing pipeline.

Acknowledgments. The work of Juampablo Heras Rivera was partially supported by the U.S. Department of Energy Computational Science Graduate Fellowship under Award Number DE-SC0024386.

Disclosure of Interests. The authors have no competing interests to declare that are relevant to the content of this article.

References

1. Alzahrani, A., Zhang, X., Albukhari, A., Wardlaw, J.M., Mair, G.: Assessing brain tissue viability on nonenhanced computed tomography after ischemic stroke. Stroke **54**(2), 558–566 (2023)
2. Bandera, E., Botteri, M., Minelli, C., et al.: Cerebral blood flow threshold of ischemic penumbra and infarct core in acute ischemic stroke. Stroke **37**(5), 1334–1339 (2006). https://doi.org/10.1161/01.STR.0000217418.29609.22
3. Brazzelli, M., Sandercock, P., Chappell, F., Celani, M., Righetti, E., Arestis, N., et al.: Magnetic resonance imaging versus computed tomography for detection of acute vascular lesions in patients presenting with stroke symptoms. Cochrane Database Syst. Rev. **4**, CD007424 (2009)
4. Chalela, J., Kidwell, C., Nentwich, L., Luby, M., Butman, J., Demchuk, A., et al.: Magnetic resonance imaging and computed tomography in emergency assessment of patients with suspected acute stroke: a prospective comparison. Lancet **369**, 293–298 (2007)
5. Czap, A.L., Sheth, S.A.: Overview of imaging modalities in stroke. Neurology **97**(20_Supplement_2), S42–S51 (2021)
6. Fiebach, J., Schellinger, P., Jansen, O., Meyer, M., Wilde, P., Bender, J., et al.: CT and diffusion-weighted MR imaging in randomized order: diffusion-weighted imaging results in higher accuracy and lower interrater variability in the diagnosis of hyperacute ischemic stroke. Stroke **33**, 2206–2210 (2002)
7. González, R., Schaefer, P., Buonanno, F., Schwamm, L., Budzik, R., Rordorf, G., et al.: Diffusion-weighted MR imaging: diagnostic accuracy in patients imaged within 6 hours of stroke symptom onset. Radiology **210**, 155–162 (1999)
8. Hasford, F., Mumuni, A.N., Trauernicht, C., et al.: A review of MRI studies in Africa with special focus on quantitative MRI: Historical development, current status and the role of medical physicists. Phys. Med. **103**, 46–58 (2022)
9. Hatamizadeh, A., Nath, V., Tang, Y., Yang, D., Roth, H.R., Xu, D.: Swin UNETR: Swin transformers for semantic segmentation of brain tumors in MRI images. In: Crimi, A., Bakas, S. (eds.) International MICCAI Brainlesion Workshop. pp. 272–284. Springer, Cham (2021). https://doi.org/10.1007/978-3-031-08999-2_22
10. Heras Rivera, J.E., et al.: An ensemble approach for brain tumor segmentation and synthesis. arXiv preprint arXiv:2411.17617 (2024)
11. Hernandez Petzsche, M.R., et al.: ISLES 2022: a multi-center magnetic resonance imaging stroke lesion segmentation dataset. Sci. data **9**(1), 762 (2022)
12. Hoopes, A., Mora, J.S., Dalca, A.V., Fischl, B., Hoffmann, M.: Synthstrip: skull-stripping for any brain image. Neuroimage **260**, 119474 (2022). https://doi.org/10.1016/j.neuroimage.2022.119474

13. Isensee, F., Jaeger, P.F., Kohl, S.A.A., Petersen, J., Maier-Hein, K.H.: nnu-net: a self-configuring method for deep learning-based biomedical image segmentation. Nat. Methods **18**(2), 203–211 (2021). https://doi.org/10.1038/s41592-020-01008-z
14. Isensee, F., Wald, T., Ulrich, C., et al.: nnU-Net revisited: a call for rigorous validation in 3D medical image segmentation. In: Proceeding of MICCAI, pp. 488–498 (2024)
15. Knauth, M., von Kummer, R., Jansen, O., Hähnel, S., Dörfler, A., Sartor, K.: Potential of CT angiography in acute ischemic stroke. Am. J. Neuroradiol. **18**(6), 1001–1010 (1997)
16. Nukovic, J.J., et al.: Neuroimaging modalities used for ischemic stroke diagnosis and monitoring. Medicina **59**(11), 1908 (2023)
17. Olivot, J., Mlynash, M., Thijs, V.N., et al.: Optimal Tmax threshold for predicting penumbral tissue in acute stroke. Stroke **40**(2), 469–475 (2009). https://doi.org/10.1161/STROKEAHA.108.526954
18. Pantano, P., Caramia, F., Bozzao, L., Dieler, C., von Kummer, R.: Delayed increase in infarct volume after cerebral ischemia: correlations with thrombolytic treatment and clinical outcome. Stroke **30**, 502–507 (1999)
19. Pulli, B., Schaefer, P.W., Hakimelahi, R., et al.: Acute ischemic stroke: infarct core estimation on CT angiography source images depends on CT angiography protocol. Radiology **262**(2), 593–604 (2012)
20. Ren, T., et al.: How we won the ISLES'24 challenge by preprocessing (2025). https://arxiv.org/abs/2505.18424
21. Ren, T., et al.: Re-diffinet: Modeling discrepancy in tumor segmentation using diffusion models. In: Medical Imaging with Deep Learning (2024)
22. Ronneberger, O., Fischer, P., Brox, T.: U-Net: Convolutional networks for biomedical image segmentation, arXiv:1505.04597 (2015)
23. de la Rosa, E., Su, R., Reyes, M., et al.: ISLES'24: Improving final infarct prediction in ischemic stroke using multimodal imaging and clinical data, arXiv:2408.10966 (2024)
24. Tong, E., Hou, Q., Fiebach, J.B., Wintermark, M.: The role of imaging in acute ischemic stroke. Neurosurg. Focus **36**(1), E3 (2014)
25. Wintermark, M., Flanders, A.E., Velthuis, B., et al.: Perfusion-CT assessment of infarct core and penumbra. Stroke **37**(4), 979–985 (2006). https://doi.org/10.1161/01.STR.0000209238.61459.39
26. World Health Organization: Medical devices: magnetic resonance imaging units per million population, total density. https://www.who.int/data/gho/data/indicators/indicator-details/GHO/total-density-per-million-population-magnetic-resonance-imaging, Accessed 02 July 2025
27. Yousufuddin, M., Young, N.: Aging and ischemic stroke. Aging (Albany NY) **11**(9), 2542–2544 (2019). https://doi.org/10.18632/aging.101931

Deep Ensemble Approach for Enhancing Brain Tumor Segmentation in Resource-Limited Settings

Jeremiah Fadugba[1(✉)], Isabel Lieberman[3], Olabode Ajayi[5], Mansour Osman[2], Solomon Oluwole Akinola[4], Tinashe Mutsvangwa[3], Farouk Dako[7], Dong Zhang[8], Udunna C. Anazondo[3,6,9], and Raymond Confidence[6,9]

[1] University of Ibadan, Ibadan, Nigeria
jfadugba@aimsric.org
[2] Carnegie Mellon University Africa, Kigali, Rwanda
[3] University of Cape Town, Cape Town, South Africa
[4] Institute for Intelligent Systems, University of Johannesburg, Johannesburg, South Africa
[5] South African National Bioinformatics Institute (SANBI), University of Western Cape, Cape Town, South Africa
[6] Medical Artificial Intelligence Laboratory (MAI Lab), Lagos, Nigeria
[7] Department of Radiology, University of Pennsylvania, Philadelphia, USA
[8] Department of Electrical and Computer Engineering, University of British Columbia, Vancouver, Canada
[9] Montreal Neurological Institute, McGill University, Montréal, Canada

Abstract. Segmentation of brain tumors is a critical step in treatment planning, yet manual segmentation is both time-consuming and subjective, relying heavily on the expertise of radiologists. In Sub-Saharan Africa, this challenge is magnified by overburdened medical systems and limited access to advanced imaging modalities and expert radiologists. Automating brain tumor segmentation (BraTS) using deep learning offers a promising solution. Convolutional Neural Networks (CNNs), especially the U-Net architecture, have shown significant potential. However, achieving generalizability across different datasets remains a challenge. This study addresses this gap by developing a deep learning ensemble that integrates three CNN models (3DUNet, V-Net, and MSA-VNet) for the semantic segmentation of gliomas. By initially training on the larger BraTS glioma dataset (n = 1251) and fine-tuning with the BraTS-Africa dataset (n = 60), we enhance model performance. Our ensemble approach significantly outperforms individual models, achieving DICE scores of 0.84 for Tumor Core, 0.85 for Whole Tumor, and 0.82 for Enhancing Tumor. Individually, MSA-VNet had the top performing score of 0.81 for TC and 0.80 for ET while V-Net achieved 0.83 for WT. These results underscore the potential of ensemble methods in improving the accuracy and reliability of automated brain tumor segmentation, particularly in resource-limited settings.

U. Anazodo et al. (Eds.): MIRASOL 2025, LNCS 16398, pp. 260–269, 2026.
https://doi.org/10.1007/978-3-032-13654-1_26

Keywords: Deep Learning · Image Segmentation · Brain Tumor Segmentation · BraTS2024 · Deep Ensemble

1 Introduction

Brain-tumour research and care in low- and middle-income countries (LMICs) have long been hampered by the sparse availability of MRI scanners [17]. In Sub-Saharan Africa (SSA) this shortage is reflected in disproportionately high glioma mortality, underscoring an urgent need for wider access to advanced imaging and multidisciplinary treatment. Brain tumours—abnormal proliferations of brain cells, remain among the most severe and life-threatening neurological conditions, and their management hinges on precise medical imaging [9].

Manual tumor delineation on MRI is time-consuming and prone to variability due to its slice-by-slice nature across 3D volumes [18]. These difficulties are magnified in resource-constrained SSA health systems that already struggle with workforce shortages [1]. Although manual tracing may achieve high accuracy, the wide variation in lesion shape, size and location, as well as the burden of inspecting several MRI contrasts—renders it impractical for large-scale studies and routine care [15]. Computer-aided diagnosis through machine-learning methods therefore offers an avenue to improve diagnostic accuracy, accelerate early detection and stratify patient prognosis.

Traditional automated techniques such as thresholding and region growing [6] cannot reliably cope with heterogeneous tumour appearance; they often miss subtle borders or fail under changing imaging conditions. Deep-learning approaches, especially convolutional neural networks (CNNs), have emerged as state-of-the-art alternatives. Architectures derived from U-Net [20] and, more recently, prompt-guided models such as the Segment Anything Model (SAM) [4], learn rich hierarchical features and have achieved impressive segmentation accuracy across diverse datasets. Yet important gaps remain.

Models trained on data from high-income settings often generalise poorly to SSA cohorts, where scanner hardware, imaging protocols and patient demographics differ markedly [2]. Moreover, most cutting-edge architectures assume abundant GPU resources, limiting their deployment in LMIC clinics with modest computational infrastructure [5,8]. Addressing these challenges calls for segmentation methods that are accurate, robust and efficient in real-world low-resource environments. The present study therefore develops and validates a deep-learning pipeline tailored to SSA imaging data, with the broader aim of strengthening clinical decision-making and improving survival outcomes in this underserved region.

2 Related Works

Deep-learning work on glioma segmentation has moved quickly from generic CNN pipelines to specialised architectures that confront three hard realities: a chronic shortage of annotated data, pronounced tumour heterogeneity, and tight computational budgets. Canonical encoder–decoders 2D U-Net, its volumetric cousins 3D U-Net and V-Net, and the dual-path DeepMedic—still set the reference bar because their skip connections protect spatial detail while the contracting path captures global context [16,20,24]. They dominate BraTS scoreboards [2,10], yet pure convolutions stumble when lesions span multiple anatomical compartments or scans carry site-specific artefacts [10].

Recent variants tackle those limits head-on. Attention gates, dense residual blocks, and full self-attention have been grafted onto the U-Net scaffold [6,13]. Multiscale-Attention U-Nets tighten indistinct borders, while Transformer-CNN hybrids such as TransBTS and UNETR marry long-range self-attention with 3D convolutions, driving whole-tumour Dice beyond 0.85 on the latest BraTS releases [22,23]. Self-configuring nnU-Net pipelines push post-operative performance further by auto-tuning pre-processing, patch size, and learning schedules [10]. Ensembling adds another layer of resilience: fusing U-Net, V-Net, and attention-augmented backbones via majority voting or uncertainty weighting routinely lifts Dice by two to four points for whole tumour (WT), tumour core (TC), and enhancing tumour (ET) masks [12].

Because expertly labelled MR volumes are scarce—particularly in Sub-Saharan Africa—research has pivoted toward label-efficient learning [7]. Semi-supervised workflows harvest information from vast unlabelled repositories, GAN- or diffusion-based augmentation plugs subtype gaps, and knowledge-distillation shrinks heavyweight ensembles into lightweight students that still run on CPU-only clinics [7]. Pruned or quantised backbones such as MobileNet-U-Net and Efficient U-Net push inference below 30 ms per slice on modest hardware, edging real-time surgical guidance into reach [21]. Clinically, automated segmentation now saves 8–15 min per case and trims inter-rater variance by roughly a third in radiotherapy planning and longitudinal follow-up [14]. This evolutionary arc—from classical CNNs through attention-rich hybrids to resource-aware distilled models, defines the challenge we take up: delivering state-of-the-art accuracy under the data and compute constraints that characterise low and middle income settings [3,15,17].

3 Methodology

3.1 Dataset Description

Sixty brain MRI scans from the Brain Tumor Segmentation (BraTS)-Africa 2024 data set [2], supplemented by the labelled training cohort ($n = 1251$) of the BraTS-GLI 2024 post-treatment glioma data set, were used in this study. For further validation of our method, the BraTS-Africa validation set ($n = 35$) was used. The BraTS-Africa subset was retrospectively acquired from patients with gliomas in Sub-Saharan Africa (SSA) and includes T1-weighted (T1), post-gadolinium contrast T1-weighted (T1Gd), T2-weighted (T2), and T2-Fluid-Attenuated Inversion Recovery (T2-FLAIR) MRI volumes, with expert annotations of necrotic core, enhancing tumour, and oedema. Both data sets were collected under standard-of-care imaging protocols, then pre-processed and annotated with the BraTS pipeline and annotation protocol, ensuring label consistency with previous BraTS Challenge releases, as detailed in the BraTS-Africa data descriptor paper [2].

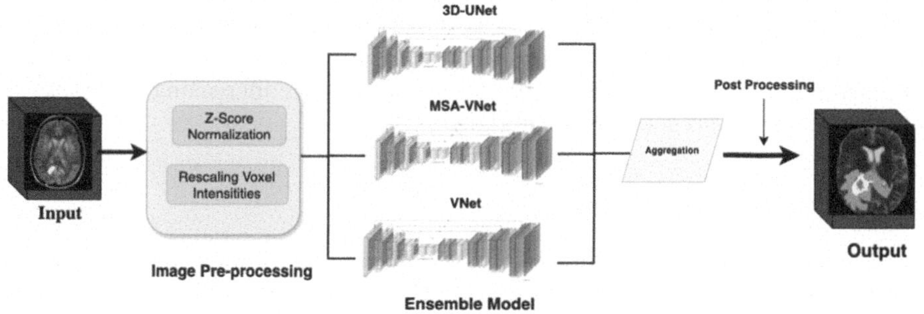

Fig. 1. Pipeline for the Deep Ensemble Learning.

3.2 Deep Ensemble Learning

Deep Ensemble learning is a common technique in machine learning and in particular, brain tumor segmentation where multiple deep learning models are trained and combined to make a single prediction [11,12,19].

In this work, our approach combined the 3DUNet [24], VNet [16], and MSA-VNet [23] to achieve superior segmentation performance. As seen in the pipeline, the final results were obtained by averaging the results of the individual predictions of the models to produce a final segmentation, achieving a higher performance compared to the single model's prediction. By integrating the predictions from these diverse models, the ensemble method reduces individual model biases and variances, leading to more robust and accurate segmentation outcomes. This pipeline is seen in Fig. 1.

We divided the official BraTS-GLI training dataset into a 60:20:20 split, holding 20% of the dataset for validation and 20% for testing. After every epoch, we iterate through this validation hold, calculating the Dice score between model prediction and the ground truth for the regions of ET, TC, and WT and then averaging these scores. For fine-tuning on the BraTS-Africa dataset we freeze the decoder for UNet3D to avoid overfitting since it has more parameters than other models while for V-Net and MSA-VNEt, we continue training on the BraTS-Africa dataset.

3.3 Image Pre-processing

In addition to the standard pre-processing pipeline from the BraTS Challenge organizers [2], we made further pre-processing of the scans following [19]. The pre-processing pipeline adopted is widely established in prior work [19], and we observed improved stability and convergence during training, which suggests a positive contribution to model performance. First, **Z-score normalization** was used to standardize voxel intensities across patients by subtracting the mean and dividing by the standard deviation, enabling consistent comparisons and reducing the impact of outliers. Second, **voxel intensity rescaling** was employed to enhance feature visibility by stretching the intensity range between the 2nd and 98th percentiles to span the full spectrum, thereby increasing contrast and highlighting subtle structures in the data.

3.4 Network Architecture

The proposed network architecture combines three model architecture to enhance automated tumor segmentation of the BraTS-Africa dataset. The architectures are:

1. 3DUNet [24] is an extension of the two dimensional (2D) U-Net architecture, specifically designed for volumetric medical image segmentation. It leverages three dimensional (3D) convolutions to capture spatial context across all three dimensions, thereby improving the accuracy of segmentation for complex structures within medical images. The architecture consists of an encoder-decoder structure with skip connections that allow for the preservation of fine-grained details while progressively capturing higher-level features. This makes 3DUNet particularly effective for tasks such as brain tumor segmentation, where understanding the spatial relationships within the volume is crucial.
2. V-Net [16] is a similar network to 3DUNet and is also designed for volumetric medical image segmentation. It features an encoder-decoder structure resembling a 'V' shape, hence its name. With a major similarity to 3DUNet, V-Net differs in the kernel sizes by using a volumetric kernels having size $5 \times 5 \times 5$ voxels within its residual blocks to enhance feature extraction. Skip connections in V-Net sums the input features with the output of the convolutional layers, allowing the network to combine low-level spatial information with high-level semantic information. In this work, We modify the original V-Net architecture [16] by first increasing the number of feature maps from 16 to 32 in contrast to the original implementation.
3. Multi-Scale Attention VNet (MSA-VNet) [23] is an advanced 3D CNN for volumetric medical image segmentation that builds upon V-Net principles. It incorporates multi-scale attention mechanisms and consists of an encoder-decoder structure, where the encoder compresses input through convolutional and pooling layers, capturing multi-scale features. A central convolutional block refines deep features before the decoder path, which uses transposed convolutions for up-sampling and integrates encoder information via skip connections. Multi-scale attention blocks at each skip connection allow the model to focus on relevant features from both paths, improving its ability to

distinguish subtle input differences. The network concludes with a final convolutional layer outputting the segmented map, resulting in a powerful tool for precise medical image segmentation.

3.5 Image Post-processing

Post-processing is an essential step in glioma segmentation, focused on refining initial results to improve accuracy and reduce errors. Common techniques include morphological operations, region growing, and level set evolution. Following the approach in [19] connected ET voxel components smaller than 50 voxels are removed and relabeled as background. Additionally, the Tumor Core (TC), comprising ET and NCR voxels, is adjusted to ensure that ET removal does not introduce holes, maintaining segmentation integrity.

4 Experiments and Results

4.1 Training Details

All model training and development were carried out using PyTorch, a deep-learning framework. The training experimental setup on the BraTS-GLI data includes four NVIDIA GeForce RTX 2080 Ti GPUs while a single GPU was used for finetuning on BraTS-Africa dataset. The models were trained using the Dice Loss function available on MONAI for 40 epochs with a batch size of 4. We employed the AdamW optimizer, incorporating a weight decay factor of 1×10^{-5}, an initial learning rate set to 6×10^{-5}, and a cosine annealing strategy for learning rate scheduling. For finetuning the same hyperparameters were used however, for cross validation experiments, we reduce the epochs to 30 for each folds.

4.2 Model Results

The results in Table 1 shows the DSC and the Hausdorff distance (95%) (HD95) values for a hold out test set of 12 cases from the BraTS-Africa training set. The use of this post-processing technique saw just a bit of an increase in performance for each of the model (see Table 1).

4.3 Online Evaluation

The performance of 3DUNet, V-Net, and MSA-VNet was assessed using 5-fold cross-validation on the BraTS-Africa 2024 Validation set via the Synapse platform[1]. Results (in Table 2) showed that V-Net outperformed the others, achieving the highest mean Dice scores across all tumor regions ET, TC, and WT. MSA-VNet showed improvement over the baseline 3DUNet, which had the lowest performance. All models performed best on WT segmentation, followed by TC, with ET being the most challenging. The strong performance of V-Net

Table 1. Dice and HD95 scores for TC, WT, and ET for each models on the hold-out test cases with and without post-processing.

Model	Post-Processing	DSC ↑			HD95 ↓		
		ET	TC	WT	ET	TC	WT
UNet	Without	0.73	0.65	0.82	19.85	20.36	14.76
	With	0.74	0.65	0.82	20.11	21.77	15.84
V-Net	Without	0.74	0.77	0.80	9.13	10.14	17.62
	With	0.74	0.76	0.81	9.56	11.52	17.34
MSA-VNet	Without	0.56	0.59	0.75	11.91	14.22	21.28
	With	0.64	0.59	0.73	10.89	14.96	20.32

Table 2. Results of 5 Fold cross-validation of each model on the BraTS-Africa 2024 validation dataset. The average of DSC and HD95 scores are computed via the Synapse online platform

Model	DSC ↑			HD95 ↓		
	ET	TC	WT	ET	TC	WT
UNet	0.79	0.79	0.83	21.13	22.73	12.05
V-Net	0.84	0.86	0.85	15.23	17.38	21.32
MSA-VNet	0.81	0.83	0.81	26.12	27.94	22.06

highlights the effectiveness of its simple but improved architecture over UNEt for segmenting complex brain tumor structures.

The online validation of the deep ensemble approach of these models is presented in Table 3. It summarizes the results of each model on the BraTS-Africa validation set computed on the Synapse Platform. The Dice score and Hausdorff distance 95% (HD95) were computed via the platform. Our results for the three evaluated tumor sub-regions ET, WT and TC regions with the post-processing technique shows that the ensemble model gained substantially for the WT region and not so much improvement in the other regions.

These results underscore the effectiveness of the proposed ensemble learning approach in improving segmentation accuracy for brain tumor segmentation, particularly on the Sub-Saharan Africa dataset.

5 Limitations

This study has some limitations. First, due to computational and time constraints, we did not conduct ablation experiments to isolate the impact of image pre-processing. Such tests would compare training with and without pre-processing, and while their absence limits definitive attribution of performance gains, prior work [19] suggests a positive effect.

[1] https://www.synapse.org.

Table 3. Further evaluation of each model's performance on the BraTS-Africa 2024 validation dataset. The average of DSC and HD95 scores were computed via the Synapse online platform

Model	DSC ↑			HD95 ↓		
	ET	TC	WT	ET	TC	WT
UNet	0.77	0.77	0.82	22.85	24.21	**11.93**
V-Net	0.80	0.81	0.83	**16.76**	19.38	19.278
MSA-VNet	0.80	0.81	0.73	29.05	31.59	37.55
Ensemble	**0.82**	**0.84**	**0.85**	16.86	**18.65**	13.44

Second, the model was not evaluated on external datasets. This limits claims of broad generalisability, though internal cross-validation helped mitigate overfitting. Future work will extend validation to independent datasets to confirm robustness.

Third, we did not benchmark the ensemble against fully trained single models, nor explore advanced fusion strategies beyond simple averaging. These omissions limit conclusions about ensemble superiority, but future studies will investigate alternative strategies expected to further enhance performance.

6 Conclusion

This paper presented our contribution to the brain tumor segmentation task on data from low resource settings with significantly different resolutions. This is a unique opportunity provided by the BraTS-Africa Challenge. This also is essential in developing and evaluating deep learning methods for brain tumors management in resource-limited settings. With the numerous challenges posed by the peculiarity of the BraTS-Africa dataset; such as the image quality and the number of cases, better performance was achieved with individual deep learning model. Moreover our study demonstrates the effectiveness of ensemble models. It is also noteworthy to mention that 3DUNet which served as a baseline competes favourably with other models. However, our investigation into a different architecture; V-Net showed a 4.5%, 5.4% and 1.8% increase in Dice score for ET, TC, and WT, respectively. We considered how attention mechanism can help push the performance using MSA-VNet but this architecture did not give any improvement to the baseline 3DUNet and the V-Net model. Overall, we see that the ensemble approach gave a 6.4 %, 8.6% and 3.9% performance increase in Dice score for ET, TC, and WT, respectively.

Acknowledgments. The authors would like to thank the instructors of the Sprint AI Training for African Medical Imaging Knowledge Translation (SPARK) Academy 2024 summer school on deep learning in medical imaging for providing insightful background knowledge on brain tumors that informed the research presented here. The authors would also like to thank Linshan Liu for administrative assistance in supporting the SPARK Academy training and capacity building activities which the authors immensely benefited from. The authors acknowledge the computational infrastructure support from the Digital Research Alliance of Canada (The Alliance) and knowledge translation support from the McGill University Doctoral Internship program through student exchange program for the SPARK Academy. The authors are grateful to McMedHacks for providing foundational information on python programming for medical image analysis as part of the 2023 SPARK Academy program. This research was funded by the Lacuna Fund for Health and Equity (PI: Udunna Anazodo, grant number 0508-S-001), National Science and Engineering Research Council of Canada (NSERC) Discovery Launch Supplement (PI: Udunna Anazodo, grant number DGECR-2022-00136), and Radiological Society of North America (RSNA) Research and Education Foundation Derek Harwood-Nash International Education Scholar Grant (PI: Farouk Dako and Udunna Anazodo, grant number ESCH24?284). This work was partly supported by the Italian Ministry of University and Research (MUR) under project PE0000013 – Future of Artificial Intelligence Research (FAIR).

Disclosure of Interests. The authors have no competing interests to declare that are relevant to the content of this article.

References

1. Adewole, M., et al.: Status of magnetic resonance imaging systems and quality control programs in Nigeria. medRxiv (2023)
2. Adewole, M., et al.: The brats-Africa dataset: expanding the brain tumor segmentation data to capture African populations. Radiol. Artif. Intell. **7**(4), e240528 (2025). https://doi.org/10.1148/ryai.240528, pMID: 40237600
3. Anazodo, U.C., Adewole, M., Dako, F.: AI for population and global health in radiology (2022)
4. Barakat, M., et al.: Towards samba: segment anything model for brain tumor segmentation in sub-Sharan African populations. arXiv preprint arXiv:2312.11775 (2023)
5. Barakat, M., et al.: Towards SAMBA: Segment Anything Model for Brain Tumor Segmentation in Sub-Sharan African Populations (2023). http://arxiv.org/abs/2312.11775. arXiv:2312.11775
6. Biratu, E.S., Schwenker, F., Debelee, T.G., Kebede, S.R., Negera, W.G., Molla, H.T.: Enhanced region growing for brain tumor MR image segmentation. J. Imaging **7**(2), 22 (2021). https://doi.org/10.3390/jimaging7020022. https://www.mdpi.com/2313-433X/7/2/22
7. Cho, W., et al.: Hollowed Net for On-Device Personalization of Text-to-Image Diffusion Models (2024). https://arxiv.org/abs/2411.01179v1
8. Futrega, M., Milesi, A., Marcinkiewicz, M., Ribalta, P.: Optimized U-Net for Brain Tumor Segmentation (2021). http://arxiv.org/abs/2110.03352. arXiv:2110.03352
9. Havaei, M., et al.: Brain tumor segmentation with deep neural networks. Med. Image Anal. **35**, 18–31 (2017)
10. Isensee, F., Jaeger, P.F., Kohl, S.A.A., Petersen, J., Maier-Hein, K.H.: nnU-Net: a self-configuring method for deep learning-based biomedical image segmentation. Nat. Methods **18**(2), 203–211 (2021). https://doi.org/10.1038/s41592-020-01008-z. https://www.nature.com/articles/s41592-020-01008-z

11. Işín, A., Direkoğlu, C., Şah, M.: Review of MRI-based brain tumor image segmentation using deep learning methods. Procedia Comput. Sci. **102**, 317–324 (2016). https://doi.org/10.1016/j.procs.2016.09.407. https://linkinghub.elsevier.com/retrieve/pii/S187705091632587X
12. Jiang, Z., Zhao, C., Liu, X., Linguraru, M.G.: Brain tumor segmentation in multi-parametric magnetic resonance imaging using model ensembling and super-resolution (2022). https://doi.org/10.1007/978-3-031-09002-8_12
13. Khan, W.R., et al.: A hybrid attention-based residual unet for semantic segmentation of brain tumor. Comput. Mater. Continua **76**(1), 647–664 (2023)
14. Kickingereder, P., et al.: Automated quantitative tumour response assessment of MRI in neuro-oncology with artificial neural networks: a multicentre, retrospective study. Lancet Oncol. **20**(5), 728–740 (2019)
15. Magadza, T., Viriri, S.: Deep learning for brain tumor segmentation: a survey of state-of-the-art. J. Imaging **7**(2), 19 (2021). https://doi.org/10.3390/jimaging7020019. https://www.mdpi.com/2313-433X/7/2/19
16. Milletari, F., Navab, N., Ahmadi, S.A.: V-net: fully convolutional neural networks for volumetric medical image segmentation (2016). https://arxiv.org/abs/1606.04797
17. Murali, S., et al.: Bringing MRI to low-and middle-income countries: directions, challenges and potential solutions. NMR Biomed. e4992 (2023)
18. Razzak, M.I., Imran, M., Xu, G.: Efficient brain tumor segmentation with multi-scale two-pathway-group conventional neural networks. IEEE J. Biomed. Health Inform. **23**(5), 1911–1919 (2018)
19. Ren, T., Honey, E., Rebala, H., Sharma, A., Chopra, A., Kurt, M.: An Optimization Framework for Processing and Transfer Learning for the Brain Tumor Segmentation (2024). http://arxiv.org/abs/2402.07008. arXiv:2402.07008
20. Ronneberger, O., Fischer, P., Brox, T.: U-net: convolutional networks for biomedical image segmentation. In: Medical Image Computing and Computer-Assisted Intervention–MICCAI 2015: 18th International Conference, Munich, Germany, 5–9 October 2015, Proceedings, Part III 18, pp. 234–241. Springer (2015)
21. Tan, M., Le, Q.V.: EfficientNet: Rethinking Model Scaling for Convolutional Neural Networks (2020). https://doi.org/10.48550/arXiv.1905.11946. http://arxiv.org/abs/1905.11946. arXiv:1905.11946
22. Wang, W., Chen, C., Ding, M., Li, J., Yu, H., Zha, S.: TransBTS: Multimodal Brain Tumor Segmentation Using Transformer (2021). https://arxiv.org/abs/2103.04430v2
23. Xu, C.Y., Sang, Z.J., Shao, Y.Q.: MSA-VNet: multi-scale attention-based V-Net for DCE-MRI lesion segmentation. In: 2022 International Conference on Image Processing, Computer Vision and Machine Learning (ICICML), pp. 309–312 (2022). https://doi.org/10.1109/ICICML57342.2022.10009647. https://ieeexplore.ieee.org/document/10009647
24. Çiçek, Ö., Abdulkadir, A., Lienkamp, S.S., Brox, T., Ronneberger, O.: 3D U-Net: learning dense volumetric segmentation from sparse annotation (2016). https://arxiv.org/abs/1606.06650

Non-invasive Mean Pulmonary Artery Pressure Prediction Using Multi-modal Feature Fusion of Chest X-Ray and ECG

Siyeop Yoon[1], Jerome Charton[1], Michal Grzeszczyk[1], Arkadiusz Sitek[1], Hui Ren[1], Quangzheng Li[1], and Kei Nakata[2](✉)

[1] Department of Radiology and Center for Advanced Medical Computing and Analysis, Massachusetts General Hospital and Harvard Medical School, Boston, MA, USA

[2] Sapporo Medical University, Sapporo, Japan
k.nakata@sapmed.ac.jp

Abstract. Pulmonary hypertension (PH) is a life-threatening condition marked by elevated mean pulmonary arterial pressure (mPAP), with high morbidity and mortality. Right heart catheterization (RHC) is the gold standard for mPAP measurement because it provides direct and accurate hemodynamic assessment. However, RHC necessitates specialized facilities and continuous monitoring, which limits its accessibility, especially in community hospitals. This study introduces a deep learning model that leverages DINOv2 to estimate mPAP from chest X-ray and ECG images. The DINOv2-based chest X-ray encoder is fine-tuned to extract high-dimensional representations followed by feature fusion with those extracted from ECG images using a light-weight convolutional neural network, enabling the model to generate accurate mPAP predictions. The model was trained on 290 RHC invasive mPAP measurements from 163 patients and subsequently tested on 71 measurements from 38 patients at a town-based hospital. Performance evaluation using Bland-Altman analysis and regression correlation with invasive mPAP measurements showed low bias (−1.55 mmHg, limits of agreement = [−21.34, 18.24]), and moderate agreement (R^2 = 0.43). Moreover, the model demonstrates the potential for tracking long-term disease progression trajectories by correlating longitudinal changes in imaging features with mPAP variations. The model is deployable as a web tool, enabling scalable, non-invasive PH screening and monitoring with routine CXR and ECG, particularly in settings with limited RHC access.

Keywords: Pulmonary Hypertension · Multi-Modal Deep Learning · Vision Foundation Models · Low-Resource Healthcare

1 Introduction

Pulmonary hypertension (PH) is a progressive and life-threatening cardiovascular disorder defined by elevated mean pulmonary artery pressure (mPAP) equal

U. Anazodo et al. (Eds.): MIRASOL 2025, LNCS 16398, pp. 270–278, 2026.
https://doi.org/10.1007/978-3-032-13654-1_27

or higher than 20 mmHg at rest. PH proceeds to right ventricular failure and mortality if undiagnosed or untreated [10]. Early identification of PH is critical since targeted therapies can substantially improve patient outcomes. Nevertheless, diagnosis is frequently delayed due to non-specific symptoms and limited access to advanced diagnostic tools in many healthcare settings [6].

The current gold standard for PH diagnosis and mPAP measurement is right heart catheterization (RHC). Although RHC provides a direct hemodynamic assessment, its invasive nature, associated procedural risks, high costs, and limited availability—-especially in low-resource environments—-pose significant challenges [7]. In non-invasive clinical practice, transthoracic echocardiography (TTE) is commonly employed to estimate pulmonary pressures via Doppler-derived tricuspid regurgitation velocity. However, TTE suffers from operator dependency, unreliability in up to 30% of cases, and a tendency to overestimate pulmonary pressures in patients with concurrent lung disease [4,11].

Recent advances in deep learning have demonstrated the feasibility of predicting PH and elevated pulmonary pressures from non-invasive data. For instance, machine learning models applied to 12-lead ECGs have identified latent markers of pulmonary vascular pathology, while deep learning models trained on cardiac magnetic resonance imaging (CMR) have achieved accurate PH classification that slightly outperforms standard diagnostic metrics, delivering results within seconds [5,8]. Although resource intensive CMR was used, these studies underscore the potential of AI-driven, non-invasive PH detection. Moreover, previous models have typically relied on a single imaging modality, which limits their ability to capture complementary information that could be obtained by integrating multiple data sources. A promising solution lies in multi-modal AI approaches that integrate both structural and electrophysiological markers, such as those derived from chest X-ray (CXR) and ECG data. Although CXR and ECG lack the specificity of traditional diagnostics, they can capture secondary disease markers—including pulmonary artery enlargement, right ventricular hypertrophy, and electrical conduction abnormalities—that are often challenging to quantify manually. We hypothesize that deep learning models can extract and interpret these subtle patterns, correlating them with hemodynamic abnormalities and disease progression [11].

The field of computer vision has been transformed by foundation models—large-scale, self-supervised models that learn general-purpose visual representations. Among these, DINOv2, a self-supervised Vision Transformer trained on 142 million images, has exhibited strong generalization across diverse tasks without the need for labeled data during pre-training [2,3,9]. The robustness and scalability of DINOv2 render it particularly attractive for medical imaging applications, where labeled data are frequently limited.

In this study, we introduce a multi-modal deep learning framework for PH screening and monitoring that combines chest X-rays and images of ECG signals to predict continuous mPAP values non-invasively. Our model is designed for deployment in low-computation environments, utilizing standard X-ray and scanned ECG images, and it can be accessed publicly via web platforms and open-source repositories, ensuring its availability in resource-limited settings (Fig. 1). We propose a DINOv2-based foundation model for chest X-ray analysis, integrated with a convolutional ECG encoder, to directly estimate mPAP. Model performance is validated using Bland-Altman analysis, linear regression, and subgroup analyses (e.g., smokers versus non-smokers) to assess predictive consistency. Additionally, we demonstrate the model's capacity to track estimated mPAP trajectory over time and compare with repeated invasive measurements in PH patients. We assess the feasibility of deploying the model via a web-based prototype, enabling broader testing and feedback from the various clinical and research settings.

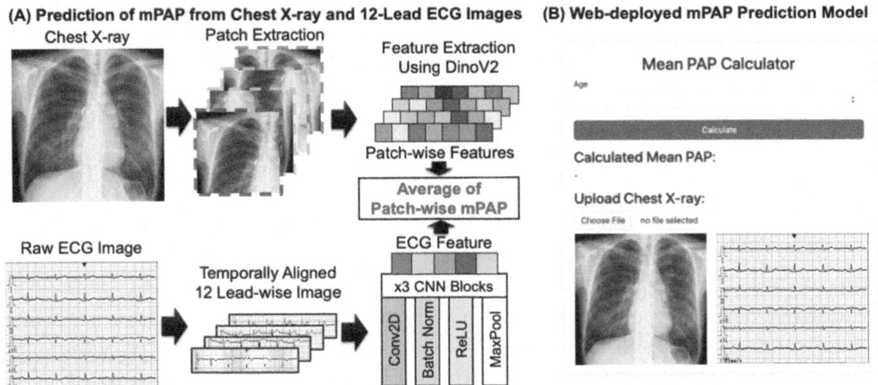

Fig. 1. Multi-Modal Deep Learning Framework for mPAP Prediction and Web-Based Deployment. (A) The proposed pipeline for estimating mean pulmonary arterial pressure (mPAP) from chest X-ray and 12-lead electrocardiogram (ECG) images. A pretrained foundation model (DINOv2) extracts high-level features from the chest X-ray, while a convolutional neural network (CNN) processes temporally aligned ECG lead images. The extracted feature representations are combined to predict mPAP. (B) A web-based mPAP prediction tool that allows users to upload chest X-ray and ECG images for real-time estimation of mPAP, demonstrating the feasibility of deploying the model for potential clinical use in resource-limited settings.

2 Methods

2.1 Data Description

This study utilized a retrospective dataset from a Sapporo Medical University Hospital. This study was approved by the Institutional Review Board (IRB) (Protocol: 2358678) and utilized retrospectively collected data. Our dataset consists of patients clinically diagnosed with PH who underwent both chest X-ray and ECG examinations close to the time of invasive hemodynamic testing. Inclusion criteria required that a frontal chest radiograph (posteroanterior view) and a standard 12-lead ECG were available within a 7-days of a right heart catheterization measuring mPAP. The final dataset included 201 PH patients (age 65 ± 16 years, 54% female), with a total of 361 records of mPAP measurements, X-rays, and ECG. PH etiologies in the cohort were mixed, including Group 1 PAH, Group 2 PH (due to left heart disease), and Group 3 PH (due to lung disease) [1], reflecting a real-world case mix.

For each patient, we obtained the digital chest X-ray image and a digitized ECG recording. The 12-lead ECGs (originally recorded on paper or electronically) were converted into a scanned image format for input into the vision model. For training, the images underwent preprocessing: chest X-rays were rescaled to a uniform size (512×512 pixels) and normalized, and 12-lead ECG images were similarly resized to a uniform size (224×48) per lead, and concatenated to channel dimension. We paired each X-ray with its corresponding ECG image and the ground-truth mPAP value (measured by RHC). The 361 records were split into training, validation, and test sets (70% train, 10% val, 20% test) ensuring that each patient appears in only one set.

2.2 Multi-modal mPAP Regression Model

The proposed model estimates mPAP by integrating information from chest X-rays and 12-lead ECG images. The chest X-ray branch employs a DINOv2-based Vision Transformer, pre-trained on 142 million images, as a robust feature extractor for pulmonary and cardiovascular structures. In our implementation, the DINOv2 backbone is wrapped within a custom fine-tuning module that reshapes its output into a 1024-dimensional embedding.

During **training**, the DINOv2 is optimized on 224×224 chest patches obtained by randomly cropping from 512×512 resized images. When ECG images are provided, the ECG feature extractor branch extracted a 128-dimensional vector via a series of convolutional layers. The first layer accepts an input with 36 channels (representing the 12-lead ECG image with RGB channels) and outputs 64 feature maps using a kernel size of 3×3, stride 1, and padding 1. This is followed by batch normalization, ReLU activation, and a 2×2 max-pooling operation. The channel depth increases to 128 and 256 for the second and third blocks, respectively.

The final ECG-features are projected with an adaptive average pooling layer reducing the spatial dimensions to 1×1, and a fully connected layer projects the 256-dimensional output to a 128-dimensional feature vector. This branch, therefore, yields a 128-dimensional embedding that captures key waveform features indicative of right ventricular strain and electrical conduction abnormalities.

For each 224×224 chest patch, the DINOv2 backbone produces a 1024-dimensional feature, which is concatenated with the ECG feature. The individual feature vectors of the patch are concatenated into a single fused feature vector with a total dimensionality of $d_{\text{fused}} = d_{\text{chest}} + d_{\text{ECG}} = 1024 + 128 = 1152$. This fused representation is processed by a regression head that first reduces the dimensionality to 592 through a fully connected layer, applies a ReLU activation, and finally maps the features to a single scalar output representing the mPAP (in mmHg).

The networks are trained at the same time using the Mean Squared Error (MSE) loss defined as $L = \frac{1}{n}\sum_{i=1}^{n}(\hat{y}_i - y_i)^2$, where $\hat{y}_i$ is the predicted mPAP and y_i is the ground-truth measurement obtained from right heart catheterization. Optimization is performed for 20 epochs using the Adam optimizer with a learning rate of 5×10^{-5}.

In **validation/testing**, the full 512×512 chest X-ray is partitioned into a 3×3 grid of 224×224 patches. A patch-level regressor then predicts an mPAP value for each patch; the final chest-based mPAP estimate is computed as the average of patch-wise predictions.

For the **model deployment**, we developed a web-based mPAP prediction tool that enables users to upload chest X-ray and 12-lead ECG images for real-time estimation of mPAP. The web-based mPAP prediction tool leverages a multi-modal deep learning model that integrates DINOv2-based Vision Transformer with complementary ECG-derived features, estimating mPAP. This tool utilizes the Hugging Face API for deployment in resource-limited settings, thereby offering a cost-effective, rapid, and accessible solution for clinical decision-making and patient management (https://github.com/siyeopyoon/PulmoFusion-mPAP and https://jcharton-mean-pap-api.hf.space).

Evaluation Metrics. To assess the impact of different chest X-ray feature extractors on mPAP prediction accuracy, we conducted experiments comparing DINOv2, ResNet34, and VGG16 backbones. Moreover, we evaluated the performance of the multi-modal framework both with and without the inclusion of the ECG encoder. Evaluation of the model was performed using a suite of quantitative analyses to assess the agreement and accuracy of the predicted mPAP values against the reference measurements obtained via RHC. A Bland-Altman plot was generated to identify systematic biases and to define the limits of agreement between the two methods. In addition, the Mean Absolute Error (MAE) was computed as $\text{MAE} = \frac{1}{n}\sum_{i=1}^{n}|\hat{y}_i - y_i|$, providing a direct measure of the average discrepancy (in mmHg) between predictions and true values. Finally, a linear regression analysis was conducted between the predicted and catheterized mPAP measurements, yielding key statistical parameters such as the coefficient of determination (R^2) and the regression slope.

3 Results

Figure 2 presents a Bland-Altman plot that compares the differences between predicted and measured mPAP values against their averages. The plot includes a bias line representing the mean difference (−1.55 mmHg) and dashed lines that indicate the 95% limits of agreement (from −21.34 to 18.24 mmHg). Table 1 details the performance of different deep learning models for mPAP prediction. Standalone imaging models (VGG16, Resnet34, DinoV2) achieved regression slopes of 0.15, 0.03, and 0.09 with correlation coefficients (R^2) of 0.13, 0.02, and

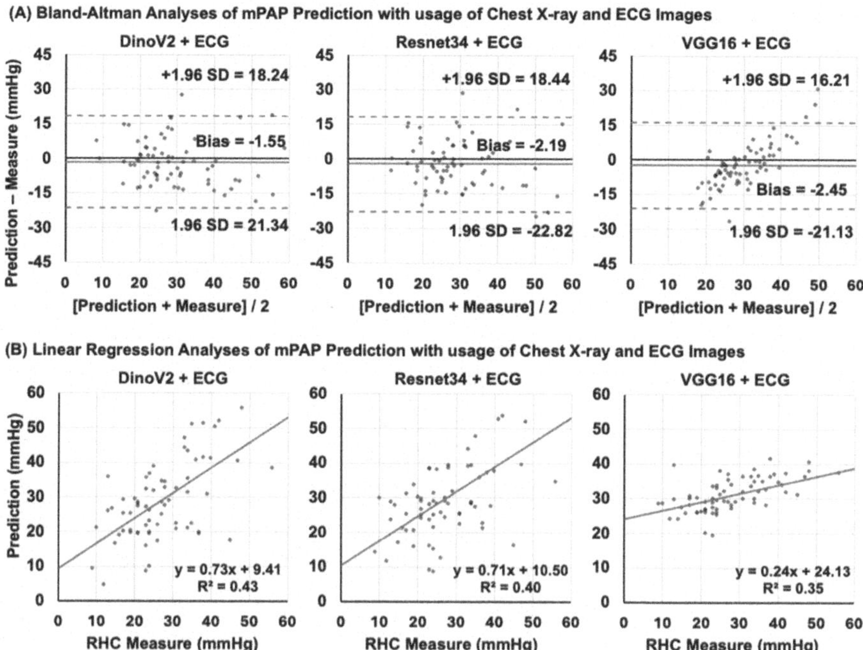

Fig. 2. The measurement agreement of mean pulmonary artery pressure between right heart catheterization and AI-based prediction with the usage of Chest X-ray and ECG image. (A) Bland-Altman plots show bias and limits of agreement ($\pm$1.96 SDs), the solid red line and red dotted lines, respectively. In (B) linear regression the correlation coefficient (R2) and slope (y) are shown. (Color figure online)

0.06, respectively. The incorporation of ECG data led to improved performance; notably, the DinoV2+ECG model reached a regression slope of 0.73 and an R^2 of 0.43, along with a mean absolute error of 8.20 ± 6.03 and a bias of -1.55 mmHg (limits of agreement: -21.34 to 18.24 mmHg). These numerical results demonstrate that the integration of ECG with chest X-ray data enhances the accuracy of non-invasive mPAP estimation.

To assess the model's ability to monitor changes over time, we selected a patient from our cohort who underwent multiple imaging studies and corresponding RHC measurements over a period from Day 0 to Day 4513. For this case, our trained model was applied to each paired chest X-ray and ECG image, and predicted mPAP values were recorded at each time point. As illustrated in Fig. 3, the patient initially showed high mPAP values together with clear ECG indicators of right ventricular strain (highlighted by the red circle in panel (B)). Over time, as the ECG features indicative of RV strain diminished, the predicted mPAP values decreased correspondingly. This observation provides a concrete example of how changes in RV strain may be associated with alterations in pulmonary arterial pressure.

Table 1. Performance evaluation of deep learning models for mPAP prediction. Reported are mean absolute error (MAE ± SD), regression slope, R^2, and bias with limits of agreement for standalone imaging models (VGG16, Resnet34, DinoV2) and their counterparts augmented with ECG data.

Method	MAE ↓	Regression Slope	Correlation R^2	Bias [LoA%]
VGG16	8.77 ± 6.63	0.15	0.13	−2.41 [−26.52, 18.69]
Resnet34	10.14 ± 6.46	0.03	0.02	−3.92 [−26.52, 18.69]
DinoV2	9.05 ± 6.70	0.09	0.06	−1.28 [−23.31, 20.75]
VGG16+ECG	**7.53 ± 6.26**	0.24	0.35	−2.45 [−21.13, 16.21]
Resnet34+ECG	8.62 ± 6.35	0.71	0.40	−2.19 [−22.82, 18.44]
DinoV2+ECG	8.20 ± 6.03	0.73	0.43	**−1.55** [−21.34, 18.24]

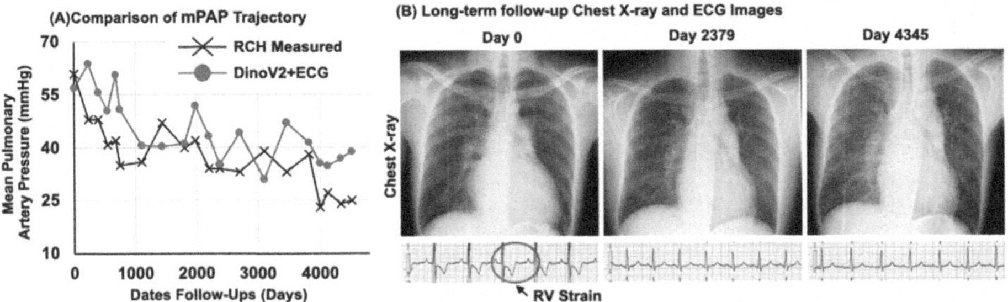

Fig. 3. Comparison of measured mean pulmonary artery pressure (mPAP) from right heart catheterization (black crosses) and predicted mPAP from the proposed DINOv2+ECG model (red circles) over 4513 days, illustrating the framework's ability to track disease progression. Chest X-ray and ECG images at Day 0, 2379, and 4345 reveal progressive cardiopulmonary changes, including right ventricular strain (red circle) captured by ECG morphology. These findings demonstrate how our proposed model can non-invasively estimate mPAP from routine clinical imaging and follow pulmonary hypertension trajectories using chest X-ray and ECG.

4 Conclusion

We presented a novel approach for non-invasive pulmonary hypertension assessment using a DINOv2 vision foundation model to analyze chest X-ray and ECG images for mPAP prediction. The proposed model demonstrated high accuracy in estimating mean pulmonary arterial pressure, with strong correlation and close agreement to invasive measurements. Through Bland-Altman analysis, we showed minimal bias and clinically acceptable limits of agreement between the model's predictions and right heart catheterization values. We further illustrated the model's value by tracking mPAP changes in longitudinal case studies, highlighting potential use in monitoring disease progression or treatment response. It could also be applied retrospectively to prior images to see trends (for example, reading old X-rays to estimate what mPAP might have been before). Importantly, the method is non-invasive and repeatable, making it safe for serial use. Our model could be integrated into hospital PACS systems or web-based access, as shown in the example website, Fig. 1(B).

While the results are encouraging, our study has several limitations. The dataset size (201 patients) is relatively modest, drawn from a single center. This raises concerns about overfitting to specific population characteristics or image acquisition protocols. Although using DINOv2 should improve generalizability, the model should be validated on external cohorts (multi-center data) to ensure it works broadly. Future studies will focus on validating our model across broader populations and integrating additional data modalities to further improve robustness and precision. With additional refinement, this foundation model-based approach has the potential to be implemented as a low-cost, widely accessible solution for improving pulmonary hypertension care. In essence, our study highlights the promise of cross-domain AI models in bridging the gap between simple diagnostic tests and complex invasive measurements for PH management.

Acknowledgments. This study was partially supported by NIH R01HL159183.

Disclosure of Interests. The authors have no competing interests to declare that are relevant to the content of this article.

References

1. Anderson, J.J., Lau, E.M.: Pulmonary hypertension definition, classification, and epidemiology in Asia. JACC Asia **2**(5), 538–546 (2022)
2. Caron, M., et al.: Emerging properties in self-supervised vision transformers. In: Proceedings of the IEEE/CVF International Conference on Computer Vision, pp. 9650–9660 (2021)
3. Dosovitskiy, A., et al.: An image is worth 16x16 words: transformers for image recognition at scale. arXiv preprint arXiv:2010.11929 (2020)
4. Frea, S., et al.: Noninvasive assessment of hemodynamic status in heartware left ventricular assist device patients: validation of an echocardiographic approach. JACC Cardiovasc. Imaging **12**(7 Part 1), 1121–1131 (2019)
5. Grzeszczyk, M.K., Korzeniowski, P., Alabed, S., Swift, A.J., Trzciński, T., Sitek, A.: Tabmixer: noninvasive estimation of the mean pulmonary artery pressure via imaging and tabular data mixing. In: International Conference on Medical Image Computing and Computer-Assisted Intervention, pp. 670–680. Springer (2024)
6. Hambly, N., Alawfi, F., Mehta, S.: Pulmonary hypertension: diagnostic approach and optimal management. CMAJ **188**(11), 804–812 (2016)
7. Hasan, B., et al.: Challenges and special aspects of pulmonary hypertension in middle-to low-income regions: JACC state-of-the-art review. J. Am. Coll. Cardiol. **75**(19), 2463–2477 (2020)
8. Johns, C.S., et al.: Diagnosis of pulmonary hypertension with cardiac MRI: derivation and validation of regression models. Radiology **290**(1), 61–68 (2019)
9. Oquab, M., et al.: Dinov2: learning robust visual features without supervision. arXiv preprint arXiv:2304.07193 (2023)

10. Simonneau, G., et al.: Haemodynamic definitions and updated clinical classification of pulmonary hypertension. Eur. Respir. J. **53**(1) (2019)
11. Thenappan, T., Prins, K.W., Pritzker, M.R., Scandurra, J., Volmers, K., Weir, E.K.: The critical role of pulmonary arterial compliance in pulmonary hypertension. Ann. Am. Thorac. Soc. **13**(2), 276–284 (2016)

Large Scale DICOM Compliance Evaluation of Medical Image Data Elements in Low-Resource Settings

Elijah Chileshe[1,2](✉) and Lighton Phiri[1,2](✉)

[1] Department of Computing and Informatics, University of Zambia, Lusaka, Zambia
[2] DataLab Research Group, University of Zambia, Lusaka, Zambia
{elijah.chileshe,lighton.phiri}@cs.unza.zm

Abstract. The Medical Imaging Technology Association (MITA) has highlighted the significance of Artificial Intelligence (AI), from an interoperability perspective, during the interaction between systems within the medical imaging ecosystem. MITA has, in part, recommended conformance to specifications, such as Digital Imaging and Communication in Medicine (DICOM) standard, as being vital for DICOM producers. While the significance of DICOM metadata during the implementation of AI models has been extensively documented, the DICOM compliance levels of medical images generated is largely unknown. This paper presents an empirical study conducted at a large University Teaching Hospital in order to assess the DICOM compliance levels associated with medical images generated. In order to determine the relative levels of importance of each of the 5,078 DICOM Standard 2024a Data Elements, interviews were conducted with modality operators and, additionally, document analysis of Radiology Examination Request Forms was conducted. A total of 7,632,544 DICOM files from The University Teaching Hospitals (UTHs), were analysed in order to determine the availability of DICOM metadata element values and, additionally, the correctness of the values. The results revealed consistently low DICOM compliance levels, with an overall average DICOM compliance rate of 3.62%. In addition, the results indicate that DICOM Data Elements vital for facilitating discoverability of records is platforms such as Picture Archiving and Communication Systems had missing values. This study demonstrates how DICOM compliance could potentially be accomplished in ensuring adherence to international standards.

Keywords: DICOM Data Elements · Medical Images · Metadata

1 Introduction

The Digital Imaging and Communications in Medicine (DICOM) standard [1] is a widely used standard for storing and transmitting medical images. A crucial component of medical images generated with modalities is the metadata that is embedded within the medical image.

The Medical Imaging Technology Association (MITA) has highlighted a set of criteria to consider, in medical imaging workflows, in order to effectively integrate AI in

U. Anazodo et al. (Eds.): MIRASOL 2025, LNCS 16398, pp. 279–288, 2026.
https://doi.org/10.1007/978-3-032-13654-1_28

medical imaging, with conformance to the DICOM specification cited as an important point to consider [2].

The UTHs presently does not have the technological infrastructure for managing medical images; specifically, there is no Picture Archiving and Communication System (PACS) platform. In addition, most radiological workflow activities are conducted using manual processes, as outlined in our prior work [3]. Our previous work highlighted the importance of policies and procedures and the work presented in this paper is motivated by the need to put in place policies and standard operating procedures that will ensure that medical images and radiological reports are effectively stored and managed.

The study was driven by the following research questions: (i) How can DICOM Data Element compliance be assessed in a resource constrained environment? (ii) What is the current DICOM Data Element compliance level associated with medical images at UTHs?

The main contribution of this work are as follows:

- Large scale DICOM compliance evaluation on real-world medical images, providing insights into data quality challenges
- Methodological approach and techniques for determining DICOM compliance for medical images
- DICOM compliance dataset with all DICOM Data Elements associated with the DICOM 2024a standard
- Experimental results for analyses conducted at UTHs, highlighting specific areas on non-compliance

The remainder of this paper is organised as follows: Sect. 2 outlines studies involving DICOM Data Elements analyses and, additionally, general metadata analysis; Sect. 3 describes the methodological approach employed when conducting the study; Sect. 4 presents and discusses the study results and; Sect. 5 presents concluding remarks.

2 Related Work

2.1 Metadata in Medical Imaging

DICOM Data Elements. Digital objects such as DICOM files consist of bitstreams, representing the digital content and; the metadata, providing auxiliary information about the digital content. The metadata is associated with the digital content by embedding the metadata elements into the file consisting of the bitstreams or in a separate file; DICOM file metadata—commonly referred to as DICOM Data Elements—are embedded in the same file as the bitstreams.

Numerous literature has extensively studied the crucial role of metadata in the management of digital content, preservation of digital content and also discoverability of digital content. Riley has categorised metadata elements into three broad categories—Administrative Metadata, Structural Metadata and Descriptive Metadata—based on their contextual usage when handling digital content [4]. The usage of metadata elements is generally done using prescribed standards which dictate how the metadata is encoded and used.

DICOM Data Elements are embedded in the same DICOM file containing the medical image data, and encoded based on the DICOM standard; with the current DICOM 2024a standard [1] comprising 5,078 Data Elements.

Similarly to other metadata schemes, DICOM Data Elements are crucial in facilitating discoverability of medical images in platforms such as PACS platforms. In addition, patient-specific DICOM Data Elements generally help Radiologists and other consumers of medical images in understanding the medical images being interpreted, using DICOM Viewers.

Metadata Quality Assessment. Quality metadata generally facilitates effective administration, management and discoverability of digital content. Existing literature on metadata quality have emphasised the use of three key metrics: accuracy of metadata element values; completeness of metadata element values and consistency of metadata element values. Park states that these three metrics are some of the most commonly used during metadata assessment [5].

Extensive studies on analysis of metadata quality have been conducted on scholarly data. For instance, Phiri proposes the use of supervised machine learning for the automatic classification of digital objects, in order to improve metadata quality [6]. Bertha et al. demonstrate the adverse effects and potential machine learning centric solutions in improving descriptive metadata [7]. Metadata quality assessments have also been done on learning objects, with Currier et al. analysing learning object repositories in order to detect errors by untrained creators and the lack of use of authority control [8].

This work focuses on evaluating DICOM Data Elements completeness, using the approach of assessing metadata completeness proposed by Kasonde and Phiri in which they propose an approach to assessing the compliance of Electronic Theses and Dissertation metadata [9].

2.2 DICOM Data Element Analysis

In addressing the need to analyse DICOM metadata in Mammography studies to characterise radiation exposure, Santos et al. investigate the quality of DICOM metadata and its role in assessing radiation exposure for patients undergoing mammography [10]. By focusing on the analysis of DICOM data elements, the study highlights the importance of DICOM data elements in enhancing the quality and effectiveness of medical imaging procedures.

Additionally, Prieto, et al. present a methodology for automatically detecting potential image retakes in digital radiography by analysing DICOM header information. By analysing key DICOM Data elements such as patient identification number, modality, description, projection, date, cassette orientation, and image comments, the system identifies images that may require retakes [11]. The analysis of DICOM Data Elements is crucial for quality assurance in radiology departments, as it helps in identifying image faults, sources of error, and inappropriate practices as well as the detection of deficiencies in department performance, including wrong identifications, positioning errors, wrong radiographic techniques and equipment malfunctions, potentially improving the accuracy of identifying images that require retakes [11].

The critical role of analysing DICOM metadata in radiology for population characterisation is also emphasised by the work by Santos et al. [12], where they categorise patient populations based on age, gender and imaging modality, providing valuable insights into the population accessing radiology services. Through this research, the importance of leveraging DICOM metadata for informed decision-making and healthcare improvement is underscored.

Our work differs from the state-of-the-art as it is arguably the first attempt in empirically analysing DICOM compliance on a large scale.

3 Methodology

A quantitative approach was employed when conducting this study, with the CRoss Industry Standard Process for Data Mining (CRISP-DM) [13] methodology used to guide the overall experimentation process. Ethical clearance was granted by The University of Zambia Biomedical Research Ethics Committee (Reference Number: 2731–2022) and The National Health Research Authority (Reference Number: NHRA000024/10/05/2022), to conduct this study. In addition, formal permission was granted from UTHs. The study was conducted at a single institution, with the approach designed to fit existing workflows using the compliance score to assess DICOM metadata.

3.1 Experimental Setting

The experiments were conducted on a standalone Lenovo® ThinkPad T480, with an Intel® Core™ i5-8350U (CPU @ 1.70 GHz), using 16 GB RAM and running Ubuntu 20.04.6 LTS. The Python programming language was primarily used to implement all scripts used during data collection, pre-processing and analysis. Specifically, Pydicom [14], Concurrent.Futures [15], Pandas [16] and Numpy [17] Python libraries were used.

3.2 Data Collection and Preparation

Our most recent work highlighted the unavailability of Enterprise Medical Imaging infrastructure, in particular, a Picture Archiving and Communication System platform at UTHs. In order to collect data required for the planned analyses, three (3) Research Assistants were recruited and trained to digitalise medical images stored on Compact Discs and Digital Versatile Disc; in addition, medical images archived on External Hard Disk Drives were digitalised forming the dataset used in the study[1]. Additional technical details are available in our work outlining the large-scale study that was conducted [18].

The embedded DICOM Data Elements were extracted from each DICOM file and the data pre-processed by deduplicating DICOM files associated with the same DICOM Series and Study [18] since the DICOM Data Elements are shared by DICOM files associated with the same Series and Study.

[1] Datasets, scripts and Jupyter Notebooks used during analysis are publicly available.

3.3 DICOM Compliance

The DICOM compliance assessment was performed by determining the relative level of importance for all the 5,078 DICOM Data Elements associated with the DICOM standard 2024a [19]. Weights for each of the 5,078 DICOM Data Elements were subsequently assigned and the DICOM compliance scores for each DICOM record computed.

DICOM Data Elements Analysis. The relative importance of DICOM Data Elements was conducted by interviewing modality operators and through systematic content analysis. Semi-structured interview sessions were conducted with two (2) modality operators as they are primarily involved with processing DICOM worklists at the UTHs. In addition, a systematic content analysis was conducted by analysing Radiology Examination Request Forms[2] used at UTHs. Essentially, form fields on Request Forms were mapped on to DICOM Data Elements.

DICOM Compliance Scoring. The DICOM compliance scoring was performed by initially assigning weights to each of the 5,078 DICOM Data Elements for each DICOM record, summing up the weights and subsequently computing the compliance score relative to the total weighting score.

Three categories—Mandatory, Optional and Retired—of DICOM Data Elements were identified and their weightings ascribed, as outlined in Table 1. The weightings for the elements were informed by interviews conducted with operators and, additionally, the resulting UTHs request form content analysis. Specifically, the weighting was performed as outlined below.

1. Input: DICOM Record dicom_record consisting of multiple DICOM Data Elements
2. Output: Total Weight total_weight of the DICOM Record based on the weights of its Data Elements.
3. Function calculate_weight (data_element):
    - Input: DICOM Data Element data_element
    - Output: Weight weight assigned to the Data Element.
    - Procedure:
        - If data_element is mandatory; Set weight to 1
        - Else if data_element is optional; Set weight to 0.5
        - Else if data_element is retired; Set weight to 0
4. Function process_dicom_record(dicom_record):
    - Input: DICOM Record dicom_record
    - Output: Total Weight total_weight of the DICOM Record.
    - Procedure:
        - Initialise total_weight to 0.
        - For each data_element in dicom_record:
        - Calculate the weight of data_element
        - Add the calculated weight to total_weight.

[2] See supplementary data for request form.

Table 1. DICOM Data Element weighting.

Category	Description	Weight
Mandatory Data Elements	DICOM Data Elements identified as being important by modality operators and also available on Request Forms. Example elements include Patient ID and Patient Name	1
Optional Data Elements	DICOM Data Elements with no corresponding mapping on Request Forms. Example elements include Patient Address and Patient's Size	0.5
Retired Data Elements	DICOM Data Elements no longer supported	0

4 Results and Discussion

Table 2. UTHs DICOM compliance dataset[1].

		DICOM Files	Unique Instances
Radiographs	Computed Radiography (CR)	4,071	1,385
	Computed Tomography (CT)	7,575,236	4,595
	Diagnostic Radiography (DX)	48,455	18,806
	Radiographic Fluoroscopy (RF)	8	4
	Structured Report (SR)	4,773	72
	X-ray Angiography (XA)	1	1

[1]DICOM compliance dataset publicly available but omitted as part of blind review

Table 2 shows the characteristics of the dataset used in the study. A total of 7,632,544 DICOM files were included in the dataset for analysis. The deduplication conducted during pre-processing resulted in a total of 24,863 study instances. As outlined in Sect. 3, DICOM records with the same Study ID (0020,0010) and Study Instance UID (0020,000D), were deduplicated since the DICOM Data Elements are shared amongst all files linked to the same Study.

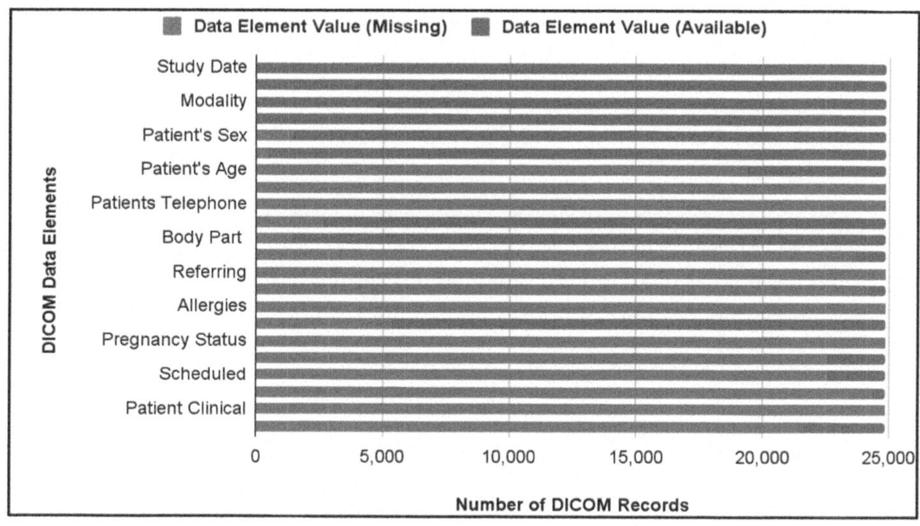

Fig. 1. Distribution of selected DICOM Data Elements with NULL entries.

It is evident from Table 2 that a large proportion of the DICOM files are CT and, additionally, were produced in 2020. It was mentioned in Sect. 3 that UTHs currently uses a manual workflow and has yet to implement an Enterprise Medical Imaging strategy and as such, there is presently no infrastructure to support the storage of medical images. The medical images used in the analysis were compiled from CDs and External Hard Disk Drives. However, works are underway to implement infrastructure aimed at digitising the storage, management, and retrieval of medical images.

4.1 Exploratory Analysis

An exploratory data analysis exercise was conducted in order to explore the DICOM Data Element values. Figure 1 shows the distribution of DICOM Data Elements with missing values, a key metric for computing the completeness score, as outlined in Sect. 3. While the DICOM 2024a Standard comprises 5,078 Data Elements, the focus of the analyses was on Data Elements appearing on UTHs request forms and, additionally, Data Elements identified as being important in a typical radiological workflow.

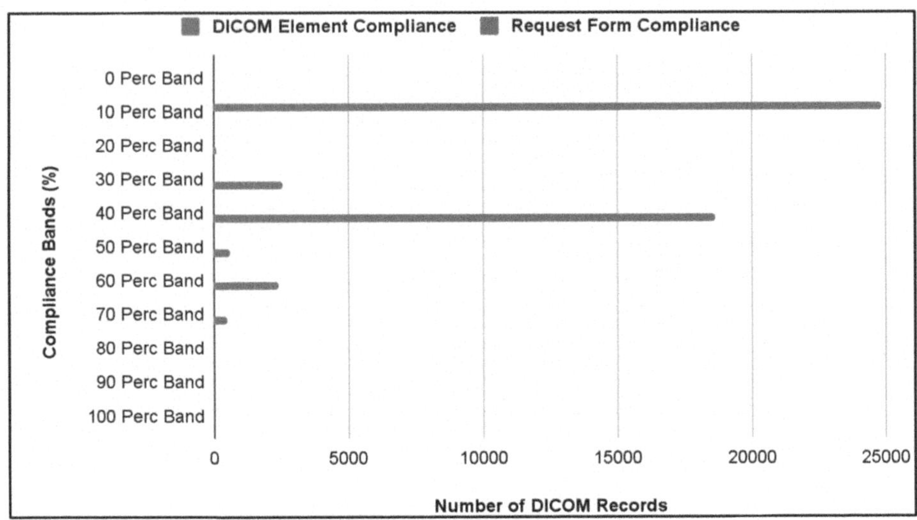

Fig. 2. Distribution of DICOM Element completeness scores by percentage bands.

Figure 1 clearly indicates that most Data Elements have missing values, with the exception of those whose values are automatically generated (such as Study Date and Modality) and, arguably, those perceived to be required to properly identify examinations. The proportion of missing values associated with Data Elements necessary to facilitate the discoverability of medical images, and in some instances, the augmentation of data in AI pipelines, signals the importance of addressing the issue of metadata completeness. It is important to note that the presence and relevance of specific DICOM tags, such as "Pregnancy Status," depend on the protocols and practices of each hospital according to their specific needs.

4.2 DICOM Data Element Completeness

The average DICOM compliance score relative to all the 5,078 DICOM 2024a Data Elements was 3.62%, while the average compliance score for DICOM Request Form fields for UTHs was 38.88%.

Figure 2 shows the distribution of DICOM completeness compliance using predefined compliance bands. The DICOM compliance scores were computed relative to all the DICOM Data Elements and relative to the Data Elements available on UTHs request forms. In both instances, it is evident that the DICOM compliance is significantly low, averaging 10% and 40% for all the Data Elements and Request Form, respectively.

Figure 3 illustrates the average completeness scores for various modalities included in the analyses. In addition, an analysis of the annual distribution of DICOM compliance scores indicates that historically, the problem with DICOM compliance has consistently been an issue for UTHs.

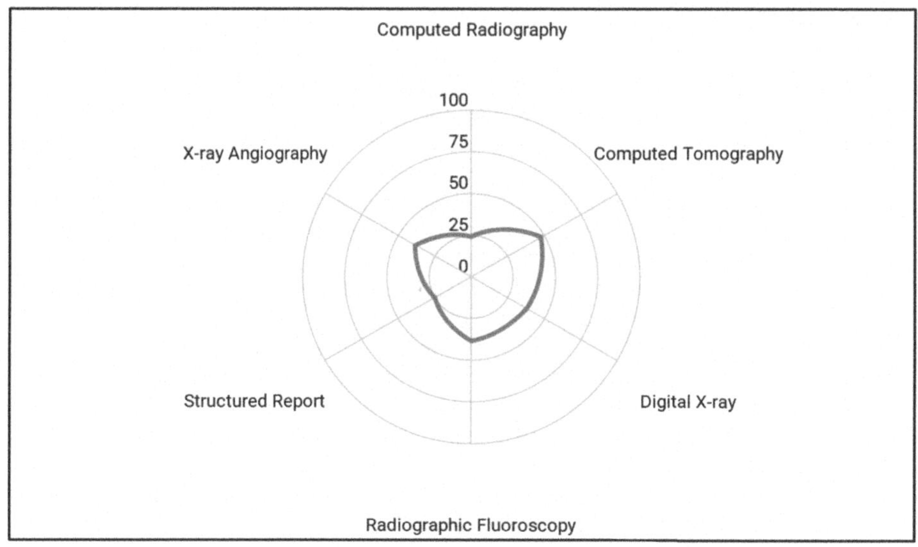

Fig. 3. Average Request Form DICOM completeness scores for Modalities at UTHs.

5 Conclusion

This paper outlined an effective approach for assessing DICOM Data Element compliance in medical imaging. It introduced a technique for evaluating compliance and illustrated its application through a case study of UTHs. The approach has the potential to guide policy direction regarding the comprehensive description of DICOM files and, additionally, provides a way of helping institutions uncover anomalies associated with DICOM data elements. As the volume of medical images grows and their integration into AI solutions expands, ensuring the quality and consistency of DICOM Data Elements becomes paramount. This not only enhances the accuracy and reliability of medical diagnoses but also supports interoperability across healthcare systems. Ultimately, standardised DICOM compliance contributes to improved patient care, streamlined workflows in hospitals, and advances in medical research, benefiting both healthcare providers and society at large.

Using the findings from this study, ongoing work is focused on developing processes and procedure documents aimed at addressing challenges associated with DICOM completeness. Potential future work could involve applying the approach to multiple sites and analysing the impact on DICOM completeness on AI and workflows.

Acknowledgement. This study was conducted as part of a much larger "Enterprise Medical Imaging in Zambia" project. The project is funded through generous grants from Google Research, Data Science Africa and The University of Zambia. In addition, this work was partly supported by the Italian Ministry of University and Research (MUR) under project PE0000013 – Future of Artificial Intelligence Research (FAIR). We are very appreciative of this funding.

References

1. DICOM Standards Committee: DICOM PS3.1 2023c - Introduction and Overview, https://dicom.nema.org/medical/dicom/current/output/html/part01.html, Accessed 10 Jul 2023
2. Medical Imaging Technology Association: AI and DICOM, https://www.dicomstandard.org/ai, Accessed 10 Jul 2023
3. Zulu, E.O., Phiri, L.: Enterprise medical imaging in the global south: challenges and opportunities. In: 2022 IST-Africa Conference (IST-Africa), pp. 1–9. IEEE Explore (2022). https://doi.org/10.23919/ist-africa56635.2022.9845508
4. Riley, J.: Understanding Metadata: What is Metadata, and What is it For?: A Primer. National Information Standards Organization, Baltimore (2017)
5. Park, J.-R.: Metadata quality in digital repositories: a survey of the current state of the art. Cataloging Classif. Q. **47**, 213–228 (2009). https://doi.org/10.1080/01639370902737240
6. Phiri, L.: Automatic classification of digital objects for improved metadata quality of electronic theses and dissertations in institutional repositories. Int. J. Metadata Semant. Ontol. **14**, 234–248 (2021). https://doi.org/10.1504/IJMSO.2020.112804
7. Chipangila, B., et al.: Controlled vocabularies in digital libraries: challenges and solutions for increased discoverability of digital objects. Int. J. Digit. Libr. **25**, 139–155 (2023). https://doi.org/10.1007/s00799-023-00374-1
8. Currier, S., Barton, J., O'Beirne, R., Ryan, B.: Quality assurance for digital learning object repositories: issues for the metadata creation process. Res. Learn. Technol. **12** (2004). https://doi.org/10.3402/rlt.v12i1.11223
9. Kasonde, C.C., Phiri, L.: Assessing and Promoting Metadata Quality for Electronic Theses and Dissertations in Institutional Repositories Using a Policy-Driven Approach. INFLIBNET Centre, Gandhinagar (2023)
10. Santos, M., Silva, A., Rocha, N.P.: DICOM metadata quality analysis for mammography radiation exposure characterization. In: Rocha, Á., Adeli, H., Reis, L.P., Costanzo, S., Orovic, I., Moreira, F. (eds.) Advances in Intelligent Systems and Computing. Springer Nature, Switzerland AG (2020). https://doi.org/10.1007/978-3-030-45688-7_16
11. Prieto, C., et al.: Image retake analysis in digital radiography using DICOM header information. J. Digit. Imaging **22**, 393–399 (2009). https://doi.org/10.1007/s10278-008-9135-y
12. DICOM metadata analysis for population characterization: a feasibility study. Proc. Comput. Sci. **100**, 355–361 (2016). https://doi.org/10.1016/j.procs.2016.09.169
13. Wirth, R., Hipp, J.: CRISP-DM: towards a standard process model for data mining. In: Proceedings of the Fourth International Conference on the Practical Application of Knowledge Discovery and Data Mining, pp. 29–39 (2000)
14. Mason, D., et al.: pydicom/pydicom: Pydicom v2.4.0 (v2.4.0). https://doi.org/10.5281/zenodo.8034250, Accessed 05 Mar 2024
15. concurrent.futures — Launching parallel tasks, https://docs.python.org/3/library/concurrent.futures.html, last accessed 2025/08/23
16. The pandas Development Team: pandas-dev/pandas: Pandas (v2.2.1), https://zenodo.org/records/10697587, Accessed 23 Aug 2025. https://doi.org/10.5281/zenodo.10697587
17. Harris, C.R., et al.: Array programming with NumPy. Nature **585**, 357–362 (2020). https://doi.org/10.1038/s41586-020-2649-2
18. Chileshe, E., Phiri, L.: Large-scale analysis of medical image metadata. Zambia ICT J. **5**, 44–48 (2023)
19. DICOM Standards Committee: PS3.4, https://dicom.nema.org/medical/dicom/current/output/chtml/part04/PS3.4.html, Accessed 23 Aug 2025

EDGE-KD: Explainability-Driven Guidance for Efficient Knowledge Distillation in Chest X-Ray Classification

Houda El Mohamadi[1,3](✉), Mohammed El Hassouni[2], and Rachid Jennane[3]

[1] LRIT, FS, Mohammed V University in Rabat, Rabat, Morocco
houda.elmohamadi@um5r.ac.ma

[2] FLSH, Mohammed V University in Rabat, Rabat, Morocco
mohamed.elhassouni@flsh.um5.ac.ma

[3] IDP, University of Orleans, Orléans, France
rachid.jennane@univ-orleans.fr

Abstract. Deep learning models have demonstrated remarkable results in the automatic classification of chest X-ray images. However, their adoption is often limited by high computational costs and a lack of interpretability, which are critical for clinical trust and deployment. Knowledge Distillation (KD) offers an effective means of model compression, but standard KD techniques primarily focus on predictive performance, which does not cover the important aspect of explainability. In this paper, we propose a novel Knowledge Distillation framework (EDGE-KD) that integrates explainability directly into the distillation process, ensuring the transfer of both predictive and interpretable knowledge. Our approach leverages a texture-shape explainability method, using saliency maps that encode texture and shape features to guide the training of the student model. This dual-feature alignment guides the student model to not only mimic the teacher's predictions but also inherit its interpretable patterns, enhancing the performance and interpretability. To validate the robustness of our approach, we tested several teacher/student network architectures on a complex chest X-ray dataset containing multiple pathologies. Experimental results demonstrate that our EDGE-KD approach consistently outperforms baseline KD methods, improving classification accuracy by (0.5–3%) while enhancing training stability and preserving model explainability.

Keywords: Knowledge distillation (KD) · Explainable Methods · Texture-Shape features · Saliency maps · CXR imaging

1 Introduction

Chest X-ray (CXR) image analysis has seen transformative advances through deep learning (DL) models [1], achieving state-of-the-art performance in multi-class classification of pulmonary pathologies including pneumonia, tuberculosis, and other thoracic infections [2]. These DL-based systems have been extensively deployed for detecting lung abnormalities and diagnosing thoracic diseases

U. Anazodo et al. (Eds.): MIRASOL 2025, LNCS 16398, pp. 289–298, 2026.
https://doi.org/10.1007/978-3-032-13654-1_29

[3]. Despite their diagnostic prowess, DL models remain constrained by two fundamental limitations: computational complexity and inadequate interpretability. To address the first challenge, knowledge distillation (KD) has emerged as an effective model compression strategy. First proposed by [4], KD transfers knowledge from a computationally intensive teacher model to a compact student network while preserving predictive accuracy [5]. Recent innovations include Progressive Knowledge Distillation (PKD) [6], which incrementally refines teacher knowledge transfer, and hybrid frameworks [7] that combine soft targets with feature representations to achieve 7.2% accuracy gains. Further studies [8,9] confirm KD's effectiveness in medical imaging while maintaining diagnostic performance. However, conventional KD methods neglect explainability, a critical requirement for clinical adoption. Prior research [10–12] documents several explainability methods (EXMs), with saliency maps being among the most widely used techniques for visualizing decision-influential regions [13]. While typically applied a posteriori, recent works [14,15] have integrated EXMs into KD frameworks to enhance both performance and transparency. Nevertheless, existing approaches rely on unimodal saliency maps that inadequately capture the textural and structural features essential for medical diagnosis.

Building on these foundations, we propose EDGE-KD (Explainability-Driven Guidance for Efficient Knowledge Distillation), a novel framework that synergizes dual-feature saliency maps with knowledge transfer. Our method incorporates the Texture-Shape Explainability Method (TS-EXM) [16], which fuses textural and structural saliency maps to provide richer feature representations. Unlike traditional KD that transfers only logits, we explicitly incorporate explainability as an additional knowledge stream through:

- Dual saliency map extraction from teacher/student networks
- Explainability-aligned loss functions enforcing feature consistency

The remainder of this paper is organized as follows: Sect. 2 details our methodology. Section 3 describes experimental setups, and presents comparative results. Section 4 discusses implications and future directions.

2 Methodology

The proposed method, as depicted in Fig. 1, follows a structured process: first, standard training is conducted to establish baseline model performance. Next, an explainable dual-feature extraction mechanism is applied, capturing both texture and shape saliency maps to enhance interpretability. Finally, explainability-driven knowledge distillation is performed, ensuring that the student model inherits both predictive accuracy and interpretable feature representations from the teacher model.

2.1 Standard Training

In the initial stage of our methodology, both teacher and student models are trained on a chest X-ray dataset using deep learning models. A variety of teacher-

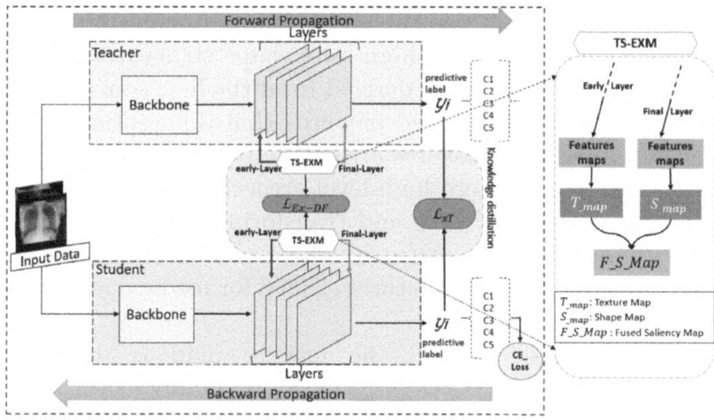

Fig. 1. Illustration of the proposed EDGE-KD framework, highlighting the stages of standard training, explainable dual-feature extraction, and explainability-driven knowledge distillation.

student model pairs are considered, incorporating both convolutional neural network (CNN) architectures and transformer-based architectures. The pretrained models are fine-tuned to adapt to the specific characteristics of the dataset.

During this phase, models are optimized using the standard Cross-Entropy loss function, as defined in Eq. (1). For each training sample x_i with its corresponding ground truth label $y_i \in [0,1]^C$ (where C is the number of classes), the model produces a probability distribution over the classes:

$$\mathcal{L}_{CE} = -\frac{1}{N}\sum_{i=1}^{N}\sum_{c=1}^{C} y_{i,c} \log P_c(x_i;\theta), \tag{1}$$

where $P_c(x_i;\theta)$ represents the predicted probability of class c, and θ denotes the model parameters.

This step is crucial for two main reasons: (1) it ensures that the teacher network achieves high performance by learning relevant patterns and subtle variations present in CXR images, which is essential for subsequent knowledge transfer; and (2) it establishes a performance baseline for the student model, facilitating meaningful comparison in later stages of training.

2.2 Explainable Dual-Features Extraction

To integrate explainability into knowledge distillation, we employ a dual-feature extraction mechanism that captures both texture and shape characteristics. This ensures that the student model learns not only the predictive patterns but also the interpretable representations of the teacher model. This process generates saliency maps using the Texture-Shape Explainability Method (TS-EXM), strategically integrated into the optimization pipeline to serve as both an interpretable representation and a mechanism for enforcing dual-feature alignment.

TS-EXM is designed to extract two types of features: fine-grained texture details from early convolutional layers and high-level shape structures from deeper layers. In practice, texture features are derived from the first convolutional block of each architecture, capturing localized patterns that differentiate fine structural details in images, while shape features are extracted from the final block preceding the classification head, where high-level structural information is encoded. This consistent choice across CNNs and transformer-based models reflects the hierarchical nature of deep architectures, where early layers emphasize texture and deeper layers encode global structures crucial for medical abnormality detection.

Given a convolutional layer l, the feature map is defined as $F^l \in \mathbb{R}^{B \times D_l \times H \times W}$, where B is the batch size, D_l is the number of feature channels, and $H \times W$ represents the spatial dimensions. The saliency map is generated using a thresholding function:

$$B_j(b,h,w) = \begin{cases} 1, & \text{if } F^l_{b,j,h,w} > \mu_j + \alpha \cdot \sigma_j \\ 0, & \text{otherwise} \end{cases} \tag{2}$$

where μ_j and σ_j represent the mean and standard deviation of the feature activations, respectively, and α is a sensitivity parameter.

For transformer-based architectures, we utilize Multi-Head Self-Attention (MSA) mechanisms to extract important features by aggregating attention weights across multiple layers:

$$A_{\text{global}} = \frac{1}{L} \sum_{l=1}^{L} A^l \tag{3}$$

where A^l denotes the attention weights at transformer layer l, and L is the total number of attention layers.

After generating texture and shape saliency maps, a guided filter is applied to refine and merge these maps, ensuring that both fine-grained details and high-level structural features are preserved. The final fused saliency map, F_{SalM}, serves as the primary knowledge representation in the explainability-driven knowledge distillation process, ensuring that the student model learns from the most relevant regions.

2.3 Incorporating Explainability into Knowledge Distillation

As mentioned above, the fused saliency maps extracted from both the teacher and student networks play a central role in our explainability-driven knowledge distillation framework. This ensures that the student model not only replicates the teacher's predictions but also aligns its internal feature representations, enforcing spatial consistency in the learned features. By guiding the student model to focus on the most relevant regions, similar to the teacher model, our approach enhances both predictive performance and interpretability.

To achieve this, we introduce a composite loss function that balances soft-target matching (4), dual-feature alignment using the fused saliency maps (5), and predictive accuracy. The dual-feature alignment loss is enforced using the Mean Squared Error (MSE), capturing the pixel-wise intensity differences between the saliency maps of the teacher and student models:

$$\mathcal{L}_{\text{ST}} = \frac{1}{N}\sum_{i=1}^{N} \left\| \sigma(y_i^{(T)}) - \sigma(y_i^{(S)}) \right\|^2 \tag{4}$$

where $\sigma(y_i^{(T)})$ and $\sigma(y_i^{(S)})$ represent the softmax-normalized probability distributions of the teacher and student models.

$$\mathcal{L}_{\text{ExDf}} = \frac{1}{N}\sum_{i=1}^{N} \left\| F_{\text{SalM}}^{(T)}(x_i) - F_{\text{SalM}}^{(S)}(x_i) \right\|^2 \tag{5}$$

where $F_{\text{SalM}}^{(T)}(x_i)$ and $F_{\text{SalM}}^{(S)}(x_i)$ denote the fused saliency maps extracted from the teacher and student models, respectively.

The loss formulations in (4) and (5) ensure that the student model learns not only *what* to predict but also *where* to focus, making the knowledge transfer process more transparent. The total loss function is defined as:

$$\mathcal{L}_{\text{total}} = \alpha \cdot (\mathcal{L}_{\text{ExDf}} + \mathcal{L}_{\text{ST}}) + (1 - \alpha) \cdot \mathcal{L}_{\text{CE}}^{(S)} \tag{6}$$

where $\alpha \in [0, 1]$ is a weighting factor that balances the contributions of the explainability-driven alignment loss $\mathcal{L}_{\text{ExDf}}$ and the soft-target loss $\mathcal{L}_{\text{ST}}$ with the standard cross-entropy loss $\mathcal{L}_{\text{CE}}^{(S)}$. By integrating explainability constraints into the knowledge transfer process, EDGE-KD ensures that the student model inherits both the diagnostic relevance and predictive quality of the teacher model.

3 Experimental Results

3.1 Dataset Description

The dataset used in this study is an extended version of publicly available CXR data designed for lung disease research. The original dataset [17,18] comprised four classes: Covid-19 (3,616 images), Normal (10,192 images), Lung Opacity (6,012 images), and Viral Pneumonia (1,345 images). Each class was provided with corresponding masks. However, in our work, only the images were used, and an additional class, Tuberculosis (TB), was included, consisting of 2,494 images sourced from public repositories [19]. All images, originally in PNG format, were resized from their original resolution of 299×299 to 224×224 pixels. The dataset was divided into three subsets: training, validation, and testing with 70% of the images used for training and the remaining 30% for validation and test.

3.2 Implementation Details

Our experiments were implemented using the PyTorch deep learning framework on Colab, utilizing an NVIDIA Tesla T4 GPU. For all models, we initialized the weights using pretrained parameters where applicable, and the models were fine-tuned for our specific task. We used the Stochastic Gradient Descent (SGD) optimizer with a fixed learning rate of 0.001, a momentum value of 0.9, and a weight decay of 1×10^{-3}. Training was conducted for 50 epochs with a batch size of 64.

3.3 Teacher-Student Configurations

Our study investigates various teacher-student configurations, grouped into two categories: *same-family* (architectures sharing a common lineage) and *cross-family* (heterogeneous architectures).

In the *same-family pairs*, we have: `EfficientNetV2S` (Teacher), a 23-layer network with paired with `EfficientNetB0` (Student), a 16-layer variant from the same efficiency-optimized family; `MobileNetV3Large` (Teacher), a 16-layer depthwise separable convolution network, paired with `MobileNetV3Small` (Student), an 11-layer lightweight counterpart; `ResNet101` (Teacher), a 101-layer residual network, paired with `ResNet34` (Student), a 34-layer simplified variant; and `ViT-Base/16` (Teacher), a Vision Transformer with 12 blocks and 16×16 patch embedding, paired with `DeiT-Small` (Student), a data-efficient transformer with distilled attention.

In the *cross-family pairs*, we examine: `EfficientNetV2S` (Teacher) paired with `DenseNet121` (Student), a 121-layer densely connected CNN; `MobileNetV3 Large` (Teacher) paired with `ShuffleNetV2×0.5` (Student), an 11-layer channel-shuffling network; and `ResNet101` (Teacher) paired with `DenseNet169` (Student), a 169-layer densely connected architecture.

3.4 Results

The results in Table 1 provide a comprehensive comparison of various teacher-student configurations across different knowledge distillation (KD) methods, including our proposed EDGE-KD framework, traditional KD approaches, and baseline performance. The evaluation is based on three key performance metrics: accuracy, loss, and F1-score.

We start by evaluating the baseline performance and find that teacher models generally achieve higher accuracy and F1-score while maintaining lower loss compared to their corresponding student models. An exception is observed with `ResNet34`, which surpasses its teacher, `ResNet101`, in raw accuracy (0.945 vs. 0.921). Despite this result, `ResNet34` exhibits strong instability, as illustrated in Fig. 2b, with sharp accuracy fluctuations and poor validation loss convergence. In contrast, `ResNet101` follows a smoother and more stable learning curve, as shown in Fig. 2a, indicating superior structured feature extraction and better

Table 1. Performance Comparison of Explainability-Driven Knowledge Distillation (EDGE-KD) Against Baseline and Traditional KD Methods

Models		Params (M)		Baseline Performance						Comparative KD Methods								
				T-Performance			S-Baseline (W/o)			S KL-Div			S I-KD			S EDGE-KD (Our)		
Teacher(T)	**Student(S)**	**Teacher**	**Student**	**Acc**	**Loss**	**F1-Score**	**Acc**	**Loss**	**F1-Score**	**Acc**	**Loss**	**F1-Score**	**Acc**	**Loss**	**F1-Score**	**Acc**	**Loss**	**F1-Score**
EfficientNetV2S	EfficientNetB0	21.5	5.23	0.955	0.106	0.965	0.925	0.123	0.939	0.946	0.118	0.955	0.950	0.109	0.957	0.954	0.085	0.964
MobileNetV3Large	MobileNetV3Small	5.48	2.5	0.957	0.107	0.959	0.920	0.121	0.933	0.868	0.260	0.874	0.932	0.186	0.945	0.934	0.111	0.947
ResNet101	ResNet-34	44.5	21.0	0.921	0.171	0.933	0.945	0.218	0.901	0.936	0.119	0.948	0.939	0.201	0.954	0.948	0.126	0.959
ViT-Base16	DeiT-Small	84.6	5.71	0.948	0.142	0.955	0.940	0.159	0.935	0.941	0.184	**0.954**	0.922	0.166	0.926	0.938	0.132	0.954
EfficientNetV2S	DenseNet121	21.5	8.2	0.955	0.106	0.965	0.884	0.155	0.910	0.938	0.146	0.944	0.907	0.195	0.917	0.941	0.141	0.953
MobileNetV3	ShuffleNetV2*0.5	5.48	1.6	0.957	0.107	0.959	0.912	0.125	0.921	0.940	0.168	0.942	0.923	0.194	0.934	0.935	0.117	0.949
ResNet101	denseNet169	44.5	14.5	0.921	0.171	0.933	0.872	0.186	0.891	0.928	0.172	0.941	0.922	0.168	0.938	0.943	0.177	0.942

generalization. This initial evaluation provides a reference point to quantify the performance limitations before applying knowledge transfer.

Then, we compare three knowledge distillation (KD) methods: KL Divergence (KL-Div) [4], which transfers knowledge through soft target probabilities; Intermediate Knowledge Distillation (I-KD) [7], which distills knowledge from intermediate feature activations; and our proposed EDGE-KD framework. To ensure a fair comparison, we set α to 0.5 for all KD methods, while for KL-Div, the temperature parameter T is fixed at 2. As shown in Table 1, EDGE-KD consistently outperforms both KL-Div and I-KD, achieving higher accuracy and F1-scores while maintaining a lower validation loss. Additionally, we evaluate the model reduction rate, highlighting efficiency improvements.

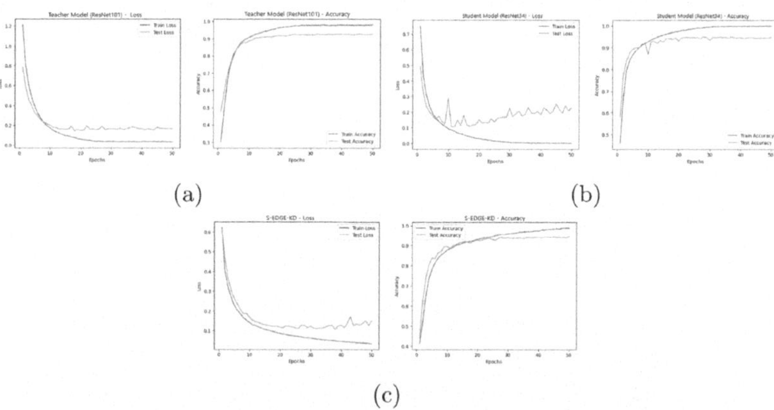

Fig. 2. Improving ResNet34 Training Stability and Convergence: (a) Teacher (ResNet101), (b) Baseline Student (ResNet34), (c) EDGE-KD (Our).

For same-family Teacher-Student, EDGE-KD improves accuracy from 0.925 to 0.954 with a model reduction rate of 75.7% for the `EfficientNetV2S` $\rightarrow$ `EfficientNetB0` configuration. In `MobileNetV3Large` $\rightarrow$ `MobileNetV3Small`, accuracy increases from 0.920 to 0.934 with a reduction rate of 54.4%. For the pair `ResNet101` $\rightarrow$ `ResNet34`, as shown in Fig. 2c, EDGE-KD enhances classification accuracy while stabilizing training, achieving a 52.8% reduction and ensuring better generalization. In `ViT-Base/16` → `DeiT-Small`, improvements are limited due to the fundamental differences between CNNs and Transformers, where self-attention mechanisms make saliency-based explainability alignment less effective. The reduction rate for this configuration is 93.3%.

For the cross-family Teacher-Student, in `EfficientNetV2S` → `DenseNet121`, EDGE-KD improves accuracy from 0.884 to 0.941 with a reduction rate of 61.9%. Similarly, in `ResNet101` → `DenseNet169`, accuracy increases from 0.872 to 0.943 with a reduction rate of 67.4%. Although KL-Div attains slightly higher accuracy in some cases, such as `MobileNetV3` → `ShuffleNetV2`×0.5 (0.940 vs. 0.935), EDGE-KD provides superior training stability, achieving a significantly lower validation loss (0.117 vs. 0.168) and a reduction rate of 70.8%.

To further assess the explainability-preserving capabilities of EDGE-KD, we employ the Feature Map Extraction Method (FEM) [20] to visualize the most relevant regions for model predictions. Table 2 presents an example of explainability maps for `EfficientNetV2S` (Teacher) → `EfficientNetB0` (Student), comparing importance maps generated across pathology categories: Covid-19, Normal, Viral Pneumonia (V_P), Lung Opacity (L_O), and Tuberculosis (TB). Based on this figure, we show that EDGE-KD closely aligns the student's explanations with those of the teacher, enhancing the focus on diagnostically critical regions.

These results confirm the generalizability of EDGE-KD, demonstrating consistent improvements in classification accuracy and explainability across both same-family and cross-family configurations. By effectively preserving key features from the baseline student model while integrating structured knowledge from the teacher, EDGE-KD ensures stable knowledge transfer.

Table 2. Example of explainability maps generated using FEM method for the `EfficientNetV2S` (Teacher) → `EfficientNetB0` (Student) configuration

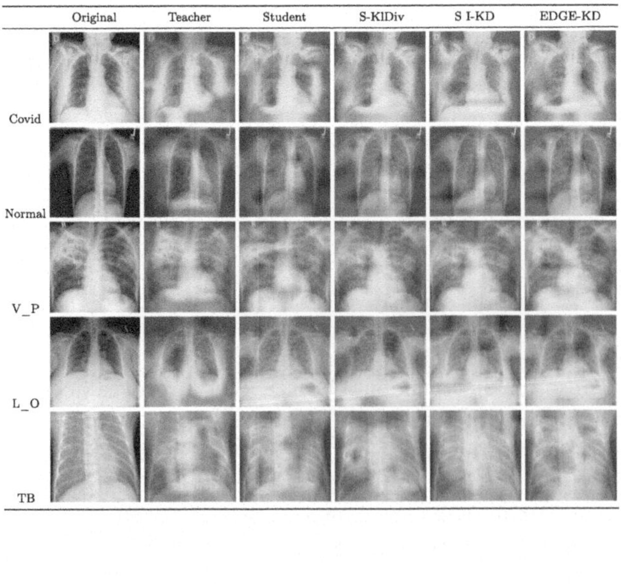

	Original	Teacher	Student	S-KlDiv	S I-KD	EDGE-KD
Covid						
Normal						
V_P						
L_O						
TB						

4 Conclusion

In this paper, we introduced EDGE-KD, a Explainability-Driven Guidance for Efficient knowledge distillation framework for chest X-ray classification. By integrating saliency-based knowledge transfer, EDGE-KD enhances both classification accuracy and explainability while maintaining stable training. The results demonstrate consistent improvements across same-family and cross-family teacherstudent configurations, achieving substantial model reduction without compromising performance or training stability. While the present evaluation already confirms the robustness of EDGE-KD, complementary directions such as component-wise analyses, cross-validated protocols with variance reporting and broader comparative baselines could provide additional granularity and further consolidate the evidence base. Future work will focus on extending EDGE-KD to other medical imaging tasks and enhance its applicability to transformer-based models by refining distillation strategies for attention-based feature alignment.

Acknowledgment. This research is supported by the project "Plateforme logicielle d'intégration de stratégies d'immunisation contre la pandémie COVID-19" funded by the grant of the Hassan II Academy of Sciences and Technology of Morocco. Additionally was conducted with the support of the National Center for Scientific and Technical Research (CNRST) under the PhD-Associate Scholarship Program (PASS).

References

1. Chehade, A.H., et al.: A systematic review: classification of lung diseases from chest x-ray images using deep learning algorithms. SN Comput. Sci. **5**(4), 405 (2024)
2. Toğaçar, M., et al.: Detection of covid-19 findings by the local interpretable model-agnostic explanations method of types-based activations extracted from CNNs. Biomed. Signal Process. Control **71**, 103128 (2022)
3. Karim, Md.R., et al.: Deepcovidexplainer: explainable covid-19 diagnosis from chest x-ray images. In: IEEE International Conference on Bioinformatics and Biomedicine (BIBM), pp. 1034–1037 (2020)
4. Hinton, G.: Distilling the knowledge in a neural network. arXiv preprint arXiv:1503.02531 (2015)
5. Termritthikun, C., et al.: Explainable knowledge distillation for on-device chest x-ray classification. IEEE/ACM Trans. Comput. Biol. Bioinform. (2023)
6. Son, M.H., et al.: Progressive knowledge distillation for automatic perfusion parameter maps generation from low temporal resolution CT perfusion images. In: International Conference on Medical Image Computing and Computer-Assisted Intervention, pp. 611–621 (2024)
7. Song, Y., et al.: Medical image classification: knowledge transfer via residual u-net and vision transformer-based teacher-student model with knowledge distillation. J. Vis. Commun. Image Represent. **102**, 104212 (2024)
8. Nguyen-Duc, T., et al.: Med-tex: transfer and explain knowledge with less data from pretrained medical imaging models. In: 2022 IEEE 19th International Symposium on Biomedical Imaging (ISBI), pp. 1–4 (2022)

9. Dapueto, J., Zini, L., Odone, F.: Knowledge distillation for efficient standard scan-plane detection of fetal ultrasound. Med. Biol. Eng. Comput. **62**(1), 73–82 (2024)
10. Zhukov, A., Benois-Pineau, J., Giot, R.: Evaluation of explanation methods of AI-CNNs in image classification tasks with reference - based and no-reference metrics. Adv. Artif. Intell. Mach. Learn. **3**(1), 620–646 (2023)
11. Mahmoudi, S.A., et al.: Explainable deep learning for covid-19 detection using chest x-ray and CT-scan images. In: Healthcare Informatics for Fighting COVID-19 and Future Epidemics, pp. 311–336 (2022)
12. Bhandari, M., et al.: Explanatory classification of CXR images into covid-19, pneumonia and tuberculosis using deep learning and XAI. Comput. Biol. Med. **150**, 106156 (2022)
13. An, J., Joe, I.: Attention map-guided visual explanations for deep neural networks. Appl. Sci. **12**(8), 3846 (2022)
14. Komodakis, N., Zagoruyko, S.: Paying more attention to attention: improving the performance of convolutional neural networks via attention transfer. In: ICLR (2017)
15. Sun, T., et al.: Explainability-based knowledge distillation. Pattern Recogn. **159**, 111095 (2025). ISSN 0031-3203
16. El Mohamadi, H., El Hassouni, M.: Enhanced deep learning explainability for covid-19 diagnosis from chest x-ray images by fusing texture and shape features. In: International Conference on Wireless Networks and Mobile Communications (WINCOM), pp. 1–6 (2023)
17. Chowdhury, M.E.H., et al.: Can AI help in screening viral and covid-19 pneumonia? IEEE Access **8**, 132665–132676 (2020)
18. Rahman, T., et al.: Exploring the effect of image enhancement techniques on covid-19 detection using chest x-ray images. Comput. Biol. Med. **132**, 104319 (2021)
19. Kiran, S., Jabeen, I.: Dataset of tuberculosis chest x-rays images (2024). https://data.mendeley.com/datasets/8j2g3csprk/2
20. Fuad, K.A.A., et al.: Features understanding in 3D CNNs for actions recognition in video. In: 2020 Tenth International Conference on Image Processing Theory, Tools and Applications (IPTA), pp. 1–6. IEEE (2020)

Contour-Guided Segmentation and X-Ray Image Validation for Pneumonia Detection Using Mobile Deep Learning in Low-Resource Settings

Gebregziabihier Nigusie[1](✉), Marawan Elbatel[2], and Yohannes Ayana[3]

[1] Mizan-Tepi University, Tepi, Ethiopia
gere@mtu.edu.et
[2] The Hong Kong University of Science and Technology, Hong Kong SAR, China
mkfmelbatel@ust.hk
[3] Amplitude Ventures, Rogaland, Norway
yohannesayana10@gmail.com

Abstract. Pneumonia is an infectious disease that affects the lungs and is commonly diagnosed using chest X-ray imaging. However, the availability of radiologists to interpret these images is often limited, especially in rural areas of developing countries. To mitigate this, research has explored computer vision models for automated pneumonia detection. Key challenges remain in accurately identifying the region of interest (ROI), deploying models on mobile devices, and verifying that input images are original and unaltered to maintain prediction reliability. In this study, we have developed a novel algorithm that validates X-ray images based on heuristic values and identifies anatomical regions through a contour-guided ROI algorithm. Experiments were conducted using both contour-guided ROI-applied images and original images on MobileNet and MobileViT architectures. Among the two models, MobileViT, a hybrid vision transformer achieved optimal performance, with an accuracy of 85.3% on contour-guided ROI images for three-class pneumonia classification, viral, bacterial, and normal cases. To enhance practical usability, we developed a mobile application for model deployment and easy access. The app automatically verifies whether the input is an original X-ray image and generates a summary of the prediction results with Processing a single image takes an average of 3 s and 1 MB of RAM on a mid-range device, making deployment feasible in resource-limited environments. This solution represents a valuable contribution toward improving pneumonia diagnosis in rural and resource-limited settings.

Keywords: Pneumonia Detection · Real-Time Diagnosis · ROI · Edge Density Analysis · MobileVit · Mobile Prediction

U. Anazodo et al. (Eds.): MIRASOL 2025, LNCS 16398, pp. 299–309, 2026.
https://doi.org/10.1007/978-3-032-13654-1_30

1 Introduction

Pneumonia is a common acute lung disease caused by bacteria or viruses leading to inflammation of the lung tissue and fluid or pus accumulation in the alveoli [1]. The disease remains a leading cause of death among children under five in developing countries [2], where limited radiologists are available for timely and accurate diagnosis [3]. Therefore, we aim to develop a mobile-based pneumonia detection that helps clinicians to perform rapid and reliable diagnosis using chest X-ray images, with a focus on supporting resource-constrained regions.

To optimize the pneumonia detection process, different studies have been conducted using machine learning and deep learning approaches such as Pneumonia Disease Detection Using Chest X-Rays and Machine Learning by employing convolutional neural networks [4] and pretrained ResNet-50 models on a dataset comprising 31.6% pneumonia and 68.4% normal chest X-ray images [5]. Their experimental results shows that CNN models achieved better accuracy and feature extraction performance [6]. Similarly, pretrained models such as InceptionV3, and VGG19 have been widely adopted for pneumonia detection due to their optimal feature learning capabilities [7]. Some studies are also integrate into mobile platforms, providing a cost-effective and portable solution for early diagnosis of pneumonia [8,9]. Although transfer learning has shown a promising and progressively better outcomes, the review emphasized the need for further optimization such as applying accurate image preprocessing and feature extraction processes which can significantly enhance performance to adapt these models for real time pneumonia detection from clinical x-ray images. Despite these advancements, previous studies are limited to support for end-to-end deployment in real-world mobile settings where ensuring both diagnostic image quality and accurate ROI detection, which are crucial for pneumonia detection from the clinical x-ray images.

Chest X-ray imaging is the most widely used diagnostic method for detecting pneumonia due to its accessibility and affordability compared to other imaging modalities. However, in many parts of Africa, the shortage of trained radiologists leads to delays in diagnosis and treatment, particularly in rural areas [10]. As highlighted by Michael G. Kawooya et al., the limited availability of qualified imaging professionals presents a significant barrier to timely medical care across the continent [11]. Based on previous studies that demonstrate the effectiveness of advanced image preprocessing and feature extraction in enhancing model accuracy. To this end, we propose an anatomically guided, mobile-based pneumonia detection framework that leverages deep learning to support clinical decision-making in resource-constrained settings to detect viral, bacterial and normal types of x-ray images. The experiment is conducted by applying advanced Contour-guided ROI detection through edge and symmetry-based X-ray image validation, along with optimized preprocessing techniques to accurately identify regions of interest, then we fine-tuned lightweight transformer and CNN-based models such as MobileViT [12] and MobileNet [13] for robust pneumonia detection. To ensure contextual relevance, we curated a dataset comprising chest X-rays from African populations and conducted ROC and per-class precisionrecall analyses to examine the effect of minor class imbalance. Finally, we integrated the trained models into a mobile application designed for ease of use and deployment in rural areas. Through rigorous evaluation, our model achieved an accuracy of 85.3%, demonstrating reasonable results for generalizability through localized ROIs, and potential to support early diagnosis in regions with limited access to radiologists.

2 Methodology

2.1 X-Ray Image Dataset

We collected a dataset of 4,600 chest X-ray images from two primary sources in Ethiopian hospitals and a publicly available online repository focused on African-based medical imaging. The dataset was annotated with diagnostic labels to facilitate supervised learning for pneumonia classification. Specifically, the dataset comprises 1,650 images labeled as bacterial pneumonia, 1,476 as viral pneumonia [14], and 1,474 as normal. As noted in previous studies, class imbalance is a common challenge that can adversely affect model performance. To address this, we ensured an approximately balanced distribution across the three classes and conducted ROC and per-class precisionrecall analyses to assess minor variations in class size.

2.2 Overall Framework

The framework that we have designed for pneumonia detection from chest X-ray images, specifically tailored for deployment in resource-constrained settings is comprise multiple stages. As shown in Fig. 1, the framework begins with image resizing, followed by a validation step to determine whether the input images are chest X-rays. This is followed by an optimized preprocessing stage, including contour-guided segmentation, to isolate the lung regions for pneumonia detection. The segmented images are then passed to the deep learning models, the first model we have used for the experiment is MobileViT [15], the architecture combines convolutional layers and vision transformers to create lightweight models optimized for mobile [9]. The second model is MobileNet which is CNN-based architecture designed for inference on mobile and embedded devices [16]. Finally, the entire framework is integrated into a mobile application for the real-time, on-device pneumonia detection. This mobile deployment is particularly valuable in rural regions where access to health care experts is limited.

2.3 Automatic X-Ray Image Validation

To ensure that the model processes only valid chest X-ray images, we developed a verification algorithm using OpenCV for mobile. This algorithm determines whether an uploaded image is a chest X-ray by analyzing key features, as outlined in the pseudo-code shown in Algorithm 1. The chest X-ray validation process is applied before ROI extraction to reduce model processing overhead.

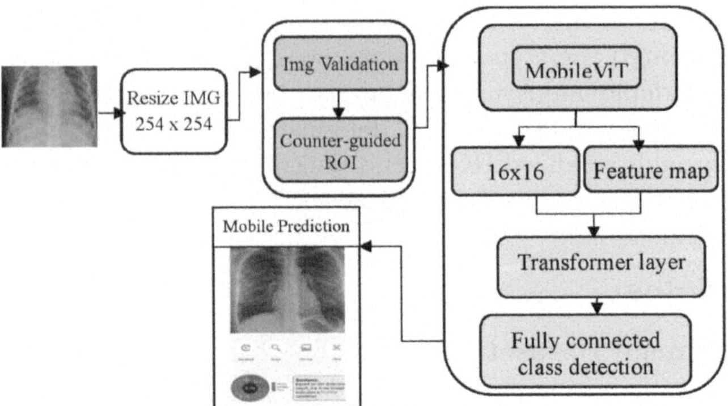

Fig. 1. Architectural Framework of the Model

Algorithm 1. Heuristic-Based Chest X-ray Image Validation

1: **Input:** Image I
2: **Output:** Boolean (True if likely chest X-ray, False otherwise)
3: $C \leftarrow$ Color variation, $S \leftarrow$ Symmetry difference, $E \leftarrow$ Edge density, $B \leftarrow$ Brightness
4: **if** $C < \tau_C$ **and** $S < \tau_S$ **and** $E > \tau_E$ **and** $B > \tau_B$ **then**
5: **return** True
6: **else**
7: **return** False
8: **end if**

Color Variation Analysis: The purpose of this color variation analysis is to filter out images that are not primarily grayscale by analyzing per-pixel color variation. The result is computed using Eq. (1), where $\sigma_{x,y}$ is the standard deviation value of the R, G, and B channels at pixel (x, y), and H and W are the height and width of the image. For the grayscale images the RGB values is approximately zero variation. Based on this the images which have Mean Color Variation less than 5 will be accepted for analyzing by the model.

$$\text{Mean Color Variation} = \frac{1}{H \times W} \sum_{x=1}^{H} \sum_{y=1}^{W} \sigma_{x,y} \tag{1}$$

Edge Density Detection: The second metric used is image edge density detection. In this stage we have measured the structural complexity of the image by applying a binary edge map value at pixel (x, y), is 1 if an edge is detected and 0 otherwise. The thresholds θ_1 and θ_2, ranging from 0.005 to 0.15 pixel intensity, are used to determine valid grayscale X-ray images, which is computed using

Eq. (2).

$$\text{Edge Density Detection} = \frac{1}{H \times W} \sum_{x=1}^{H} \sum_{y=1}^{W} E(x, y) \tag{2}$$

Symmetry Analysis: Equation (3) is used to quantify left-right symmetry as chest x-ray image is reasonably symmetrical resemblance in left and right half of the lung. To utilize this anatomical property for image validation, we perform a symmetry check by comparing the left and right halves of the image. The right half is horizontally flipped to align with the left half using the operation flip(Img_{right}, horizontal), and images with a symmetry difference of less than 90 are considered acceptable for prediction.

$$\text{Symmetry Analysis} = \frac{1}{H \times \frac{W}{2}} \sum_{x=1}^{H} \sum_{y=1}^{W/2} |Img_{\text{left}}(x, y) - Img_{\text{right-flip}}(x, y)| \tag{3}$$

Standard Deviation of Intensity: This metric measures the contrast variation of pixel intensities in the grayscale image. The regular X-ray image shows a moderate level of contrast which helps distinguish anatomical structures such as bones, soft tissues, and air spaces. Equation (4) is used to compute the standard deviation intensity and images with a standard deviation value of 20 or less are considered acceptable.

$$\text{Std Intensity} = \sqrt{\frac{1}{W \times H} \sum_{x=1}^{W} \sum_{y=1}^{H} (\text{Img}(x, y) - \mu)^2} \tag{4}$$

Center Mean Intensity and contrast characteristics of the image are also considered in this stage of image validation.

Heuristic Score: Finally, the computed values of each individual metric are combined into a heuristic score. Based on our defined acceptance ranges, an image is considered valid if the total heuristic score is greater than or equal to 4. Such images are accepted for further processing in the Contour-guided ROI and mobile based pneumonia prediction.

To assess the robustness of the image filtering stage, we conducted an ablation study by varying key thresholds used to confirm whether an image is a chest X-ray. Specifically, we tested color variation thresholds (3, 5, 7), symmetry difference thresholds (70, 90, 110), and acceptance score cutoffs (3.0, 4.0, 5.0) using 13 sample chest X-ray images. The results show that threshold combinations such as a color variation range of 5–7, symmetry difference between 90–110, and a score cutoff $\geq$ 3.0 yield the highest acceptance rate (69.2%). However, since the model is designed for medical imaging, a stricter image acceptance criterion with a score cutoff $\geq$ 5.0 was used. This indicates that the system is moderately sensitive to these thresholds. Therefore, the final metric values were selected based on this ablation study with a strict acceptance rate.

As illustrated in Fig. 2, the X-ray image is rejected by the algorithm during analysis using a mobile app for pneumonia detection with lightweight models. The image fails to meet the expected characteristics of a standard chest X-ray, as it has been modified by the addition of green color, violating the grayscale X-ray image format typically required for medical image analysis.

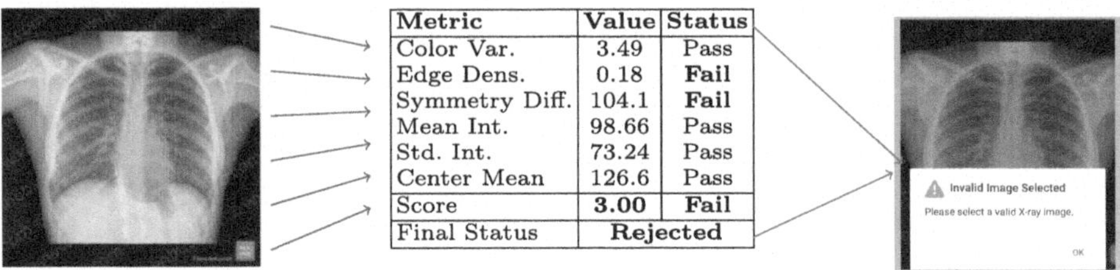

Metric	Value	Status
Color Var.	3.49	Pass
Edge Dens.	0.18	**Fail**
Symmetry Diff.	104.1	**Fail**
Mean Int.	98.66	Pass
Std. Int.	73.24	Pass
Center Mean	126.6	Pass
Score	**3.00**	**Fail**
Final Status	**Rejected**	

Fig. 2. X-ray image validation for pneumonia detection.

2.4 Contour-Guided ROI Detection

In this section, we employ an optimized algorithm to accurately segment the anatomical lung region, which helps improve model performance [17]. As illustrated in Fig. 3, image (a) shows the contour-guided ROI used to isolate the lung region [18], while image (b) presents the original X-ray image resized to 254 $\times$ 254 pixels, on which the ROI segmentation is applied. The resulting segmented image (c) highlights the extracted anatomical lung region which is subsequently analyzed for pneumonia detection.

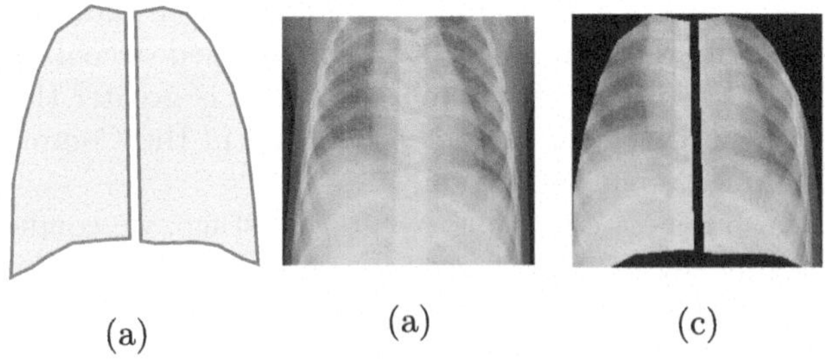

Fig. 3. The ROI extraction process based on Contour-guided algorithm

3 Experimental Result

In this study a total of 4,600 chest X-ray images were used. Experiments were conducted on both the original resized images and those processed using the advanced contour-guided ROI extraction algorithm. To evaluate the performance of the models, we used standard evaluation metrics such as precision, precision, recall, and F1 score, which were calculated based on the confusion matrix. The dataset we have used shows small size variation across classes. To evaluate the potential impact of these minor class imbalances on minority-class performance, we analyzed per-class AUC-ROC and Precision-Recall (PR) results. To ensure statistical significance and account for inherent variability, we computed 95% confidence intervals (CIs) for the AUC and Average Precision (AP) scores based on the test dataset size. For the MobileViT model, the bacterial class achieved an AUC of 96% and an AP of 0.93; the normal class had an AUC of 98% and an AP of 95%; and the virus class reached an AUC of 98% and an AP of 98%. These results provide strong evidence that our model's performance is statistically stable across all classes, despite minor variations in class distribution, as shown in Fig. 4.

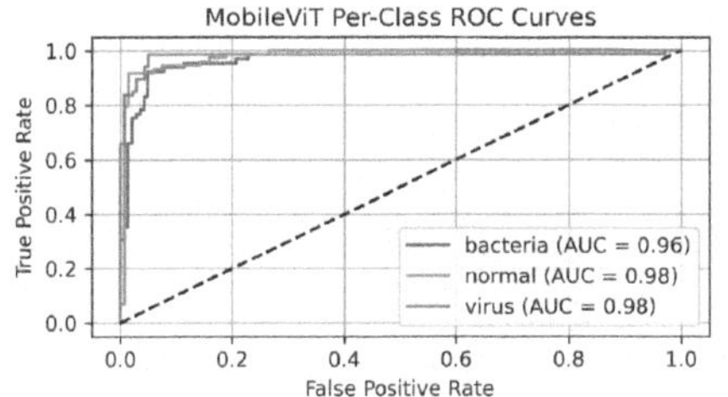

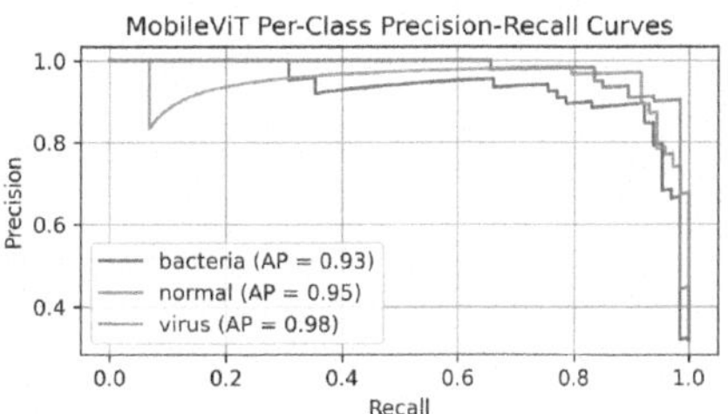

Fig. 4. Per-class ROC and Precision-Recall Curves

Contour-Based Lung Segmentation Experiment: By applying the contour-based ROI algorithm, both models shown improved classification accuracy. Based on the confusion matrix of the better-performing MobileViT model, 52 images were correctly classified as bacterial, 66 as viral, and 57 as normal. In contrast, when using the original images without the contour-based ROI algorithm, the model correctly classified only 37 of the total normal test cases. Table 1 presents the experimental results for both the contour-guided ROI and the original image inputs across the two models.

Table 1. Comparison of models performance with and without the ROI algorithm

Model	Image type	Accuracy	Precision	Recall	F1-Score
MobileNet	Original	79.5	83	80	78
	Contour ROI Based	80	84	80	80
MobileViT	Original	80	86	81	80
	Contour ROI Based	**85.3**	**87**	**85**	**85**

3.1 Deployment of Deep Learning Models on Mobile Devices

To support accessible pneumonia detection, we deployed the model in a mobile application for real-time prediction and result analysis. This mobile-based prediction is particularly beneficial in resource-limited settings. The application demonstrates low latency when analyzing a single image. To evaluate the latency, we used a mid-range device with 4GB RAM and the model is deployed using PyTorch, the average inference time was 3 to 4 s per image with peak memory usage of 1 MB. These results support the feasibility of real-time deployment of the model in resource-constrained environments. Figure 5 illustrates the prediction process and output of the mobile application, where the last screen summarizes the classification result for a valid X-ray image.

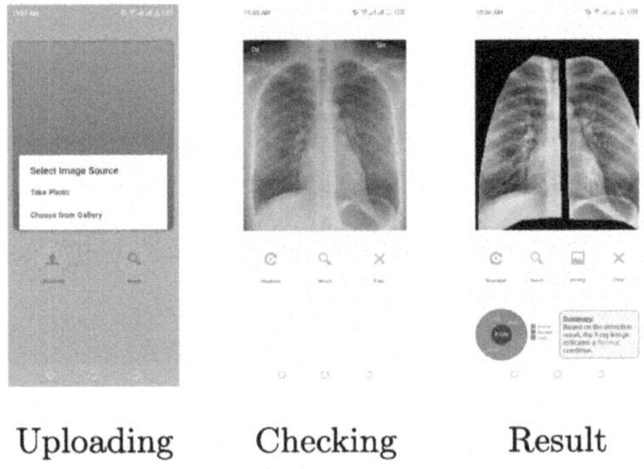

Fig. 5. Mobile based pneumonia detection process

4 Discussion

For pneumonia detection, to enhance the reliability of input data, we proposed an image preprocessing step that assesses the quality of X-ray images using key metrics including color variation, edge density, symmetry characteristics, and standard deviation. A final decision is made based on a heuristic score, where an image is accepted if its score is greater than or equal to 4.0. Furthermore, we developed a contour-guided algorithm to accurately identify the anatomical region of the image. This new ROI extraction approach improved the performance of both models. In particular, the MobileViT model, which achieved optimal accuracy, showed a 5.3% improvement when the ROI method was applied to detect the three classes of X-ray images (bacterial, viral, and normal). A total of 4,600 images were used to train the models with a 90/5/5 split for training, validation, and testing. The MobileViT model achieved an accuracy of 85.3% using the contour-guided ROI algorithm, showing better performance for pneumonia detection compared to MobileNet, which achieved an accuracy of 80%. Finally, the model was deployed as a mobile application, allowing direct use on mobile devices for pneumonia prediction. The app incorporates an initial verification step to detect and reject modified or non-chest X-ray images, ensuring that only valid images proceed. This mobile deployment facilitates real-time pneumonia detection in locations where radiological expert availability is limited.

5 Conclusion

In this work, we developed a mobile-based pneumonia detection model by integrating a novel image validation method with a contour-based ROI detection algorithm. The proposed ROI approach significantly improved the classification accuracy of the deep learning models. The main contributions of this study include the development of lightweight, mobile-based hybrid models for pneumonia detection, the introduction of a novel contour-guided X-ray preprocessing technique for accurate region of interest identification, and the implementation of new X-ray image verification algorithm to ensure that only valid images are processed. Furthermore, the successful deployment of these models on mobile platforms enables real-time, on-device pneumonia prediction. Overall, the proposed framework represents a promising step toward addressing the diagnostic gap in low-resource healthcare environments by leveraging deep learning models on widely accessible mobile technology. Future work will focus on scaling this approach with larger datasets and resource-efficient training strategies to further optimize performance and reliability.

References

1. Pates, K.M., Periselneris, J.N., Brown, J.S.: Pneumonia. Medicine **51**(11), 763–767 (2023)
2. Beletew, B., Bimerew, M., Mengesha, A., Wudu, M., Azmeraw, M.: Prevalence of pneumonia and its associated factors among under-five children in East Africa: a systematic review and meta-analysis. BMC Pediatr. **20**, 254 (2020). https://doi.org/10.1186/s12887-020-02083-z

3. Schuchat, A., Dowell, S.F.: Pneumonia in children in the developing world: new challenges, new solutions. Semin. Pediatr. Infect. Dis. **15**(3), 181–189 (2004). https://doi.org/10.1053/j.spid.2004.05.010
4. Szepesi, P., Szilágyi, L.: Detection of pneumonia using convolutional neural networks and deep learning. Biocybern. Biomed. Eng. **42**(3), 1012–1022 (2022). https://doi.org/10.1016/j.bbe.2022.08.001
5. Usman, C., Rehman, S.U., Ali, A., Khan, A.M., Ahmad, B.: Pneumonia disease detection using chest X-rays and machine learning. Algorithms **18**, 82 (2025). https://doi.org/10.3390/a18020082
6. Varshni, D., Thakral, K., Agarwal, L., Nijhawan, R., Mittal, A.: Pneumonia detection using CNN-based feature extraction. In: Proceedings of the 2019 IEEE International Conference on Electrical, Computer and Communication Technologies (ICECCT), pp. 1–7 (2019). https://doi.org/10.1109/ICECCT.2019.8869364
7. Abueed, M.A.M., et al.: Pneumonia detection using transfer learning: a systematic literature review. In: Proceedings of the 16th International Conference on Advanced Computer Science and Applications (IJACSA), pp. 1032–1040. The Science and Information Organization (2025)
8. Sait, U., Kumar, T., Shivakumar, S., Ravishankar, V.D., Lal, G.L.K.V., Bhalla, K.: A mobile application for early diagnosis of pneumonia in the rural context. In: Proceedings of the 2019 IEEE Global Humanitarian Technology Conference (GHTC), pp. 1–6. IEEE, Seattle (2019)
9. Stokes, K., et al.: A machine learning model for supporting symptom-based referral and diagnosis of bronchitis and pneumonia in limited resource settings. Biocybern. Biomed. Eng. **41**(4), 1288–1302 (2021)
10. Brady, A.P., Laoide, R.Ó., McCarthy, P., McDermott, R.: Discrepancy and error in radiology: concepts, causes and consequences. Ulster Med. J. **81**(1), 3–9 (2012). PMID: 23536732; PMCID: PMC3609674
11. Kawooya, M.G., et al.: An Africa point of view on quality and safety in imaging. Insights Imag. **13**, 59 (2022). https://doi.org/10.1186/s13244-022-01203-w
12. Mehta, S., Rastegari, M.: MobileViT: light-weight, general-purpose, and mobile-friendly vision transformer. arXiv preprint arXiv:2110.02178v2 [cs.CV] (2022). https://arxiv.org/abs/2110.02178
13. Sandler, M., Howard, A., Zhu, M., Zhmoginov, A., Chen, L.-C.: MobileNetV2: inverted residuals and linear bottlenecks. In: Proceedings of the IEEE Conference on Computer Vision and Pattern Recognition (CVPR), pp. 4510–4520. IEEE, Salt Lake City (2018)
14. Ho, A.: Viral pneumonia in adults and older children in sub-Saharan Africa: epidemiology, aetiology, diagnosis and management. Pneumonia (Nathan) **5**(Suppl 1), 18–29 (2014)
15. Boukhari, D.E., Chemsa, A., Baarir, Z.E.: MobileViT architecture for facial beauty prediction. In: 2024 International Conference on Telecommunications and Intelligent Systems (ICTIS), pp. 1–6. IEEE (2024). https://doi.org/10.1109/ICTIS62692.2024.10894508
16. Howard, A., et al.: Searching for MobileNetV3. In: Proceedings of the IEEE/CVF International Conference on Computer Vision (ICCV), pp. 1314–1324. IEEE, Seoul (2019)

17. Zhou, Z., Siddiquee, M.M.R., Tajbakhsh, N., Liang, J.: UNet++: a nested U-Net architecture for medical image segmentation. In: Deep Learning in Medical Image Analysis and Multimodal Learning for Clinical Decision Support, pp. 3–11. Springer (2018)
18. Sun, S., Zhang, R.: Region of interest extraction of medical image based on improved region growing algorithm. In: Advances in Engineering Research, vol. 125, International Conference on Material Science, Energy and Environmental Engineering (MSEEE 2017), pp. 1–5. Atlantis Press, Beijing (2017). https://doi.org/10.2991/mseee-17.2017.1

From Development to Deployment of AI-Assisted Telehealth and Screening for Vision and Hearing-Threatening Diseases in Resource-Constrained Settings: Field Observations, Challenges and Way Forward

Mahesh Shakya[1(✉)], Bijay Adhikari[2], Nirsara Shrestha[2], Bipin Koirala[2], Arun Adhikari[2], Prasanta Poudyal[2], Luna Mathema[2], Sarbagya Buddhacharya[2], Bijay Khatri[2], and Bishesh Khanal[1]

[1] Nepal Applied Mathematics and Informatics Institute for research (NAAMII), Lalitpur, Nepal
{mahesh.shakya,bishesh.khanal}@naamii.org.np
[2] Hospital for Children, Eye, ENT and Rehabilitation Services (CHEERS), BP Eye Foundation, Bhaktapur, Nepal

Abstract. Vision- and hearing-threatening diseases cause preventable disability, especially in resource-constrained settings(RCS) with few specialists and limited screening setup. Large scale AI-assisted screening and telehealth has potential to expand early detection, but practical deployment is challenging in paper-based workflows and limited documented field experience exist to build upon.

We provide insights on challenges and ways forward in development to adoption of scalable AI-assisted Telehealth and screening in such settings. Specifically, we find that iterative, interdisciplinary collaboration through early prototyping, shadow deployment and continuous feedback is important to build shared understanding as well as reduce usability hurdles when transitioning from paper-based to AI-ready workflows. We find public datasets and AI models highly useful despite poor performance due to domain shift. In addition, we find the need for automated AI-based image quality check to capture gradable images for robust screening in high-volume camps.

Our field learning stress the importance of treating AI development and workflow digitization as an end-to-end, iterative co-design process. By documenting these practical challenges and lessons learned, we aim to address the gap in contextual, actionable field knowledge for building real-world AI-assisted telehealth and mass-screening programs in RCS.

Keywords: AI-assisted screening · clinical deployment · end-to-end workflow · field experience

U. Anazodo et al. (Eds.): MIRASOL 2025, LNCS 16398, pp. 310–320, 2026.
https://doi.org/10.1007/978-3-032-13654-1_31

1 Introduction

Vision loss from conditions such as Diabetic Retinopathy (DR), Glaucoma, and Age-related Macular Degeneration (AMD) as well as preventable hearing loss due to middle ear diseases like Otitis Media is often avoidable if these conditions are detected early through systematic screening [1–4]. Early identification and timely intervention can significantly reduce the risk of irreversible vision or hearing impairment, especially in settings where patients otherwise present late in the disease course.

However, resource-constrained settings lack ophthalmologists and otolaryngologists for diagnosis and treatment of such conditions, and do not have robust, structured screening infrastructure [5,6]. Such settings may consist of Primary Care Centers (PCC) where community health workers (CHW) screen and treat common eye and ear conditions and are the first point of contact for health services for large section of the population in Low and Middle Income Countries(LMICs). These centers are often linked with Hub Hospitals providing specialist services and act as referral destination for these centers [7]. Basic tools (often analog) such as ophthalmoscope, slit lamp, fundus camera, audiometer, otoscope etc. are available to observe structural abnormalities but lack advanced confirmatory tools like visual field analyzers, optical coherence tomography (OCT), or audiometry booths and required expertise for comprehensive vision and hearing tests.

While these extended services help deliver care beyond urban hospitals and yearly short-term trainings are provided to improve screening capabilities, they are often constrained by limited operator skill and varying levels of diagnostic expertise in screening [8,9].

Artificial Intelligence (AI) offers a promising solution to bridge these capacity gaps [10]. When embedded within telehealth and screening workflows, AI could assist non-specialist health workers in rapidly identifying referable cases as well as assist clinicians in triaging referred cases, expanding screening reach, enabling timely referrals to higher-level centers and scaling services. A possible model may involve AI model flagging suspected referable cases, which are then reviewed remotely by specialists at a tertiary hospital, creating a tiered, efficient workflow.

Despite these opportunities, developing and deploying AI models at the point of care is still highly challenging-especially in low-resource [10–12] settings where workflow remain paper-based and use of analog image taking devices are common (For e.g. analog instead of digital otoscopes, analog slit lamp instead of digital portable fundus camera) with limited experience of Healthcare Professionals in AI-based Projects and that of AI researchers limited experience outside of lab settings resulting in trust barriers and terminology gaps. There is a need to co-design and mutual learning among interdisciplinary teams towards shared understanding for meaningful AI-assisted workflow that health workers can own up.

We started AI-assisted Telehealth Project for the very first time at the Hospital site with the mandate to apply AI to improve service delivery to PCCs, outreach and screening camps in telehealth setting. Except the Lead Computer

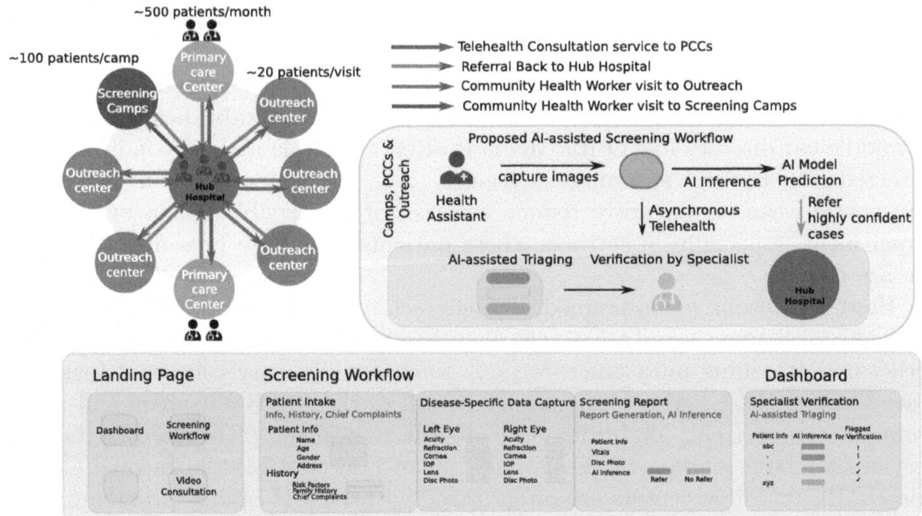

Fig. 1. *Left*: Hub-and-Spoke Model of Healthcare Delivery which represents our current clinical context. *Right & Bottom*: Prototype AI-assisted Screening, Telehealth and Referral Workflow and WebApp being tested in the current study.

Scientist, this was the first time being involved in building a real world AI-based workflow from scratch for most of the members of the Clinical team as well as the AI Researchers. In this unique scenario of multidisciplinary collaboration in AI-powered Telehealth in RCS, useful insights were gained with limited documentation available in the literature.

Hence, this paper contributes practical experience and documentation working toward AI-ready and AI-assisted screening and telehealth workflows for vision- and hearing-threatening diseases in this unique context. Specifically, We contribute

1. Field experience from screening camps and outreach centers highlighting practical usability issues such as time constraints, manual image transfer, intake form burdens requiring stakeholder-informed workflow design
2. Quantification of image acquisition challenges from digital otoscopes in high throughput screening camps and need for automated image quality assessment to support consistent, reliable screening
3. Observations on how public datasets and AI models while often showing reduced performance in new real-world settings, are highly valuable for workflow prototyping and building shared understanding between AI teams and clinicians

2 Related Work

Several studies have explored real-world AI deployment for screening vision- and hearing-threatening diseases in both well-resourced [13,14] and resource-constrained settings [15–17]. Studies in well-resourced settings have evaluated structured ML-aided triaging and grading framework showing potential benefits in prioritization and saving consultation time [13,14].

On the other hand, evidences from resource-constrained settings remain scarce. [18] validated a deep learning model for diabetic retinopathy screening in Zambian mobile clinics using Singaporean training data demonstrating good predictive performance in a new context. But these studies do not address challenges such as workflow integration and interdisciplinary collaboration required for translating from clinical gaps to AI problem formulation.

Additionally, various works [19–22], including reviews [10,12], have highlighted unique challenges and nuances of deployment in LMICs such as need for digital infrastructure, phased introduction of AI [19], inclusive design [20] and digitization of paper forms [23].

Together, these works show that while AI-assisted screening and telehealth is feasible, much of the literature still focuses on evaluation of AI model performance and pilot studies rather than the full development-to-deployment cycle, including experiences in interdisciplinary collaboration, building trust when the team is exploring AI-powered systems in clinical workflow for the first time. Our work adds to the latter discussion on development to deployment challenges specifically in the hub-and-spoke model of healthcare delivery.

3 Methods

First, we outline the context in which the project is being implemented. The initiative is led by the Ophthalmology and Otolaryngology Department of a tertiary-level hospital operating under a "hub-and-spoke" model. As shown in Fig. 1, this central hospital supports five outreach centers and two Primary Eye and Ear-Nose-Throat(ENT) Centers(PEEC) which act as "spokes" from which referrals are made back to the "hub". Health Workers from the central hub visit outreach centers weekly, whereas CHWs are stationed permanently at PEECs.

As a foundation for AI-ready workflow, we start by pre-testing/prototyping a Digitized Data Collection Workflow replicating the paper-based proforma documents used by Ophthalmic Assistants as well as ENT Assistants and then reiterating the collection workflow by field testing them in a single high-volume screening camp as well as one of the PEECs.

Additionally, to evaluate the feasibility of AI-assisted screening workflow in high throughput setting, we use AI models trained on publicly available datasets whenever feasible in parallel with data collection. *Fundus-based Ophthalmology Screening using AI Models trained on Public Datasets* - We use the publicly available EyePACS Glaucoma Justified Referral Dataset [24] to train a lightweight classifier, MobileNetv3 [25], where the focus is on ability to make inferences on low-cost hardware such as standard mid-range laptop without GPU as well as a placeholder to describe the overall workflow to clinicians. We trained the ImageNet1K-pretrained MobileNetv3 architecture [25] on fundus images square-cropped to 512 $\times$ 512 px. The model was trained till 10 epoch with Adam optimizer with learning rate $1e-3$ and batch size 4. Various Augmentations (Uniform Random Noise(0.01), Color Jitter(0.01), Random Horizontal Flipping) were done.

Additionally, we use the retrospective images stored in the Fundus Camera Devices since last 2 years in the Tertiary Hospital to test the robustness of the available multi-center large datasets and subsequent trained AI model performance in unseen context. This is described in the Datasets subsection below. *Otoscopy-based ENT Screening using AI Models trained on Public Dataset* - We find that the publicly available datasets for Otoscopy-based Screening lacked necessary metadata, low in quality as well as quantity resulting in unusable AI model. Additionally, retrospective otoscopic images from tertiary hospital were not available because of predominant use of analog otscopes in clinic and even when digital otoscopes were available, image storage was deemed too cumbersome or unnecessary.

Parallely, we also tested a web-based platform requiring internet connectivity to send the images to a on-premise server at the tertiary hospital and then run inference on the server and return the prediction to the user to evaluate feasibility of AI-assisted Triaging and Specialist Verification as shown in Fig. 1. This prototype system was shadow deployed in Primary Care Centers and screening camps as part of our workflow understanding, calibration and needs assessment phase.

Datasets. *Internal Test set - Mixed - EyePacs.* We used a subset of the Eye-PACS AIROGS Dataset as described in [26]. The dataset consists of balanced class labels (Referrable Glaucoma vs. Non Glaucoma) with 2500 images in train, and 385 images each in validation and test sets.

External Test set - Portable - Optomed AuroraIQ. This dataset consists of Fundus Images from 60 Patients taken with the portable Fundus Camera (specifically Optomed AuroraIQ). These images were taken from a Primary Care Center by an Ophthalmic Assistant. The images were taken in the context of real screening setting with multiple images of a single eye taken, if need be. These Images were labelled as Referrable Glaucoma vs. Non Glaucoma at the patient-level by the Glaucoma Specialist.

External Test set - Desktop - TopCon OCT1 Maestro and Canon CX1. The TopCon OCT1 Maestro and the Canon CX1 dataset consists of fundus images of patients suspected to have either a retinal disease or glaucoma by the clinician at the tertiary hospital, taken respectively from specific Desktop Fundus Cameras devices - TopCon OCT1 Maestro and Canon CX1. The distribution of this dataset consists of predominantly diseased fundus images. We took 100 patient with fundus images with labelled Glaucoma at the image-level by the Specialist.

4 Results

4.1 Field Experience in Workflow Digitization and Shadow Deployment

During the field testing, we embedded the AI output alongside routine clinical image review without influencing referral decisions, enabling clinicians and technicians to assess its usability in parallel with routine practice. Usability observations revealed key bottlenecks:

1. **Data Entry details** as staff spent considerable time filling paper or digital forms with patient demographics before image capture.
2. **Manual transfer frictions**, where images had to be copied from the portable fundus camera via USB wire to the laptop running the web-based platform, often adding several minutes per patient.
3. **User Feedback** Worried about the time needed to fill in the proforma document digitally. We initially had all the fields as required. User experience issues such as being able to quicken the form entry via autocompletes, dropdown menus were added after the first testing on-site at the screening camp.
4. Despite these frictions, staff appreciate the **immediate feedback** on potential disease diagnosis.

Recommendation. Despite many offerings available for digitization of the workflow, there are still barriers, one of which is perceived time consuming aspect of the digitization [27]. Despite the web app we built, frontline healthworkers were still hesitant especially in high-volume patient flow. To allow filling of the proforma details hands-free while simultaneously catering to the patient via voice-to-text based automated data entry in such noisy environment may be good future direction.

4.2 Quality Image Acquisition Challenges in Asynchronous Telehealth and Otoscopy-Based Screening

One of the barriers in deploying AI-assisted screening is inadequate image quality [28,29], which directly affects diagnostic reliability and AI model performance. In our pilot visits to Primary Care Centers and screening camps, as well as data collection site at tertiary center, we observed that non-trivial proportion of otoscopic images captured by health assistants were not of diagnostic quality, leading to failed remote grading.

Otoscopic Image-based Grading. To quantify the extent to which image quality issues occur when taken by Health Assistants for off-site consultation/screening in telehealth for ear screening, we conducted a small-scale image quality grading and screening for diseases of ear images collected with a digital otoscope in a field setting. Four otology specialists independently classified each of 100 tympanic membrane images into one of four diagnostic categories - Normal, Acute Otitis

Media (AOM), Otitis Media with Effusion (OME), Chronic Otitis Media (COM) - or selected 'Not Possible to Determine (NPD)' when the image quality or contextual information was insufficient. All specialists graded all images, enabling calculation of inter-rater reliability.

As shown in Table 1, We find that significant portion of the otoscopic images taken during screening camps by frontline Health Workers are "ungradable" mostly either due to poor image quality or obstruction due to earwax or foreign bodies. This also resulted in weak interannotator agreement with Fleiss' Kappa score of 0.37.

Table 1. Annotation results for 100 otoscopic images reviewed independently by four clinicians (A–D). The results show the number of images classified as "Possible to Determine" with specific diagnostic labels (Normal, OME, AOM, COM) and those marked as "Not Possible to Determine" further broken down by reasons such as poor image quality, obstruction (e.g., cerumen or foreign bodies), inadequate patient history, or need for additional hearing assessment.

Categories	Clinicians			
	A	B	C	D
Possible to Determine	**76**	**60**	**51**	**39**
Normal	47	17	20	15
OME	3	12	4	6
AOM	5	8	2	7
COM	21	23	25	11
Not Possible to Determine	**26**	**40**	**55**	**62**
Poor Image Quality	22	29	52	55
Obstruction	11	20	9	23
Inadequate Patient History	1	8	1	0
Hearing Assessment Required	1	7	0	2

Time pressure in high-throughput screening camps, having no time to clean or replace the speculum when contaminated with earwax, and in-general the difficulty of visualizing the full tympanic membrane in otoscopic images lead to ambiguity or unusable images for offsite diagnosis.

Recommendation. This underscores the need to integrate robust image quality assessment mechanism in both AI-assisted and asynchronous telehealth workflow through targeted training and automated AI-based feedback on gradability.

4.3 Publicly Available Datasets and AI Models Are Highly Useful for Workflow Design and Interdisciplinary Discussion

We find AI models trained on publicly available datasets are highly valuable as a starting point for designing and testing feasibility of AI-assisted screening workflows. In our project, using a glaucoma screening model trained on the widely used EyePACS Dataset [24,26] allowed us to introduce a concrete AI tool to clinical teams and frontline health workers who had no prior experience working with AI-enabled systems.

This exposure helped establish a shared understanding of how AI could be integrated into routine screening and telehealth workflows for flagging potential referable cases. By testing a working prototype, clinicians and outreach teams could see where and how AI might fit in - including how image capture, transfer and AI output could be embedded in existing screening and referral pathways. This facilitated interdisciplinary discussions, clarified expectations, and helped uncover usability and integration challenges early in the design process.

However, while these public models serve well for early-stage prototyping and workflow co-design, they do not perform reliably as drop-in solutions for local deployment. We show this by testing Zero-shot performance of an AI model trained on public dataset to classify Referrable Glaucoma(RG) vs. Non(Refereable) Glaucoma(NRG), testing their performance on datasets created from our screening camps and Outpatient Department at the tertiary hospital.

As shown in Table 2, We find that although the model performs exceptionally well (>90%) on it's internal test set (with similar population and imaging devices) across all classification metrics, it's zero-shot performance already drops to ~84% on portable fundus images taken from our screening camps. Moreover, the model's Precision reduces drastically to ~57% and ~37% respectively on images taken with TopCon OCT1 and Canon CX1 respectively. Various device-related artifacts may result in domain-shift and hence drop in performance.

Table 2. Zero-shot performance of Glaucoma Screening AI model trained on a subset of EyePACS Dataset and tested on population and devices available in our setting: The performance of the AI model on it's internal test set is shown for comparison against external test sets made from images captured from screening camps and Outpatient Department.

Dataset	Test-set Type	#RG/NRG	Accuracy	Precision	Recall	F1
EyePacs	Internal	385/385	0.9208	0.9219	0.9195	0.9207
Optomed AuroraIQ	External	28/32	0.8435	0.8367	0.8039	0.8200
TopCon OCT1 Maestro	External	100/0	–	0.5722	–	–
Canon CX1	External	100/0	–	0.3370	–	–

5 Discussion

Beyond image quality and usability, we faced challenges in translating clinical needs into robust AI tasks. Groundtruth definition is rarely straightforward, as diagnostic categories often hinge on subjective thresholds. For instance, unlike the EyePACS protocol that labels glaucoma suspects only at cup-to-disc ratios > 0.7, our team used lower thresholds prioritizing over-referral since routine screening is rare in RCS. Such choices highlight the need to explicitly document protocols, local risk preferences, and diagnostic assumptions in AI screening tool development. Additionally, the context-specific challenges of workflow digitization may require tailor-made solutions whereas challenges due to domain-shift and image quality may be broadly applicable. Advances such as Federated Learning, synthetic data augmentation techniques may benefit multisite data aggregation without compromising patient privacy as well as screening for diseases in low-data regime.

6 Conclusion

In this work, we share our practical experience in development to deployment of AI-assisted telehealth and screening workflows for vision- and hearing-threatening diseases in resource-constrained settings. Our findings highlight real-world challenges that go beyond model accuracy: specifically, image acquisition constraints, time and usability bottlenecks, and early prototyping using public datasets and models for concrete interaction between AI and Clinical team.

Our observations reinforce need for more systematic documentation of these end-to-end deployment pathways, especially in low-resource contexts with predominant paper-based high-volume workflows. By sharing these lessons learnt, we hope to inform and encourage future projects to bridge the gap between promising AI prototypes and robust screening solution to improve community health services.

Acknowledgments. This study is part of the Telehealth and AI-assisted screening Project funded by DirectRelief provided to Hospital for Children, Eye, ENT and Rehabilitation Services (CHEERS), BP Eye Foundation.

Disclosure of Interests. The authors have no competing interests to declare that are relevant to the content of this article.

References

1. Singer, D.E., Nathan, D.M., Fogel, H.A., Schachat, A.P.: Screening for diabetic retinopathy. Ann. Intern. Med. **116**(8), 660–671 (1992)
2. Topouzis, F.: Glaucoma–the importance of early detection and early treatment. J.-Glaucoma Importance Early Detect. Early Treat. (2007)
3. Bressler, N.M.: Early detection and treatment of neovascular age-related macular degeneration. J. Am. Board Fam. Pract. **15**(2), 142–152 (2002)
4. Rosenfeld, R.M., et al.: Clinical practice guideline: otitis media with effusion (update). Otolaryngol.-Head Neck Surg. **154**, S1–S41 (2016)
5. Burton, M.J., et al.: The lancet global health commission on global eye health: vision beyond 2020. Lancet Glob. Health **9**(4), e489–e551 (2021)

6. Piyasena, M.M.P.N., et al.: Systematic review on barriers and enablers for access to diabetic retinopathy screening services in different income settings. PLoS ONE **14**(4), e0198979 (2019)
7. Srivastava, S., et al.: Development of a hub and spoke model for quality improvement in rural and urban healthcare settings in India: a pilot study. BMJ Open Qual. **9**(3) (2020)
8. Sabri, K., Easterbrook, B., Khosla, N., Davis, C., Farrokhyar, F.: Paediatric vision screening by non-healthcare volunteers: evidence based practices. BMC Med. Educ. **19**(1), 65 (2019)
9. Ramkumar, V., Nagarajan, R., Shankarnarayan, V.C., Kumaravelu, S., Hall, J.W.: Implementation and evaluation of a rural community-based pediatric hearing screening program integrating in-person and tele-diagnostic auditory brainstem response (abr). BMC Health Serv. Res. **19**(1), 1 (2019)
10. Ciecierski-Holmes, T., Singh, R., Axt, M., Brenner, S., Barteit, S.: Artificial intelligence for strengthening healthcare systems in low-and middle-income countries: a systematic scoping review. NPJ Digit. Med. **5**(1), 162 (2022)
11. Krones, F., Walker, B.: From theoretical models to practical deployment: a perspective and case study of opportunities and challenges in ai-driven cardiac auscultation research for low-income settings. PLOS Digital Health **3**(12), e0000437 (2024)
12. Cabitza, F., Campagner, A., Balsano, C.: Bridging the "last mile" gap between ai implementation and operation: "data awareness" that matters. Ann. Transl. Med. **8**(7), 501 (2020)
13. Li, Y.Y.S., Vardhanabhuti, V., Tsougenis, E., Lam, W.C., Shih, K.C.: A proposed framework for machine learning-aided triage in public specialty ophthalmology clinics in Hong Kong. Ophthalmol Therapy **10**, 703–713 (2021)
14. Skevas, C., et al.: Implementing and evaluating a fully functional ai-enabled model for chronic eye disease screening in a real clinical environment. BMC Ophthalmol. **24**(1), 51 (2024)
15. Sarmiento, R.F.R., Overgaard, S.M., Gai, C., Overgaard, J.D., Ohde, J.W.: Guiding responsible ai in healthcare in the philippines. NPJ Digit. Med. **8**(1), 1–7 (2025)
16. Yim, D., Chandra, S., Sondh, R., Thottarath, S., Sivaprasad, S.: Barriers in establishing systematic diabetic retinopathy screening through telemedicine in low-and middle-income countries. Indian J. Ophthalmol. **69**(11), 2987–2992 (2021)
17. Rani, P.K., et al.: Smart (artificial intelligence enabled) drop (diabetic retinopathy outcomes and pathways): study protocol for diabetic retinopathy management. PLoS ONE **20**(5), e0324382 (2025)
18. Bellemo, V., et al.: Artificial intelligence using deep learning to screen for referable and vision-threatening diabetic retinopathy in Africa: a clinical validation study. Lancet Digital Health **1**(1), e35–e44 (2019)
19. Mollura, D.J., et al.: Artificial intelligence in low-and middle-income countries: innovating global health radiology. Radiology **297**(3), 513–520 (2020)
20. Okolo, C.T.: Ai in the "real world": examining the impact of ai deployment in low-resource contexts. arXiv preprint arXiv:2012.01165 (2020)
21. Weber, J.S., Toyama, K.: Remembering the past for meaningful ai-d. In: AAAI Spring Symposium: Artificial Intelligence for Development (2010)
22. Elahi, A., et al.: Overcoming challenges for successful pacs installation in low-resource regions: our experience in Nigeria. J. Digit. Imaging **33**(4), 996–1001 (2020)

23. Singh, G., Findlater, L., Toyama, K., Helmer, S., Gandhi, R., Balakrishnan, R.: Numeric paper forms for ngos. In: 2009 International Conference on Information and Communication Technologies and Development (ICTD), pp. 406–416. IEEE (2009)
24. Lemij, H.G., de Vente, C., Sánchez, C.I., Vermeer, K.A.: Characteristics of a large, labeled data set for the training of artificial intelligence for glaucoma screening with fundus photographs. Ophthalmol. Sci. **3**(3), 100300 (2023)
25. Howard, A., et al.: Searching for mobilenetv3. In: Proceedings of the IEEE/CVF International Conference on Computer Vision, pp. 1314–1324 (2019)
26. Steen, J., Kiefer, R., Ardali, M., Abid, M., Amjadian, E.: Standardized and open-access glaucoma dataset for artificial intelligence applications. Investig. Ophthalmol. Vis. Sci. **64**(8), 384 (2023)
27. Kabukye, J.K., et al.: Implementing smartphone-based telemedicine for cervical cancer screening in Uganda: qualitative study of stakeholders' perceptions. J. Med. Internet Res. **25**, e45132 (2023)
28. de Araujo, A.L., et al.: Ophthalmic image acquired by ophthalmologists and by allied health personnel as part of a telemedicine strategy: a comparative study of image quality. Eye **35**(5), 1398–1404 (2021)
29. Chen, J.S., et al.: Barriers to implementation of teleretinal diabetic retinopathy screening programs across the university of California. Telemedicine and e-Health **29**(12), 1810–1818 (2023)

Author Index

U. Anazodo et al. (Eds.): MIRASOL 2025, LNCS 16398, pp. 321–323, 2026.
https://doi.org/10.1007/978-3-032-13654-1

The manufacturer's authorised representative in the EU is Springer Nature Customer Service Centre GmbH, Europaplatz 3, 69115 Heidelberg, Germany. If you have any concerns regarding our products, please contact ProductSafety@springernature.com

Printed and bound by CPI Group (UK) Ltd, Croydon, CR0 4YY
07/07/2026
02160917-0010